Excel

YEAR 10

Advanced-level Mathematics Revision & Exam Workbook

ESSENTIAL skills

Get the Results You Want!

AS Kalra

Reprinted 2001, 2002, 2004, 2005, 2006, 2007, 2008, 2009, 2011, 2013
Updated in 2014 for the Australian Curriculum
Reprinted 2015 (twice), 2017, 2019, 2021, 2022, 2023

Reviewed in 2024 for the NSW Curriculum and Australian Curriculum Version 9.0 changes

ISBN 978 1 74125 567 6

Pascal Press
PO Box 250
Glebe NSW 2037
(02) 9198 1748
www.pascalpress.com.au

Publisher: Vivienne Joannou
Project editor: Michael Cole-King
Edited by Michael Cole-King
Answers checked by Peter Little
Typeset by Precision Typesetting (Barbara Nilsson) and DiZign Pty Ltd
Cover design by DiZign Pty Ltd
Printed by Vivar Printing/Green Giant Press

Dedication

This book is dedicated to the new generation of young Australians in whose hands lies the future of our nation and who by their hard work, acquired knowledge and intelligence will take Australia successfully through the 21st century.

This book is also in the loving, living and lasting memory of my dear mum, dad and uncle, who will remain a great source of inspiration and encouragement to me for times to come.

Acknowledgements
I would especially like to express my thanks and appreciation to my dear wife and my dear son, who have helped me to find the time to write this book. Without their help and support, achievement of all this work would not have been possible.

Contents

INTRODUCTION

CHAPTER 1 – Rational and irrational numbers

Unit 1 Real, rational and irrational numbers . . . 1
Unit 2 Irrational numbers and surds . . . 2
Unit 3 Simplifying surds (1) . . . 3
Unit 4 Simplifying surds (2) . . . 4
Unit 5 Pythagoras' theorem and surds . . . 5
Unit 6 Addition and subtraction of surds . . . 6
Unit 7 Multiplication and division of surds . . . 7
Unit 8 Expanding and simplifying . . . 8
Unit 9 Binomial products and conjugate surds . . . 9
Unit 10 Rationalising the denominator (1) . . . 10
Unit 11 Rationalising the denominator (2) . . . 11
Unit 12 Surds and indices . . . 12
Unit 13 Fractional indices (1) . . . 13
Unit 14 Fractional indices (2) . . . 14
Unit 15 Fractional indices and surds . . . 15
Topic Test . . . 16

CHAPTER 2 – Quadratic expressions and equations

Unit 1 Review of factorisation . . . 18
Unit 2 Non-monic, quadratic trinomials . . . 19
Unit 3 Simple quadratic equations . . . 20
Unit 4 Equations already in factorised form . . . 21
Unit 5 Equations involving the difference of two squares . . . 22
Unit 6 Equations involving a common factor . . . 23
Unit 7 Equations involving monic, quadratic trinomials . . . 24
Unit 8 Equations involving non-monic, quadratic trinomials . . . 25
Unit 9 Completing the square . . . 26
Unit 10 The quadratic formula (1) . . . 27
Unit 11 The quadratic formula (2) . . . 28
Unit 12 Mixed quadratic equations . . . 29
Unit 13 The number of solutions of quadratic equations . . . 30
Unit 14 Solving problems (1) . . . 31
Unit 15 Solving problems (2) . . . 32
Unit 16 Simultaneous equations resulting in a quadratic . . . 33
Unit 17 Equations reducible to quadratics . . . 34
Topic Test . . . 35

CHAPTER 3 – Non-linear relationships

Unit 1 Review of linear relationships (1) . . . 37
Unit 2 Review of linear relationships (2) . . . 38
Unit 3 Drawing quadratic relationships . . . 39
Unit 4 Axis of symmetry and vertex of a parabola . . . 40
Unit 5 Parabolas in the form $y = ax^2 + bx + c$ . . . 41
Unit 6 Further graphs of parabolas . . . 42
Unit 7 Determining equations of parabolas . . . 43
Unit 8 The hyperbola . . . 44
Unit 9 Sketching hyperbolic curves . . . 45
Unit 10 Further hyperbolas . . . 46
Unit 11 The circle $x^2 + y^2 = r^2$ . . . 47
Unit 12 The circle $(x - h)^2 + (y - k)^2 = r^2$ . . . 48
Unit 13 Further circles . . . 49
Unit 14 Points of intersection . . . 50
Unit 15 Exponential curves . . . 51
Unit 16 Cubic curves . . . 52
Unit 17 Further cubic and other curves . . . 53
Unit 18 Miscellaneous graphs . . . 54
Unit 19 Mixed graphs . . . 55
Topic Test . . . 56

CHAPTER 4 – Logarithms, exponentials and functions

Unit 1 Logarithms . . . 58
Unit 2 The log laws (1) . . . 59
Unit 3 The log laws (2) . . . 60
Unit 4 Logarithmic graphs . . . 61
Unit 5 Simple exponential equations . . . 62

Unit 6 Logarithmic and exponential equations ... 63
Unit 7 Functions ... 64
Unit 8 Further work with functions ... 65
Unit 9 Miscellaneous questions ... 66
Topic Test ... 67

CHAPTER 5 – Polynomials and curve sketching

Unit 1 Polynomials in general ... 69
Unit 2 Addition of polynomials ... 70
Unit 3 Subtraction of polynomials ... 71
Unit 4 Multiplication of polynomials ... 72
Unit 5 Division of polynomials ... 73
Unit 6 The remainder theorem ... 74
Unit 7 The factor theorem ... 75
Unit 8 Zeros of a polynomial ... 76
Unit 9 Solving equations involving polynomials ... 77
Unit 10 Curve sketching ... 78
Unit 11 Sketching polynomials (1) ... 79
Unit 12 Sketching polynomials (2) ... 80
Unit 13 Sketching polynomials (3) ... 81
Topic Test ... 82

CHAPTER 6 – Surface area and volume

Unit 1 Surface area of different solids ... 84
Unit 2 Perpendicular height and slant height ... 85
Unit 3 Surface area of pyramids ... 86
Unit 4 Surface area of cones ... 87
Unit 5 Surface area of a sphere ... 88
Unit 6 Surface area of composite solids ... 89
Unit 7 Volume of different solids ... 90
Unit 8 Volume of a right pyramid ... 91
Unit 9 Volume of a cone ... 92
Unit 10 Volume of a sphere ... 93
Unit 11 Volume of composite solids ... 94
Unit 12 Surface area and volume of similar figures ... 95
Unit 13 Applications of area and volume ... 96
Topic Test ... 97

CHAPTER 7 – Trigonometry

Unit 1 Review of basic trigonometry ... 99
Unit 2 Problems involving two right-angled triangles ... 100
Unit 3 Review of bearings ... 101
Unit 4 Trigonometry in three dimensions ... 102
Unit 5 Solving three-dimensional problems ... 103
Unit 6 The unit circle ... 104
Unit 7 Graphs of trigonometric ratios ... 105
Unit 8 Obtuse angles ... 106
Unit 9 Exact ratios ... 107
Unit 10 Simple trigonometric equations ... 108
Unit 11 Using the sine rule to find a side ... 109
Unit 12 Using the sine rule to find an angle ... 110
Unit 13 Further use of the sine rule ... 111
Unit 14 Using the cosine rule to find a side ... 112
Unit 15 Using the cosine rule to find an angle ... 113
Unit 16 Mixed exercises on the sine and cosine rules ... 114
Unit 17 Area of a triangle ... 115
Unit 18 Solving problems (1) ... 116
Unit 19 Solving problems (2) ... 117
Topic Test ... 118

CHAPTER 8 – Circle geometry

Unit 1 Terminology ... 120
Unit 2 Chord properties ... 121
Unit 3 Angle properties ... 122
Unit 4 Cyclic quadrilaterals ... 123
Unit 5 Tangent properties ... 124
Unit 6 Angle in the alternate segment ... 125
Unit 7 Secant properties ... 126
Unit 8 Mixed questions ... 127
Unit 9 Deductive exercises (1) ... 128
Unit 10 Deductive exercises (2) ... 129
Unit 11 Proofs involving circle geometry ... 130
Topic Test ... 131

CHAPTER 9 – Statistics and probability

Unit 1 Review of probability 133
Unit 2 Review of statistics 134
Unit 3 Standard deviation (1) 135
Unit 4 Standard deviation (2) 136
Unit 5 Comparing data sets 137
Unit 6 Comparing measures of spread 138
Unit 7 Scatter plots 139
Unit 8 Lines of best fit (1) 140
Unit 9 Lines of best fit (2) 141
Unit 10 Investigating reports 142
Topic Test 143

EXAM PAPERS

Exam Paper 1 145
Exam Paper 2 151
Exam Paper 3 157
Exam Paper 4 163

ANSWERS

Rational and irrational numbers 169
Quadratic expressions and equations 170
Non-linear relationships 172
Logarithms, exponentials and functions 176
Polynomials and curve sketching 178
Surface area and volume 180
Trigonometry 180
Circle geometry 181
Statistics and probability 182
Exam Papers 184

Introduction

This book has been written specifically for the **Year 10A Australian Curriculum and NSW Stage 5.3 Mathematics** course and forms part of a series of eight Revision & Exam Workbooks for Years 7 to 10. Each book in the series has been specifically designed to help students **revise** their work so that they can prepare for success in their **tests** during the school year and in their **half-yearly** and **yearly** exams.

This book will challenge and extend students through extensive practice. This will ensure that students are fully prepared for the advanced Mathematics courses in senior years.

The following features will help students achieve this goal:

- This book is a **workbook**. Students write in the book, ensuring that they have all their questions and working in the same place. This is invaluable when revising for exams—no lost notes or missing pages!
- Each page is a **self-contained, carefully graded unit of work**; this means students can plan their revision effectively by completing set pages of work for each section.
- **Topics are covered in depth,** providing students with practice to enable them to focus on areas of weakness or areas for extension.
- A **Topic Test** is provided at the end of each chapter. These tests are designed to help students test their knowledge of each syllabus topic. Practising tests similar to those they will sit at school will build students' confidence and help them perform well in their actual tests.
- **Four Exam Papers** have been included to test students on the Year 10A Australian Curriculum and NSW Stage 5.3 Mathematics course.
- A **marking scheme** is included in both the Topic Tests and Exam Papers to give students an idea of their progress.
- A **Topic Test and Exam Paper Feedback Chart**, found on the inside back cover, enables students to record their scores in all tests and exams.
- **Answers to all questions** are provided at the back of the book.

A note from the author

Mathematics is best learned if you have pen and paper with you and do every question in writing. Do not just read through the book—work through it and answer the questions, writing down all working. If this approach is coupled with a menu of motivation, realistic goal-setting and a positive attitude, it will lead to better marks in the examinations.

My best wishes are with you; I believe this book will help you achieve the best possible results. Good luck in your studies!

AS Kalra, MA, MEd, BSc, BEd

Chapter 1

Rational and irrational numbers

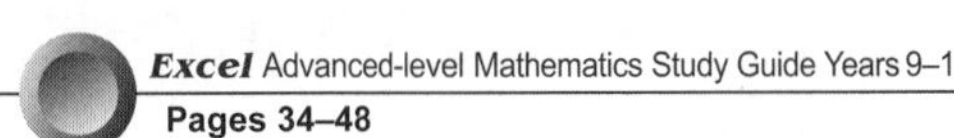

UNIT 1: Real, rational and irrational numbers

Question 1 Complete the following statements.

a A real number is any number that can be represented by a point on a ______________________.

b A rational number is any real number that can be expressed in the form $\frac{p}{q}$ where p and q are ______________ and $q \neq 0$.

c An ______________ number is any real number that is not rational.

d All integers and recurring decimals are ______________ numbers.

Question 2 Express these fractions as recurring decimals.

a $\frac{7}{9}$ ______________

b $\frac{4}{11}$ ______________

c $\frac{2}{7}$ ______________

d $\frac{5}{12}$ ______________

e $\frac{13}{33}$ ______________

f $\frac{41}{333}$ ______________

Question 3 Use a calculator to change these recurring decimals to fractions.

a $0.\dot{4}$ ______________

b $0.\dot{4}\dot{5}$ ______________

c $0.4\dot{5}$ ______________

d $0.\dot{4}\dot{5}\dot{6}$ ______________

e $0.4\dot{5}\dot{6}$ ______________

f $0.45\dot{6}$ ______________

Question 4 Without using a calculator, change each of these to fractions in simplest form.

a $0.\dot{5}$

b $0.1\dot{6}$

c $0.\dot{8}\dot{7}$

d $0.58\dot{3}$

e $0.\dot{7}\dot{5}\dot{6}$

f $0.4\dot{1}\dot{2}$

Rational and irrational numbers

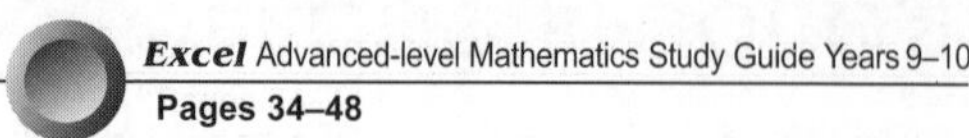

UNIT 2: Irrational numbers and surds

QUESTION 1 Write rational (Q) or irrational (Q′) for each number.

a $\frac{3}{5}$ ________ b 0.5 ________ c $\sqrt{3}$ ________ d $2\frac{1}{4}$ ________

e $\sqrt{2}$ ________ f π ________ g 12 ________ h $\sqrt{36}$ ________

i $0.\dot{3}$ ________ j $\sqrt{5}$ ________ k $\sqrt{4}+\sqrt{25}$ ________ l $\sqrt{17}$ ________

QUESTION 2 Use a calculator to find an approximation correct to 2 decimals places for:

a $\sqrt{2}$ ________ b $\sqrt{3}$ ________ c $\sqrt{6}$ ________ d $\sqrt{7}$ ________

e $\sqrt{11}$ ________ f $\sqrt{17}$ ________ g $\sqrt{29}$ ________ h $\sqrt{41}$ ________

QUESTION 3 Between which two consecutive integers is each surd?

a $\sqrt{5}$ ________ b $\sqrt{17}$ ________ c $\sqrt{39}$ ________ d $\sqrt{51}$ ________

e $\sqrt{95}$ ________ f $\sqrt{26}$ ________ g $\sqrt{79}$ ________ h $\sqrt{205}$ ________

QUESTION 4 Arrange in ascending order.

a $\sqrt{3}, 2, \sqrt{2}, 5$ b $3, \sqrt{8}, \sqrt{17}, 6$ c $\sqrt{8}, \sqrt{3}, 3, \sqrt{15}$ d $\sqrt{80}, 7, \sqrt{60}, 9$

________ ________ ________ ________

QUESTION 5

a What is: i $(+6)^2$? ________ ii $(-6)^2$? ________

b Write two possible values for the square root of 36 ________

QUESTION 6 Find.

a $\sqrt{49}$ ________ b $\sqrt{81}$ ________ c $-\sqrt{100}$ ________

d $-\sqrt{169}$ ________ e $\pm\sqrt{25}$ ________

QUESTION 7

a Which of the following do not have real solutions?

i $\sqrt{34}$ ________ ii $\sqrt{37.5}$ ________ iii $\sqrt{-51}$ ________

iv $-\sqrt{-27}$ ________ v $\sqrt{0}$ ________

b Explain why the surds you chose do not have real square roots. ________

c For what values of x in $\sqrt{x}$ are there no real solutions? ________

QUESTION 8 A right-angled isosceles triangle's two equal sides are each 1 unit long. One of these equal sides lies on the number line. Use a compass and ruler to plot the following surds on the number line. $\sqrt{2}, \sqrt{3}, \sqrt{5}, \sqrt{6}, \sqrt{7}$ and $\sqrt{10}$

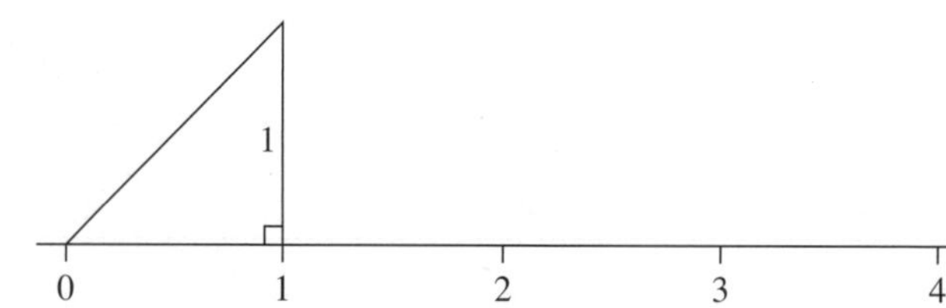

Rational and irrational numbers

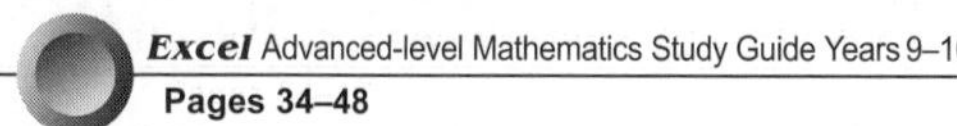

UNIT 3: Simplifying surds (1)

QUESTION 1 Simplify

a $\sqrt{6^2} =$ ______ **b** $\sqrt{10^2} =$ ______ **c** $\sqrt{2^2} =$ ______

d $(\sqrt{3})^2 =$ ______ **e** $(\sqrt{5})^2 =$ ______ **f** $(\sqrt{8})^2 =$ ______

QUESTION 2 Simplify the following surds.

a $\sqrt{8}$ **b** $\sqrt{12}$ **c** $\sqrt{18}$ **d** $\sqrt{24}$

e $\sqrt{28}$ **f** $\sqrt{50}$ **g** $\sqrt{40}$ **h** $\sqrt{44}$

i $\sqrt{75}$ **j** $\sqrt{48}$ **k** $\sqrt{27}$ **l** $\sqrt{32}$

m $\sqrt{72}$ **n** $\sqrt{98}$ **o** $\sqrt{242}$ **p** $\sqrt{200}$

QUESTION 3 Write the following surds in simplest form.

a $\sqrt{54}$ **b** $\sqrt{150}$ **c** $\sqrt{96}$ **d** $\sqrt{160}$

e $\sqrt{450}$ **f** $\sqrt{125}$ **g** $\sqrt{63}$ **h** $\sqrt{108}$

i $\sqrt{192}$ **j** $\sqrt{288}$ **k** $\sqrt{162}$ **l** $\sqrt{243}$

QUESTION 4 Simplify the following.

a $3\sqrt{12}$ **b** $4\sqrt{18}$ **c** $5\sqrt{27}$ **d** $3\sqrt{18}$

e $4\sqrt{48}$ **f** $2\sqrt{54}$ **g** $6\sqrt{72}$ **h** $7\sqrt{56}$

i $5\sqrt{32}$ **j** $3\sqrt{162}$ **k** $-\sqrt{12}$ **l** $-5\sqrt{300}$

Rational and irrational numbers

Excel Advanced-level Mathematics Study Guide Years 9–10
Pages 34–48

UNIT 4: Simplifying surds (2)

QUESTION 1 Write true or false for:

a $\sqrt{4} \times \sqrt{9} = \sqrt{36}$ ________ **b** $\sqrt{25} - \sqrt{9} = \sqrt{16}$ ________ **c** $\sqrt{100} - \sqrt{64} = \sqrt{36}$ ________

d $\sqrt{81} + \sqrt{49} = \sqrt{121}$ ________ **e** $\sqrt{18} \times \sqrt{100} = \sqrt{400}$ ________ **f** $\sqrt{144} \div \sqrt{4} = \sqrt{36}$ ________

QUESTION 2 Simplify.

a $\sqrt{3} \times \sqrt{2}$ ________ **b** $\sqrt{5} \times \sqrt{3}$ ________ **c** $\sqrt{5} \times \sqrt{6}$ ________

d $\sqrt{6} \times \sqrt{6}$ ________ **e** $\sqrt{8} \times \sqrt{8}$ ________ **f** $\sqrt{3} \times \sqrt{7}$ ________

g $\sqrt{18} \div \sqrt{3}$ ________ **h** $\sqrt{20} \div \sqrt{5}$ ________ **i** $\sqrt{50} \div \sqrt{5}$ ________

j $\sqrt{5} \times 4\sqrt{5}$ ________ **k** $\sqrt{6} \div \sqrt{2}$ ________ **l** $\sqrt{72} \div \sqrt{18}$ ________

QUESTION 3 Find the square of:

a $\sqrt{3}$ ________ **b** $\sqrt{7}$ ________ **c** $\sqrt{9}$ ________ **d** $\sqrt{5}$ ________

e $\sqrt{10}$ ________ **f** $2\sqrt{2}$ ________ **g** $3\sqrt{5}$ ________ **h** $2\sqrt{6}$ ________

i $3\sqrt{10}$ ________ **j** $4\sqrt{11}$ ________ **k** $5\sqrt{7}$ ________ **l** $4\sqrt{2}$ ________

QUESTION 4 Simplify.

a $\sqrt{20}$ ________ **b** $\sqrt{45}$ ________ **c** $\sqrt{90}$ ________

d $\sqrt{112}$ ________ **e** $\sqrt{128}$ ________ **f** $\sqrt{147}$ ________

g $\sqrt{500}$ ________ **h** $\sqrt{363}$ ________ **i** $\sqrt{135}$ ________

QUESTION 5 Simplify.

a $2\sqrt{18}$ **b** $3\sqrt{8}$ **c** $5\sqrt{125}$ **d** $3\sqrt{96}$ **e** $5\sqrt{72}$

f $8\sqrt{75}$ **g** $3\sqrt{54}$ **h** $8\sqrt{24}$ **i** $7\sqrt{32}$ **j** $8\sqrt{18}$

k $3\sqrt{128}$ **l** $3\sqrt{245}$ **m** $3\sqrt{50}$ **n** $3\sqrt{300}$ **o** $9\sqrt{63}$

QUESTION 6 Express as an entire surd.

a $2\sqrt{7}$ ________ **b** $3\sqrt{5}$ ________ **c** $5\sqrt{3}$ ________

d $5\sqrt{10}$ ________ **e** $4\sqrt{6}$ ________ **f** $8\sqrt{10}$ ________

Rational and irrational numbers

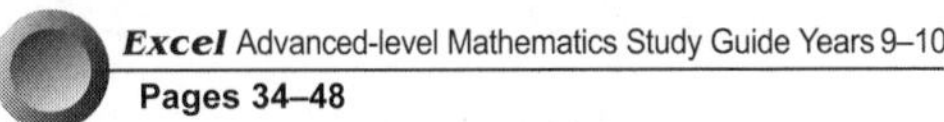

UNIT 5: Pythagoras' theorem and surds

QUESTION 1 Use Pythagoras' theorem to find the value of x. Give your answer as a surd in its simplest form.

a

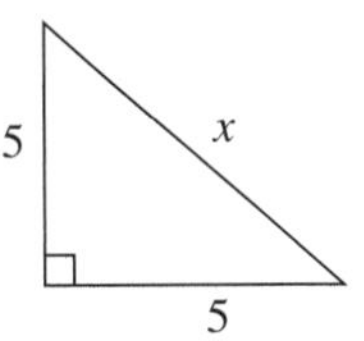

b

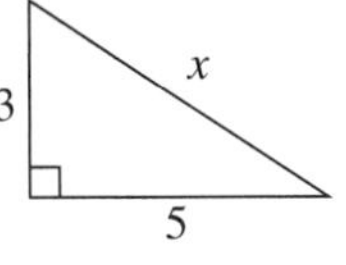

c

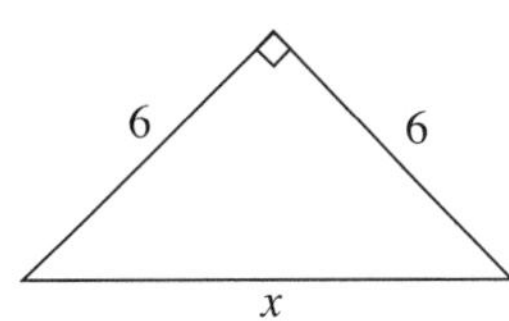

d

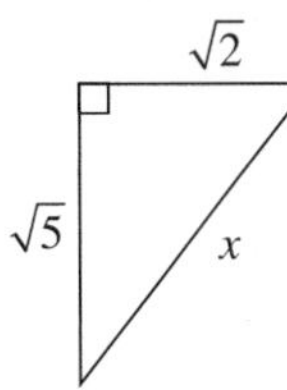

e

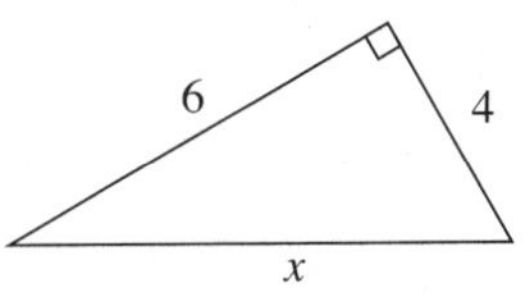

f

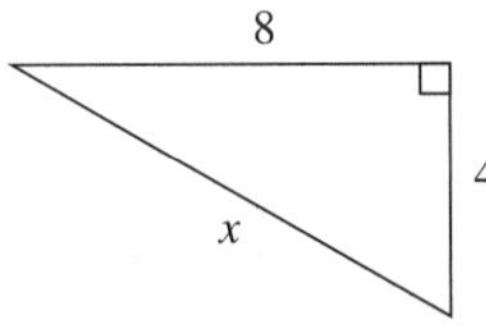

g

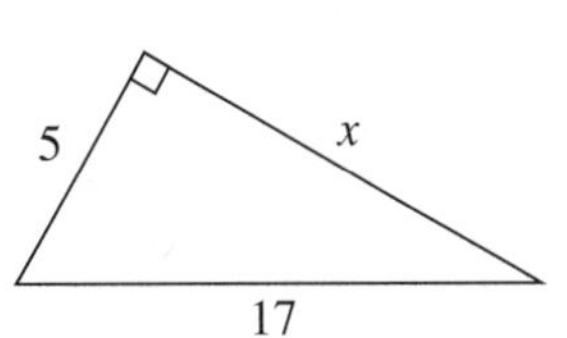

h

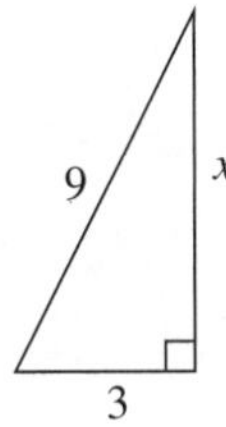

i

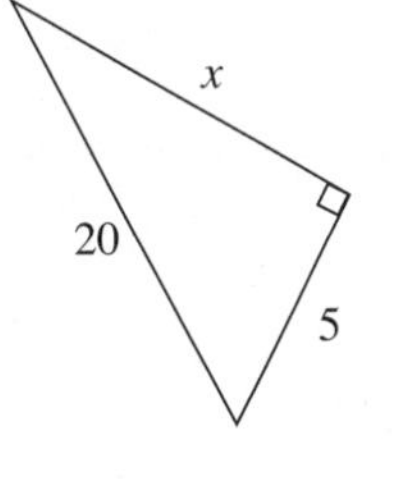

j

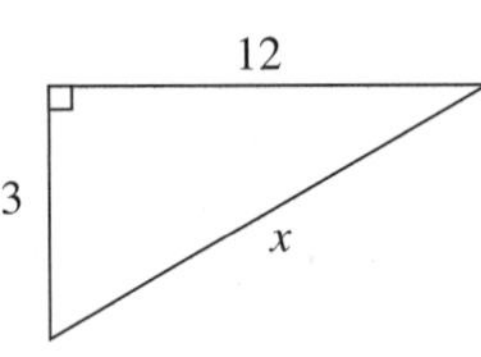

k

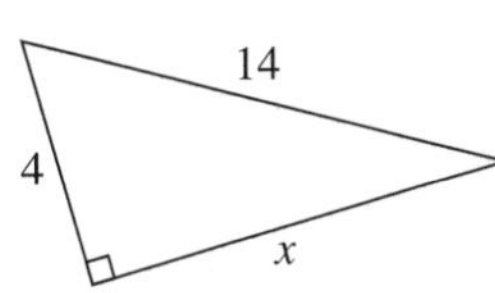

l

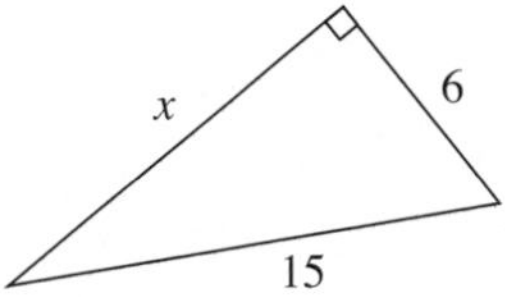

QUESTION 2 Find the length of the diagonal BD in the given rectangle $ABCD$.

QUESTION 3 $PQRS$ is a square with side length $\sqrt{7}$ units. Find the:

a area of the square

b length of the diagonal

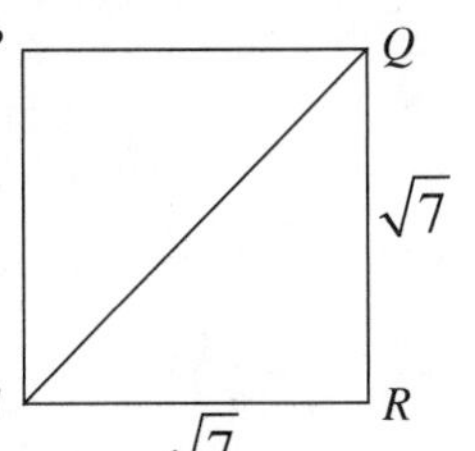

Rational and irrational numbers

Excel Advanced-level Mathematics Study Guide Years 9–10
Pages 34–48

UNIT 6: Addition and subtraction of surds

QUESTION 1 Simplify the following by collecting like surds.

a $5\sqrt{2} + 6\sqrt{2} =$ ____________

b $8\sqrt{5} + 7\sqrt{5} =$ ____________

c $6\sqrt{3} + 9\sqrt{3} =$ ____________

d $8\sqrt{5} - 4\sqrt{5} =$ ____________

e $9\sqrt{3} + 4\sqrt{3} - 2\sqrt{3} =$ ____________

f $12\sqrt{2} + 6\sqrt{2} - 5\sqrt{2} =$ ____________

g $2\sqrt{x} + 3\sqrt{x} =$ ____________

h $5\sqrt{y} + 6\sqrt{y} - 3\sqrt{y} =$ ____________

i $8\sqrt{m} + 15\sqrt{m} - 7\sqrt{m} =$ ____________

j $5\sqrt{2} + 6\sqrt{2} =$ ____________

k $11\sqrt{3} + 15\sqrt{3} - 16\sqrt{3} =$ ____________

l $3a\sqrt{x} + 2a\sqrt{x} =$ ____________

QUESTION 2 Simplify the following.

a $2\sqrt{3} + 5\sqrt{2} + 8\sqrt{3} =$ ____________

b $3\sqrt{7} + 5\sqrt{7} - 2\sqrt{11} =$ ____________

c $3\sqrt{3} + 6\sqrt{2} - 7\sqrt{3} - 4\sqrt{2} =$ ____________

d $8\sqrt{5} - 2\sqrt{3} - 5\sqrt{5} + 8\sqrt{3} =$ ____________

e $9\sqrt{2} - 6\sqrt{2} + 5\sqrt{2} =$ ____________

f $6\sqrt{3} - 2\sqrt{3} + \sqrt{3} =$ ____________

g $2\sqrt{5} - 2\sqrt{5} =$ ____________

h $8\sqrt{3} + 7\sqrt{3} =$ ____________

i $5\sqrt{10} - 3\sqrt{10} + \sqrt{3} =$ ____________

j $4\sqrt{5} - 8\sqrt{5} + 7\sqrt{5} =$ ____________

k $9\sqrt{2} + 8\sqrt{2} - 6\sqrt{3} =$ ____________

l $4\sqrt{7} + 6\sqrt{7} - 3\sqrt{7} =$ ____________

m $\sqrt{2} + 12\sqrt{2} - \sqrt{2} =$ ____________

n $8\sqrt{3} + 11\sqrt{3} - 3\sqrt{5} =$ ____________

QUESTION 3 Simplify.

a $\sqrt{8} + \sqrt{18} =$ ____________

b $\sqrt{90} + \sqrt{160} =$ ____________

c $\sqrt{27} + \sqrt{75} =$ ____________

d $\sqrt{54} - \sqrt{24} =$ ____________

e $\sqrt{48} - \sqrt{27} =$ ____________

f $\sqrt{81} - \sqrt{96} =$ ____________

g $\sqrt{18} + \sqrt{50} =$ ____________

h $3\sqrt{24} - 2\sqrt{96} =$ ____________

i $\sqrt{12} + 3\sqrt{72} =$ ____________

j $\sqrt{5} + \sqrt{125} - \sqrt{20} =$ ____________

k $\sqrt{12} + 4\sqrt{48} =$ ____________

l $\sqrt{8} + \sqrt{32} - \sqrt{18} =$ ____________

m $\sqrt{20} + \sqrt{45} - \sqrt{75} =$ ____________

n $\sqrt{242} - \sqrt{162} + \sqrt{18}$ ____________

Rational and irrational numbers

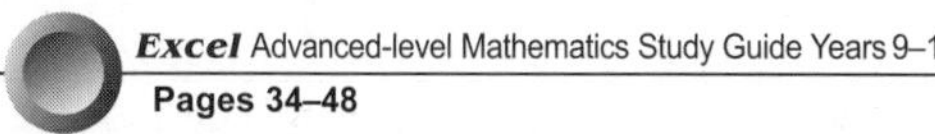

UNIT 7: Multiplication and division of surds

QUESTION 1 Find these products.

a $\sqrt{2} \times \sqrt{3} =$ ______ **b** $\sqrt{5} \times \sqrt{6} =$ ______ **c** $\sqrt{2} \times \sqrt{7} =$ ______

d $\sqrt{3} \times \sqrt{11} =$ ______ **e** $\sqrt{6} \times \sqrt{7} =$ ______ **f** $\sqrt{7} \times \sqrt{10} =$ ______

g $5\sqrt{2} \times 6\sqrt{3} =$ ______ **h** $2\sqrt{3} \times 3\sqrt{2} =$ ______ **i** $4\sqrt{7} \times 5\sqrt{5} =$ ______

j $6\sqrt{5} \times 4\sqrt{3} =$ ______ **k** $3\sqrt{11} \times 5\sqrt{7} =$ ______ **l** $10\sqrt{10} \times 3\sqrt{3} =$ ______

QUESTION 2 Find these products, giving the answers in simplest form.

a $\sqrt{5} \times \sqrt{20}$ **b** $\sqrt{8} \times \sqrt{2}$ **c** $\sqrt{12} \times \sqrt{3}$ **d** $\sqrt{10} \times \sqrt{6}$ **e** $\sqrt{15} \times \sqrt{6}$

f $2\sqrt{2} \times 5\sqrt{8}$ **g** $7\sqrt{10} \times 4\sqrt{20}$ **h** $2\sqrt{3} \times 4\sqrt{3}$ **i** $3\sqrt{2} \times 7\sqrt{2}$ **j** $5\sqrt{6} \times 2\sqrt{8}$

QUESTION 3 Find.

a $\sqrt{15} \div \sqrt{5} =$ ______ **b** $\sqrt{35} \div \sqrt{7} =$ ______ **c** $\sqrt{70} \div \sqrt{10} =$ ______

d $12\sqrt{14} \div 2\sqrt{2} =$ ______ **e** $20\sqrt{10} \div 10\sqrt{2} =$ ______ **f** $15\sqrt{6} \div 3\sqrt{3} =$ ______

QUESTION 4 Find, giving the answer in simplest form.

a $\sqrt{12} \div \sqrt{3}$ **b** $\sqrt{8} \div \sqrt{2}$ **c** $\sqrt{48} \div \sqrt{3}$ **d** $\sqrt{45} \div \sqrt{5}$

e $10\sqrt{18} \div 5\sqrt{2}$ **f** $14\sqrt{14} \div 2\sqrt{2}$ **g** $36\sqrt{60} \div 9\sqrt{6}$ **h** $21\sqrt{55} \div 7\sqrt{11}$

i $\sqrt{50} \div 5$ **j** $\sqrt{63} \div 3$ **k** $\sqrt{98} \div 7$ **l** $\sqrt{32} \div 2$

UNIT 8: Expanding and simplifying

QUESTION 1 Expand.

a $2(\sqrt{3}+2\sqrt{2}) =$ ______

b $5(3\sqrt{6}+5) =$ ______

c $\sqrt{2}(\sqrt{7}+\sqrt{3}) =$ ______

d $\sqrt{5}(\sqrt{2}+\sqrt{3}) =$ ______

e $\sqrt{5}(5-\sqrt{3}) =$ ______

f $\sqrt{3}(2\sqrt{5}-\sqrt{7}) =$ ______

g $\sqrt{2}(\sqrt{11}+2\sqrt{7}) =$ ______

h $\sqrt{6}(2\sqrt{5}+3\sqrt{7}) =$ ______

i $3\sqrt{2}(\sqrt{11}+\sqrt{5}) =$ ______

j $2\sqrt{2}(\sqrt{3}-3\sqrt{5}) =$ ______

k $3\sqrt{7}(2\sqrt{2}+3\sqrt{3}) =$ ______

l $4\sqrt{3}(5\sqrt{2}-3\sqrt{5}+1) =$ ______

QUESTION 2 Expand, giving the answer in simplest form.

a $\sqrt{3}(\sqrt{5}+\sqrt{3})$

b $\sqrt{5}(\sqrt{5}+\sqrt{3})$

c $\sqrt{7}(3\sqrt{7}-2)$

d $\sqrt{2}(3\sqrt{2}-\sqrt{2})$

e $\sqrt{3}(5\sqrt{2}-\sqrt{3})$

f $\sqrt{11}(2\sqrt{3}-6\sqrt{11})$

g $\sqrt{2}(\sqrt{6}+3)$

h $\sqrt{3}(\sqrt{15}-\sqrt{6})$

i $2\sqrt{3}(5\sqrt{3}-\sqrt{2})$

j $8\sqrt{2}(4\sqrt{10}-3\sqrt{2})$

k $5\sqrt{5}(2\sqrt{10}-\sqrt{15})$

l $6\sqrt{3}(2\sqrt{6}+5\sqrt{21})$

QUESTION 3 Expand and simplify.

a $2(\sqrt{3}+3\sqrt{2})+5(\sqrt{3}-4\sqrt{2})$

b $\sqrt{3}(\sqrt{7}-2)-\sqrt{2}(\sqrt{5}+\sqrt{2})$

c $6\sqrt{5}(\sqrt{3}-\sqrt{2})-3\sqrt{2}(4\sqrt{5}-\sqrt{3})$

d $3\sqrt{7}(2\sqrt{2}-3\sqrt{5})+2\sqrt{5}(\sqrt{7}-4)$

UNIT 9: Binomial products and conjugate surds

QUESTION **1** Expand and simplify.

a $(\sqrt{3}+\sqrt{2})(\sqrt{3}+4)$

b $(\sqrt{5}+2)(\sqrt{3}+\sqrt{7})$

c $(2\sqrt{3}+1)(2\sqrt{3}-4)$

d $(5\sqrt{3}+2)(\sqrt{3}-4)$

e $(3\sqrt{2}+3)(\sqrt{2}+1)$

f $(\sqrt{7}-3)(\sqrt{5}+2)$

g $(5+\sqrt{2})(2\sqrt{5}-2)$

h $(2\sqrt{11}+2)(\sqrt{11}-1)$

i $(\sqrt{3}+\sqrt{2})(\sqrt{2}+1)$

j $(7\sqrt{2}+\sqrt{3})(5\sqrt{2}+2)$

k $(2\sqrt{3}+3\sqrt{2})(\sqrt{3}+1)$

l $(2\sqrt{2}+5)(\sqrt{2}-3)$

QUESTION **2** Expand and simplify.

a $(\sqrt{2}+1)^2 =$ ______

b $(2\sqrt{3}+3)^2 =$ ______

c $(\sqrt{5}+\sqrt{7})^2 =$ ______

d $(\sqrt{3}-\sqrt{5})^2 =$ ______

e $(\sqrt{3}-1)^2 =$ ______

f $(\sqrt{7}-\sqrt{2})^2 =$ ______

g $(6-\sqrt{3})^2 =$ ______

h $(2\sqrt{3}+5)^2 =$ ______

i $(3\sqrt{5}-1)^2 =$ ______

j $(\sqrt{2}+\sqrt{3})^2 =$ ______

k $(\sqrt{3}-\sqrt{2})^2 =$ ______

l $(2\sqrt{5}+3\sqrt{2})^2 =$ ______

m $(2\sqrt{5}+1)^2 =$ ______

n $(2\sqrt{7}-1)^2 =$ ______

QUESTION **3** Find the product of each pair of conjugate surds.

a $(\sqrt{3}+\sqrt{2})(\sqrt{3}-\sqrt{2}) =$ ______

b $(\sqrt{7}+\sqrt{2})(\sqrt{7}-\sqrt{2}) =$ ______

c $(9+2\sqrt{7})(9-2\sqrt{7}) =$ ______

d $(\sqrt{5}+\sqrt{3})(\sqrt{5}-\sqrt{3}) =$ ______

e $(\sqrt{11}-\sqrt{7})(\sqrt{11}+\sqrt{7}) =$ ______

f $(3\sqrt{2}+3)(3\sqrt{2}-3) =$ ______

g $(\sqrt{7}-2)(\sqrt{7}+2) =$ ______

h $(5+\sqrt{5})(5-\sqrt{5}) =$ ______

i $(8-2\sqrt{2})(8+2\sqrt{2}) =$ ______

j $(2\sqrt{11}+5)(2\sqrt{11}-5) =$ ______

k $(3\sqrt{7}+3)(3\sqrt{7}-3) =$ ______

l $(5\sqrt{2}-1)(5\sqrt{2}+1) =$ ______

m $(3\sqrt{6}+7)(3\sqrt{6}-7) =$ ______

n $(8\sqrt{3}+5)(8\sqrt{3}-5) =$ ______

Excel Advanced-level Mathematics Study Guide Years 9–10
Pages 34–48

UNIT 10: Rationalising the denominator (1)

QUESTION 1 Rationalise the denominator of the following.

a $\frac{1}{\sqrt{6}} =$ ______

b $\frac{3}{\sqrt{7}} =$ ______

c $\frac{2}{\sqrt{2}} =$ ______

d $\frac{5}{\sqrt{11}} =$ ______

e $\frac{7}{\sqrt{5}} =$ ______

f $\frac{6}{\sqrt{15}} =$ ______

g $\frac{\sqrt{3}}{\sqrt{5}} =$ ______

h $\frac{\sqrt{2}}{\sqrt{7}} =$ ______

i $\frac{\sqrt{5}}{\sqrt{11}} =$ ______

j $\frac{3\sqrt{2}}{\sqrt{5}} =$ ______

k $\frac{2\sqrt{7}}{\sqrt{3}} =$ ______

l $\frac{7\sqrt{2}}{\sqrt{5}} =$ ______

QUESTION 2 Simplify the following by rationalising the denominator.

a $\frac{3}{5\sqrt{3}} =$ ______

b $\frac{5}{2\sqrt{2}} =$ ______

c $\frac{7}{3\sqrt{3}} =$ ______

d $\frac{9}{5\sqrt{5}} =$ ______

e $\frac{4\sqrt{2}}{3\sqrt{7}} =$ ______

f $\frac{5\sqrt{3}}{2\sqrt{2}} =$ ______

g $\frac{3\sqrt{2}}{4\sqrt{5}} =$ ______

h $\frac{8\sqrt{10}}{9\sqrt{3}} =$ ______

i $\frac{2\sqrt{7}}{3\sqrt{6}} =$ ______

j $\frac{\sqrt{5}-\sqrt{3}}{\sqrt{7}} =$ ______

k $\frac{\sqrt{2}-1}{\sqrt{2}} =$ ______

l $\frac{\sqrt{3}+2}{\sqrt{3}} =$ ______

Rational and irrational numbers

UNIT 11: Rationalising the denominator (2)

QUESTIONS **1** Rationalise the denominator of the following.

a $\dfrac{1}{\sqrt{2}+1} =$

b $\dfrac{1}{\sqrt{3}-1} =$

c $\dfrac{1}{\sqrt{5}-1} =$

d $\dfrac{3}{\sqrt{2}-1} =$

e $\dfrac{5}{\sqrt{5}+2} =$

f $\dfrac{2}{\sqrt{7}-\sqrt{5}} =$

g $\dfrac{3}{\sqrt{5}-\sqrt{2}} =$

h $\dfrac{8}{\sqrt{7}-\sqrt{3}} =$

i $\dfrac{1}{2\sqrt{3}+8} =$

j $\dfrac{2}{5-\sqrt{3}} =$

k $\dfrac{3}{5+\sqrt{7}} =$

l $\dfrac{2}{\sqrt{3}+\sqrt{5}} =$

m $\dfrac{\sqrt{5}}{\sqrt{6}+2} =$

n $\dfrac{7\sqrt{3}}{3\sqrt{5}-2\sqrt{6}} =$

QUESTION **2** Rationalise the denominator, giving answers in simplest form.

a $\dfrac{\sqrt{2}+1}{\sqrt{2}-1} =$

b $\dfrac{\sqrt{3}+\sqrt{2}}{\sqrt{3}-\sqrt{2}} =$

c $\dfrac{\sqrt{10}+5}{3-\sqrt{10}} =$

d $\dfrac{2\sqrt{7}+3}{8-3\sqrt{7}} =$

Excel Advanced-level Mathematics Study Guide Years 9–10
Pages 34–48

UNIT 12: Surds and indices

QUESTION **1** Complete the following.

a **i** $5 \times 5 =$ ______ **ii** $-5 \times -5 =$ ______

b **i** $(7)^2 =$ ______ **ii** $(-7)^2 =$ ______

c **i** $(2.5)^2 =$ ______ **ii** $(-2.5)^2 =$ ______

d When two positive numbers are multiplied, the result is always ______.

e When two negative numbers are multiplied, the result is always ______.

QUESTION **2** Complete the following.

a **i** $5 \times 5 \times 5 =$ ______ **ii** $-5 \times -5 \times -5 =$ ______

b **i** $(5)^3 =$ ______ **ii** $(-5)^3 =$ ______

c **i** $(2.5)^3 =$ ______ **ii** $(-2.5)^3 =$ ______

d The product of three positive numbers is always a ______ number.

e The product of three negative numbers is always a ______ number.

QUESTION **3** Write these surds in index form.

a $\sqrt{7^3} =$ ______ **b** $\left(\sqrt{5}\right)^4 =$ ______ **c** $\sqrt[3]{10^5} =$ ______

d $\sqrt[3]{22^2} =$ ______ **e** $\left(\sqrt[5]{18}\right)^2 =$ ______ **f** $\sqrt[7]{16^3} =$ ______

QUESTION **4** Write the following as surds.

a $17^{\frac{1}{2}} =$ ______ **b** $(-10)^{\frac{1}{3}} =$ ______ **c** $18^{\frac{1}{5}} =$ ______

d $12^{\frac{1}{3}} =$ ______ **e** $24^{\frac{1}{4}} =$ ______ **f** $28^{\frac{1}{6}} =$ ______

QUESTION **5** Use your calculator to evaluate correct to one decimal place.

a $\sqrt{196} =$ ______ **b** $\sqrt[3]{1728} =$ ______ **c** $\sqrt{289} =$ ______

d $\sqrt[4]{625} =$ ______ **e** $\sqrt[5]{128} =$ ______ **f** $\sqrt[6]{542} =$ ______

QUESTION **6** Use your calculator to evaluate correct to three significant figures.

a $\sqrt[3]{146} =$ ______ **b** $\sqrt[4]{238} =$ ______ **c** $\sqrt[7]{649} =$ ______

d $\sqrt[6]{864} =$ ______ **e** $\sqrt{523} =$ ______ **f** $\sqrt[5]{988} =$ ______

QUESTION **7** Simplify.

a $x^{\frac{1}{2}} \times x^{\frac{1}{2}} =$ ______ **b** $a^{\frac{1}{3}} \times a^{\frac{2}{3}} \times a^{\frac{1}{3}} =$ ______ **c** $\left(y^{\frac{1}{5}}\right)^5 =$ ______

d $\left(m^{16}\right)^{\frac{1}{4}} =$ ______ **e** $n^{\frac{5}{6}} \div n^{\frac{5}{6}} =$ ______ **f** $(16a^4b^4)^{\frac{1}{4}} =$ ______

QUESTION **8** Use your calculator to find:

a $81^{-\frac{1}{4}} =$ ______ **b** $27^{-\frac{2}{3}} =$ ______ **c** $64^{-\frac{1}{3}} =$ ______

d $125^{-\frac{1}{3}} =$ ______ **e** $32^{-\frac{3}{5}} =$ ______ **f** $\dfrac{5^{-2}}{4^{-2}} =$ ______

UNIT 13: Fractional indices (1)

QUESTION 1 Express the following in root form.

a $5^{\frac{1}{2}} =$ ________ b $7^{\frac{1}{2}} =$ ________ c $m^{\frac{1}{2}} =$ ________

d $2^{\frac{1}{3}} =$ ________ e $6^{\frac{1}{3}} =$ ________ f $11^{\frac{1}{4}} =$ ________

g $x^{\frac{1}{6}} =$ ________ h $n^{\frac{1}{10}} =$ ________ i $a^{\frac{2}{3}} =$ ________

j $p^{\frac{3}{4}} =$ ________ k $q^{\frac{4}{3}} =$ ________ l $k^{\frac{2}{5}} =$ ________

QUESTION 2 Write the following in index form.

a $\sqrt{5} =$ ________ b $\sqrt{3} =$ ________ c $\sqrt{15} =$ ________

d $\sqrt{43} =$ ________ e $\sqrt{7} =$ ________ f $\sqrt[3]{6} =$ ________

g $\sqrt[3]{28} =$ ________ h $\sqrt[3]{240} =$ ________ i $\sqrt[4]{10} =$ ________

j $\sqrt[5]{37} =$ ________ k $\sqrt[8]{23} =$ ________ l $\sqrt[n]{14} =$ ________

QUESTION 3 Evaluate the following.

a $\sqrt{4} =$ ________ b $\sqrt{25} =$ ________ c $\sqrt{49} =$ ________

d $\sqrt{81} =$ ________ e $\sqrt[3]{27} =$ ________ f $\sqrt[3]{64} =$ ________

g $64^{\frac{1}{2}} =$ ________ h $9^{\frac{1}{2}} =$ ________ i $36^{\frac{1}{2}} =$ ________

j $8^{\frac{1}{3}} =$ ________ k $64^{\frac{1}{3}} =$ ________ l $125^{\frac{1}{3}} =$ ________

QUESTION 4 Evaluate the following.

a $16^{\frac{1}{4}} =$ ________ b $8^{\frac{2}{3}} =$ ________ c $\left(81a^{\frac{1}{2}}\right)^{\frac{1}{2}} =$ ________

d $\left(27^{\frac{2}{3}}\right)^{2} =$ ________ e $9^{\frac{3}{2}} =$ ________ f $\left(9x^{\frac{3}{2}}\right)^{2} =$ ________

g $(125)^{\frac{2}{3}} =$ ________ h $(25^{2})^{\frac{1}{2}} =$ ________ i $(x^{9})^{\frac{1}{3}} =$ ________

j $25^{\frac{3}{2}} =$ ________ k $16^{\frac{3}{4}} =$ ________ l $(y^{0})^{\frac{2}{3}} =$ ________

QUESTION 5 Evaluate the following.

a $36^{\frac{1}{2}} \times 2 \times 36^{\frac{1}{2}} =$ ________ b $8^{\frac{1}{3}} \times 8^{\frac{1}{3}} \times 25^{\frac{1}{2}} \times 25^{\frac{1}{2}} =$ ________ c $\left(7^{\frac{2}{3}}\right)^{\frac{3}{2}} =$ ________

d $\left(\frac{4}{25}\right)^{\frac{1}{2}} \times \left(\frac{8}{27}\right)^{\frac{1}{3}} =$ ________ e $\left(\sqrt{10}\right)^{2} =$ ________ f $\left(\sqrt[3]{27}\right)^{2} =$ ________

g $8^{0} \times \left(8^{\frac{2}{3}}\right)^{3} =$ ________ h $4^{\frac{1}{2}} \times 4^{\frac{1}{2}} + 8^{\frac{1}{3}} \times 27^{\frac{1}{3}} =$ ________ i $\left(5^{\frac{1}{3}}\right)^{3} + \left(\frac{4}{25}\right)^{\frac{1}{2}} =$ ________

j $\left(\frac{64}{25}\right)^{\frac{1}{2}} \times \left(\frac{8}{27}\right)^{\frac{1}{3}} =$ ________ k $25^{\frac{1}{2}} \times 25^{\frac{3}{2}} + 16^{\frac{1}{4}} \times 8^{\frac{1}{3}} =$ ________ l $\left(\sqrt{729}\right)^{\frac{1}{3}} =$ ________

QUESTION 6 Write the following in simplest index form.

a $x^{\frac{1}{3}} \times x^{\frac{1}{3}} \times x^{\frac{1}{3}} =$ ________ b $y^{\frac{1}{2}} \times y^{\frac{1}{2}} \times y^{\frac{1}{2}} =$ ________ c $m^{\frac{1}{3}} \times m^{\frac{1}{3}} \times m^{\frac{1}{3}} \times m^{\frac{1}{3}} =$ ________

d $8x^{\frac{1}{2}} \times 3x^{\frac{1}{4}} =$ ________ e $p^{\frac{2}{3}} \times p^{\frac{1}{3}} \times p^{\frac{4}{3}} =$ ________ f $\left(x^{\frac{2}{3}}\right)^{\frac{3}{4}} =$ ________

g $8^{\frac{7}{3}} =$ ________ h $2y^{\frac{1}{3}} \times 8y^{\frac{4}{3}} =$ ________ i $(a^{16}b^{8})^{\frac{1}{4}} =$ ________

j $(64a^{16})^{\frac{1}{2}} =$ ________ k $(49a^{8}b^{12})^{\frac{1}{2}} =$ ________ l $\left(\frac{y^{15}}{y^{7}}\right)^{\frac{3}{8}} =$ ________

UNIT 14: Fractional indices (2)

QUESTION 1 Express in surd form.

a $25^{\frac{1}{2}} =$ ______ **b** $64^{\frac{1}{3}} =$ ______ **c** $a^{\frac{1}{6}} =$ ______

d $8^{\frac{2}{3}} =$ ______ **e** $(5x)^{\frac{1}{3}} =$ ______ **f** $(125m)^{\frac{1}{3}} =$ ______

QUESTION 2 Express in index form.

a $\sqrt{11} =$ ______ **b** $\sqrt{x} =$ ______ **c** $\sqrt[3]{216a} =$ ______

d $\sqrt[7]{65} =$ ______ **e** $\sqrt[n]{20} =$ ______ **f** $\dfrac{1}{\sqrt{21}} =$ ______

QUESTION 3 Simplify the following.

a $64^{\frac{2}{3}} =$ ______ **b** $(512)^{\frac{4}{9}} =$ ______ **c** $(216)^{\frac{1}{3}} =$ ______

d $(49a^6)^{\frac{1}{2}} =$ ______ **e** $(a^8b^8)^{\frac{1}{4}} =$ ______ **f** $(27y^{12})^{\frac{2}{3}} =$ ______

QUESTION 4 Simplify.

a $a^{\frac{2}{3}} \times a^{\frac{7}{9}} =$ ______ **b** $5m^{\frac{1}{2}} \times 6m^{\frac{3}{4}} =$ ______ **c** $x^{\frac{1}{3}} \div x^{\frac{1}{5}} =$ ______

d $\left(y^{\frac{2}{3}}\right)^{\frac{3}{2}} =$ ______ **e** $(a^{14}b^{10})^{\frac{1}{2}} =$ ______ **f** $(243a^{20})^{\frac{1}{5}} =$ ______

QUESTION 5 Simplify the following.

a $x^{\frac{1}{3}} \times x^4 \times x^{\frac{1}{3}} =$ ______ **b** $y^{\frac{2}{3}} \times y^{\frac{1}{2}} =$ ______ **c** $\left(a^{\frac{1}{4}}\right)^{\frac{2}{3}} =$ ______

d $8a + 5a^{\frac{1}{2}} + 2a =$ ______ **e** $m^{-4} \times m^{\frac{1}{4}} =$ ______ **f** $\left(\dfrac{3}{p^2}\right)^{-\frac{1}{2}} =$ ______

QUESTION 6 Write in simplest form.

a $a^{\frac{7}{8}} \times a^{\frac{8}{7}} =$ ______ **b** $(a^{-2}b^4)^{\frac{1}{2}} =$ ______ **c** $(a^3)^{\frac{1}{3}} + \left(a^{-\frac{1}{3}}\right)^3 =$ ______

d $a^{\frac{4}{3}} \times a^{\frac{2}{3}} \times a^{-\frac{2}{3}} =$ ______ **e** $\sqrt[5]{a^4} \div \sqrt[4]{a^5} =$ ______ **f** $(x^{-4}y^3)^{-3} =$ ______

QUESTION 7 Given that $a = 8$, $b = 16$ and $c = 64$, evaluate.

a $a^{\frac{2}{3}} + b^{\frac{1}{4}} =$ ______ **b** $(bc)^{\frac{1}{2}} =$ ______ **c** $a^{\frac{2}{3}} + c^{\frac{2}{3}} =$ ______

d $a^{-\frac{1}{3}} \times c^{-\frac{1}{6}} =$ ______ **e** $a^{\frac{2}{3}} \times b^{\frac{1}{4}} =$ ______ **f** $(ab)^{-\frac{2}{7}} \times a^0 =$ ______

Rational and irrational numbers

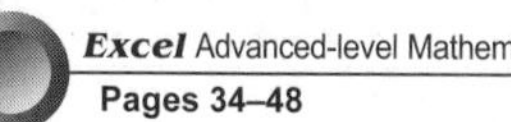

Excel Advanced-level Mathematics Study Guide Years 9–10
Pages 34–48

UNIT 15: Fractional indices and surds

QUESTION 1 Express in surd form.

a $64^{\frac{1}{2}}$ ______ **b** $100^{\frac{1}{2}}$ ______ **c** $121^{\frac{1}{2}}$ ______ **d** $144^{\frac{1}{2}}$ ______ **e** $169^{\frac{1}{2}}$ ______

f $4^{-\frac{1}{2}}$ ______ **g** $9^{-\frac{1}{2}}$ ______ **h** $16^{-\frac{1}{2}}$ ______ **i** $25^{-\frac{1}{2}}$ ______ **j** $36^{-\frac{1}{2}}$ ______

k $8^{\frac{1}{3}}$ ______ **l** $27^{\frac{1}{3}}$ ______ **m** $64^{\frac{1}{6}}$ ______ **n** $16^{\frac{1}{4}}$ ______ **o** $x^{\frac{1}{5}}$ ______

p $y^{\frac{1}{6}}$ ______ **q** $(3p)^{\frac{1}{4}}$ ______ **r** $(9x)^{\frac{1}{3}}$ ______ **s** $(5x)^{\frac{1}{2}}$ ______ **t** $(8m)^{\frac{1}{7}}$ ______

QUESTION 2 Write in index form.

a $\sqrt{5}$ ______ **b** $\sqrt{7}$ ______ **c** $\sqrt{14}$ ______ **d** $\sqrt{38}$ ______ **e** $\sqrt{46}$ ______

f $\sqrt[3]{26}$ ______ **g** $\sqrt[3]{115}$ ______ **h** $\sqrt[3]{34}$ ______ **i** $\sqrt[3]{61}$ ______ **j** $\sqrt[3]{41}$ ______

k $\sqrt[4]{11}$ ______ **l** $\sqrt[5]{37}$ ______ **m** $\sqrt[6]{82}$ ______ **n** $\sqrt[7]{19}$ ______ **o** $\sqrt[9]{123}$ ______

p $\sqrt[n]{x}$ ______ **q** $\sqrt[y]{p}$ ______ **r** $\sqrt[m]{k}$ ______ **s** $\frac{1}{\sqrt{15}}$ ______ **t** $\frac{1}{\sqrt[4]{48}}$ ______

QUESTION 3 Evaluate.

a $\sqrt{9}$ ______ **b** $\sqrt[3]{27}$ ______ **c** $\sqrt{64}$ ______ **d** $\sqrt[3]{64}$ ______ **e** $\sqrt[4]{81}$ ______

f $4^{\frac{1}{2}}$ ______ **g** $25^{\frac{1}{2}}$ ______ **h** $81^{\frac{1}{2}}$ ______ **i** $\sqrt[3]{216}$ ______ **j** $\sqrt[5]{32}$ ______

k $\sqrt[3]{343}$ ______ **l** $8^{-\frac{1}{3}}$ ______ **m** $1^{\frac{1}{5}}$ ______ **n** $\sqrt[4]{625}$ ______ **o** $\frac{1}{\sqrt{16}}$ ______

p $\frac{1}{\sqrt[4]{256}}$ ______ **q** $\sqrt[6]{64}$ ______ **r** $\sqrt[7]{128}$ ______ **s** $\sqrt[3]{125}$ ______ **t** $\sqrt[5]{243}$ ______

QUESTION 4 Evaluate.

a $8^{\frac{4}{3}}$ ______ **b** $9^{\frac{5}{2}}$ ______ **c** $16^{\frac{3}{4}}$ ______ **d** $32^{\frac{2}{5}}$ ______ **e** $100^{\frac{3}{2}}$ ______

f $625^{\frac{3}{4}}$ ______ **g** $(8^2)^{\frac{3}{2}}$ ______ **h** $\left(36^{\frac{1}{2}}\right)^3$ ______ **i** $\left(8^{\frac{2}{3}}\right)^{\frac{1}{2}}$ ______ **j** $(9^2)^{\frac{1}{2}}$ ______

k $(a^0)^{\frac{1}{7}}$ ______ **l** $\left(27^{\frac{4}{3}}\right)^{\frac{1}{4}}$ ______ **m** $(49x^0)^{\frac{1}{2}}$ ______ **n** $(81a^0)^{\frac{1}{2}}$ ______ **o** $\left(27^{\frac{2}{3}}\right)^2$ ______

p $\left(2^{\frac{5}{2}}\right)^2$ ______ **q** $\left(3^{\frac{4}{3}}\right)^3$ ______ **r** $\left(64^{\frac{2}{3}}\right)^{\frac{1}{4}}$ ______ **s** $(81)^{\frac{1}{4}}$ ______ **t** $\left(9^{\frac{3}{2}}\right)^2$ ______

QUESTION 5 Use a calculator to evaluate correct to 2 decimal places.

a $\sqrt[3]{526}$ ______ **b** $\sqrt{320}$ ______ **c** $\sqrt[4]{25 \times 16}$ ______ **d** $\sqrt[6]{485}$ ______

e $\sqrt[3]{210.3}$ ______ **f** $\sqrt{510}$ ______ **g** $\sqrt[3]{86}$ ______ **h** $\sqrt{10 \times 35}$ ______

i $\sqrt[4]{86 \times 5}$ ______ **j** $\sqrt{98}$ ______ **k** $\sqrt[3]{415}$ ______ **l** $\sqrt[3]{825}$ ______

m $\sqrt[4]{36 \times 85}$ ______ **n** $\sqrt{15}$ ______ **o** $\sqrt{59}$ ______ **p** $\sqrt[5]{328}$ ______

q $\sqrt[4]{963}$ ______ **r** $\sqrt[5]{86 \times 21}$ ______ **s** $\sqrt[6]{39}$ ______ **t** $\sqrt[7]{58 \times 38}$ ______

QUESTION 6 Without using a calculator, simplify.

a $36^{\frac{1}{2}} \times 6 \times 36^{\frac{1}{2}}$ ______ **b** $4^{\frac{1}{2}} \times 4^{\frac{1}{2}} \times 8^{\frac{1}{3}} \times 8^{\frac{1}{3}}$ ______ **c** $(32)^{\frac{2}{5}}$ ______ **d** $\left(\frac{9}{64}\right)^{\frac{1}{2}}$ ______

e $\left(\frac{27}{64}\right)^{\frac{1}{3}}$ ______ **f** $(\sqrt[3]{7})^6$ ______ **g** $(\sqrt[4]{16})^8$ ______ **h** $(\sqrt[3]{8})^6$ ______

i $9^0 \times \left(64^{\frac{1}{3}}\right)^{\frac{3}{2}}$ ______ **j** $32^{\frac{1}{2}} \times 32^{\frac{1}{2}}$ ______ **k** $27^{\frac{1}{3}} \times 27^{\frac{1}{3}}$ ______ **l** $9^{\frac{1}{2}} \times 9^{\frac{1}{2}}$ ______

m $\sqrt{16} \times \sqrt{16}$ ______ **n** $(\sqrt{25})^2$ ______ **o** $8^{\frac{1}{3}} \times 8^{\frac{2}{3}}$ ______ **p** $(\sqrt{49})^2$ ______

q $\left(9^{\frac{1}{2}}\right)^2$ ______ **r** $64^{\frac{2}{3}}$ ______ **s** $\left(81^{\frac{1}{4}}\right)^3$ ______ **t** $\left(125^{\frac{1}{3}}\right)^2$ ______

Rational and irrational numbers

TOPIC TEST PART A

Instructions
- This part consists of 10 multiple-choice questions.
- Fill in only ONE CIRCLE for each question.
- Each question is worth 1 mark.
- Calculators are NOT allowed.

Time allowed: 10 minutes **Total marks: 10**

Marks

1 $6^{\frac{1}{2}}$ equals

Ⓐ 3 Ⓑ $\sqrt{6}$ Ⓒ $\sqrt{3}$ Ⓓ $\frac{1}{6}$ **1**

2 $\sqrt{3} \times \sqrt{5}$ equals

Ⓐ $\sqrt{8}$ Ⓑ $\sqrt{15}$ Ⓒ $8\sqrt{15}$ Ⓓ $15\sqrt{8}$ **1**

3 $2\sqrt{5} - 6 + 3\sqrt{5}$ equals

Ⓐ $-\sqrt{5} - 6$ Ⓑ $-\sqrt{5}$ Ⓒ $5\sqrt{5} - 6$ Ⓓ $6\sqrt{5} - 6$ **1**

4 $5\sqrt{2} \times 3\sqrt{5}$ equals

Ⓐ $15\sqrt{10}$ Ⓑ $8\sqrt{10}$ Ⓒ $8\sqrt{7}$ Ⓓ $2\sqrt{10}$ **1**

5 $\sqrt{18}$ equals

Ⓐ $3\sqrt{2}$ Ⓑ $2\sqrt{3}$ Ⓒ $3\sqrt{6}$ Ⓓ $6\sqrt{3}$ **1**

6 $(\sqrt{a})^2$

Ⓐ $a^{\frac{1}{2}}$ Ⓑ $a^{\frac{1}{4}}$ Ⓒ a Ⓓ $\sqrt{2a}$ **1**

7 If $b = \sqrt{2}$ find the value of $(4b^2)^0$

Ⓐ $4b^2$ Ⓑ 8 Ⓒ 0 Ⓓ 1 **1**

8 Which of the following numbers is not a surd?

Ⓐ $\sqrt{5}$ Ⓑ $\sqrt{3}$ Ⓒ $\sqrt{101}$ Ⓓ $\sqrt{49}$ **1**

9 $\frac{4}{\sqrt{2}} =$

Ⓐ $2\sqrt{2}$ Ⓑ $4\sqrt{2}$ Ⓒ 2 Ⓓ $\sqrt{2}$ **1**

10 Which of the following statements is true? (i) $\sqrt{a} \times \sqrt{b} = \sqrt{ab}$, (ii) $\sqrt{a} + \sqrt{b} = \sqrt{a+b}$

Ⓐ (i) only Ⓑ (ii) only Ⓒ both (i) and (ii) Ⓓ neither **1**

Total marks achieved for PART A

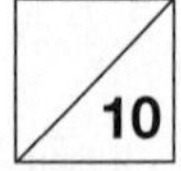

Rational and irrational numbers

TOPIC TEST

PART B

Instructions
- This part consists of 15 questions.
- Write only the answer in the answer column.
- For any working use the question column.
- Calculators are NOT allowed.

Time allowed: 20 minutes **Total marks: 15**

Questions	Answers	Marks
Simplify the following.		
1 $(2\sqrt{3})^2$		1
2 $5\sqrt{2} + 3\sqrt{2} - \sqrt{2}$		1
3 $\sqrt{8} + \sqrt{12} - 6\sqrt{2}$		1
4 Write $\sqrt[3]{x^2}$ in index form.		1
5 Find the value of $9^{\frac{5}{2}}$		1
6 Simplify. $\sqrt{16x^{16}}$		1
7 $35\sqrt{24} \div 5\sqrt{6} =$		1
8 $\sqrt{3}(2\sqrt{3} + 5\sqrt{7}) =$		1
9 $(\sqrt{7} - 5)^2 =$		1
10 Express $\frac{7\sqrt{5}}{2\sqrt{3}}$ with a rational denominator.		1
11 $2\sqrt{3} + 8\sqrt{3} - 5\sqrt{7} + \sqrt{7} =$		1
12 $\sqrt{3} \times 2\sqrt{3} =$		1
13 $\sqrt{20} + \sqrt{45} =$		1
14 Express $\sqrt{7}(\sqrt{3} + \sqrt{7})$ in its simplest surd form.		1
15 Expand and simplify $(5\sqrt{2} - 3)(5\sqrt{2} + 3)$		1

Total marks achieved for PART B /15

CHAPTER 2

Quadratic expressions and equations

Excel Advanced-level Mathematics Study Guide Years 9–10
Pages 49–68

UNIT 1: Review of factorisation

QUESTION 1 Factorise fully.

a $x^2 + 9x$

b $7x^2 + 14yz$

c $x^2 - 16$

d $4x^2y^3 - 6xy^2$

e $8a^4bc^2 + 12abc^2$

f $18xy^2 - 9xy$

QUESTION 2 Factorise these trinomials.

a $x^2 + 7x + 12$

b $x^2 - 5x - 14$

c $x^2 + 3x - 10$

d $x^2 - 9x + 20$

e $x^2 + 11x + 30$

f $x^2 - 10x + 24$

g $x^2 - 10x - 24$

h $x^2 - x - 12$

i $x^2 - 6x + 5$

j $x^2 - 10x + 25$

k $15 + 2x - x^2$

l $42 - x - x^2$

QUESTION 3 Factorise by grouping.

a $p^2 + 9p + pq + 9q$

b $2a^2 + ab - 6a - 3b$

c $3mn - 12m - 2n + 8$

d $2x^2 + 6x + x + 3$

e $10a^2 + 15a - 4a - 6$

f $2xy + 2x + y + 1$

QUESTION 4 Factorise by first taking out a common factor.

a $2x^2 + 8x - 24$

b $3x^2 + 15x + 18$

c $7x^2 + 28x + 28$

d $5x^2 - 10x - 15$

e $6x^2 - 6x - 120$

f $4x^2 - 60x + 216$

g $3x^2 - 27$

h $4x^2 + 24x + 32$

i $2x^2 - 16x + 30$

Excel Advanced-level Mathematics Study Guide Years 9–10
Pages 49–68

UNIT 2: Non-monic, quadratic trinomials

QUESTION **1** Factorise the following.

a $2x^2 + 13x + 15$

b $5t^2 + 4t - 1$

c $2x^2 + 7x + 5$

d $6y^2 + 5y - 6$

e $7x^2 + 11x - 6$

f $12x^2 + 5x - 2$

g $3a^2 + 10a + 3$

h $2a^2 + 5a + 2$

i $3y^2 - 2y - 1$

j $6x^2 + 11x + 4$

k $2a^2 + 7a + 3$

l $2x^2 - 5x + 2$

m $4x^2 + 7x + 3$

n $2y^2 - 5y + 3$

o $2m^2 - 3m - 2$

p $2y^2 + 5y - 3$

q $4x^2 - 27x + 35$

r $3t^2 + 5t - 2$

s $2x^2 - 7x + 6$

t $3m^2 + 7m - 6$

u $8a^2 - 13a - 6$

v $3y^2 - 4y - 20$

w $2n^2 - 5n - 42$

x $4x^2 + x - 5$

Quadratic expressions and equations

UNIT 3: Simple quadratic equations

QUESTION 1 Solve.

a $x^2 = 9$ **b** $x^2 = 16$ **c** $x^2 = 25$ **d** $x^2 = 1$ **e** $x^2 = 4$

f $x^2 = 64$ **g** $x^2 = 36$ **h** $x^2 = 49$ **i** $x^2 = 121$ **j** $x^2 = 144$

k $x^2 - 100 = 0$ **l** $x^2 - 81 = 0$ **m** $x^2 - 169 = 0$ **n** $2x^2 = 800$ **o** $3x^2 = 0.75$

QUESTION 2 Solve, giving the answers in simplest surd form.

a $x^2 = 7$ **b** $x^2 = 6$ **c** $x^2 = 3$ **d** $x^2 = 87$ **e** $x^2 = 93$

f $x^2 = 12$ **g** $x^2 = 72$ **h** $x^2 - 5 = 0$ **i** $x^2 = 18$ **j** $x^2 = 700$

QUESTION 3 Solve, giving each answer correct to two decimal places.

a $x^2 = 23$ **b** $m^2 = 53$ **c** $y^2 = 29$ **d** $n^2 = 67$ **e** $c^2 = 150$

f $k^2 - 19 = 0$ **g** $5y^2 = 29$ **h** $7d^2 - 15 = 0$ **i** $6(p^2 - 1) = 37$ **j** $5x^2 - 50 = 0$

QUESTION 4 Solve.

a $3x^2 - 10 = 0$ **b** $2x^2 = 16$ **c** $16x^2 - 25 = 0$ **d** $4x^2 - 24 = 0$

e $2x^2 - 11 = 0$ **f** $3x^2 = 54$ **g** $6x^2 - 216 = 0$ **h** $5(h^2 - 1) = 20$

Quadratic expressions and equations

UNIT 4: Equations already in factorised form

QUESTION 1 Solve the following quadratic equations that are already expressed in factorised form.

a $(x-2)(x-5)=0$

b $(x-1)(x+9)=0$

c $(x-2)(x-10)=0$

d $x(x+7)=0$

e $5x(x-3)=0$

f $x(4x-5)=0$

g $(x-4)(x-6)=0$

h $(x+4)(x-9)=0$

i $(x+7)(x-9)=0$

j $(x+2)(x-7)=0$

k $(x-6)(x+8)=0$

l $(x-3)(x+8)=0$

m $(x+8)(2x-3)=0$

n $(x+3)(5x-4)=0$

o $(x-2)(x-7)=0$

QUESTION 2 Solve the following quadratic equations.

a $(x-6)(x-11)=0$

b $(x+3)(x-7)=0$

c $(2x-3)(x+5)=0$

d $x(x+8)=0$

e $5x(2x-3)=0$

f $3x(x-4)=0$

g $(x+3)(x+9)=0$

h $3x(4x-5)=0$

i $-3x(x-9)=0$

j $(2x+1)x=0$

k $(x-8)x=0$

l $2x(x-4)=0$

QUESTION 3 Solve the following equations.

a $(x-7)(x-10)=0$

b $(x-8)(x+10)=0$

c $x(x-5)=0$

d $(4x-1)(x+5)=0$

e $(2x+1)(3x-5)=0$

f $3x(x-1)=0$

g $(x-3)(x-7)=0$

h $(2x+4)(3x-7)=0$

i $(x+6)(x-8)=0$

Excel Advanced-level Mathematics Study Guide Years 9–10
Pages 49–68

UNIT 5: Equations involving the difference of two squares

QUESTION 1 Solve the following quadratic equations.

a $x^2 - 9 = 0$

b $x^2 - 25 = 0$

c $x^2 - 121 = 0$

d $x^2 - 16 = 0$

e $x^2 - 64 = 0$

f $x^2 - 144 = 0$

g $x^2 - 36 = 0$

h $x^2 - 81 = 0$

i $x^2 - 169 = 0$

j $x^2 - 49 = 0$

k $x^2 - 100 = 0$

l $x^2 - 225 = 0$

QUESTION 2 Solve the following equations.

a $4x^2 - 49 = 0$

b $4x^2 - 81 = 0$

c $16x^2 - 25 = 0$

d $x^2 - 6\frac{1}{4} = 0$

e $16x^2 - 1 = 0$

f $25x^2 - 64 = 0$

g $36 - x^2 = 0$

h $3x^2 - 27 = 0$

i $36x^2 - 25 = 0$

j $9x^2 - 25 = 0$

k $4x^2 - 16 = 0$

l $(x + 3)^2 - 9 = 0$

m $x^2 - 7 = 0$

n $x^2 - 11 = 0$

o $x^2 - 8 = 0$

UNIT 6: Equations involving a common factor

QUESTION 1 Solve the following quadratic equations.

a $x^2 - 2x = 0$

b $x^2 - 3x = 0$

c $x^2 - 6x = 0$

d $x^2 + 5x = 0$

e $x^2 + 8x = 0$

f $x^2 + 3x = 0$

g $x^2 = 9x$

h $x^2 + 12x = 0$

i $3x^2 = 6x$

j $x^2 - 7x = 0$

k $6x^2 = 12x$

l $8x^2 = 32x$

m $3x^2 = 15x$

n $7x^2 = -14x$

o $3x^2 = -6x$

QUESTION 2 Solve the following equations.

a $x^2 + 13x = 0$

b $6x^2 - 18x = 0$

c $2x^2 - 18x = 0$

d $3x^2 + 9x = 0$

e $5x^2 - 20x = 0$

f $4x^2 - 28x = 0$

g $4x^2 - 12x = 0$

h $2x^2 + 14x = 0$

i $3x^2 - 15x = 0$

j $5x^2 + 10x = 0$

k $2x^2 - 5x = 0$

l $7x^2 - 21x = 0$

Excel Advanced-level Mathematics Study Guide Years 9–10
Pages 49–68

UNIT 7: Equations involving monic, quadratic trinomials

Question 1 Solve the following quadratic equations by factorising.

a $a^2 + 8a - 9 = 0$

b $y^2 - 10y + 21 = 0$

c $p^2 - 8p + 15 = 0$

d $m^2 + 5m - 24 = 0$

e $y^2 - 13y + 42 = 0$

f $x^2 + 3x - 10 = 0$

g $a^2 - 11a + 10 = 0$

h $m^2 + 2m - 35 = 0$

i $y^2 - 12y + 32 = 0$

j $y^2 + 4y - 12 = 0$

k $m^2 + 13m + 36 = 0$

l $y^2 - 13y + 40 = 0$

Question 2 Factorise and solve the following quadratic equations.

a $x^2 - 3x - 18 = 0$

b $x^2 - 15x + 54 = 0$

c $m^2 + 5m - 36 = 0$

d $a^2 + 3a - 28 = 0$

e $x^2 - 11x + 24 = 0$

f $x^2 - 7x + 12 = 0$

g $x^2 - 17x + 72 = 0$

h $x^2 - 3x = 40$

i $x^2 + 4x = 21$

j $x^2 - 9x + 18 = 0$

k $x^2 - 19x = -88$

l $x^2 + 4x = 12$

UNIT 8: Equations involving non-monic, quadratic trinomials

QUESTION **1** Solve the following quadratic equations by factorising.

a $2x^2 + x - 3 = 0$

b $3x^2 + 5x - 2 = 0$

c $2m^2 + 9m + 4 = 0$

d $2p^2 + 9p - 5 = 0$

e $2a^2 + 3a + 1 = 0$

f $2t^2 - 5t - 12 = 0$

g $3x^2 + 13x - 10 = 0$

h $2a^2 + 7a + 6 = 0$

i $3x^2 + 2x - 5 = 0$

j $3x^2 + 11x + 6 = 0$

k $5x^2 - 6x + 1 = 0$

l $4x^2 + 11x + 6 = 0$

QUESTION **2** Factorise and solve the following quadratic equations.

a $3y^2 + 4y - 7 = 0$

b $2m^2 + m - 10 = 0$

c $3p^2 - 16p + 20 = 0$

d $2y^2 = 11y + 21$

e $3t^2 + 2t = 8$

f $2x^2 = -10 + 9x$

g $x(2x + 7) = -6$

h $2y^2 = 3(y + 3)$

i $2y(y - 3) = -4$

j $4x(x + 2) = 5$

k $x(2x + 5) = 42$

l $x(6x + 5) = 6$

UNIT 9: Completing the square

QUESTION 1 Complete.

a $x^2 + 8x + 16 = (x + \_\_\_\_)^2$

b $x^2 - 18x + 81 = (x - \_\_\_\_)^2$

c $a^2 + 22a + \_\_\_\_ = (a + 11)^2$

d $4x^2 + 20x + 25 = (\_\_\_\_ + 5)^2$

e $9n^2 - 12n + \_\_\_\_ = (3n - \_\_\_\_)^2$

f $16p^2 + 8p + \_\_\_\_ = (\_\_\_\_\_\_\_\_)^2$

QUESTION 2 What number must be added to make each of the following a perfect square?

a $x^2 + 12x$ ________

b $x^2 - 20x$ ________

c $x^2 + 16x$ ________

d $x^2 - 24x$ ________

e $x^2 + 2x$ ________

f $x^2 - 5x$ ________

g $4x^2 + 12x$ ________

h $9x^2 + 30x$ ________

i $16x^2 - 56x$ ________

j $9x^2 - 66x$ ________

k $25x^2 + 10x$ ________

l $4x^2 - 36x$ ________

QUESTION 3 Solve the following quadratic equations by completing the square. Where applicable, leave answers in simplest surd form.

a $x^2 + 4x = 12$

b $x^2 = 12x - 27$

c $x^2 - 6x - 4 = 0$

d $x^2 - 8x = 5$

e $x^2 = 10x - 3$

f $x^2 + 14x = 7$

g $x^2 + 20x - 25 = 0$

h $4x^2 + 20x + 22 = 0$

i $9x^2 = 24x - 9$

j $100x^2 - 60x + 1 = 0$

k $25x^2 + 30x = 7$

l $4x^2 - 28x = -37$

UNIT 10: The quadratic formula (1)

QUESTION 1 Use the quadratic formula to solve:

a $x^2 + 40x + 391 = 0$

b $4x^2 + 7x - 2 = 0$

c $6x^2 + 19x - 36 = 0$

QUESTION 2 Use the quadratic formula to solve these equations. Leave each answer in simplest surd form.

a $x^2 + 5x - 2 = 0$

b $x^2 + 7x + 3 = 0$

c $2x^2 - 3x - 4 = 0$

d $x^2 - x - 1 = 0$

e $3x^2 - 7x + 1 = 0$

f $5x^2 - 11x + 5 = 0$

g $x^2 + 4x - 7 = 0$

h $x^2 - 10x - 19 = 0$

i $3x^2 + 2x - 7 = 0$

Quadratic expressions and equations

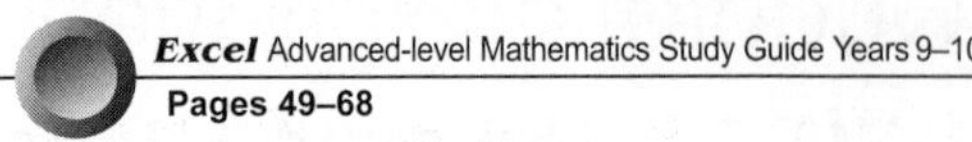

Excel Advanced-level Mathematics Study Guide Years 9–10
Pages 49–68

UNIT 11: The quadratic formula (2)

QUESTION 1 Solve, giving each answer in simplest surd form.

a $2x^2 + 6x - 5 = 0$

b $6x^2 - 10x + 1 = 0$

c $4x^2 + 9x - 18 = 0$

QUESTION 2 Solve, giving each answer correct to three decimal places.

a $3x^2 - 8x + 3 = 0$

b $2x^2 + 7x + 1 = 0$

c $x^2 - 11x + 5 = 0$

d $2x^2 + 9x - 3 = 0$

e $3x^2 - 4x - 5 = 0$

f $5x^2 + 2x - 9 = 0$

g $x^2 + 13x - 7 = 0$

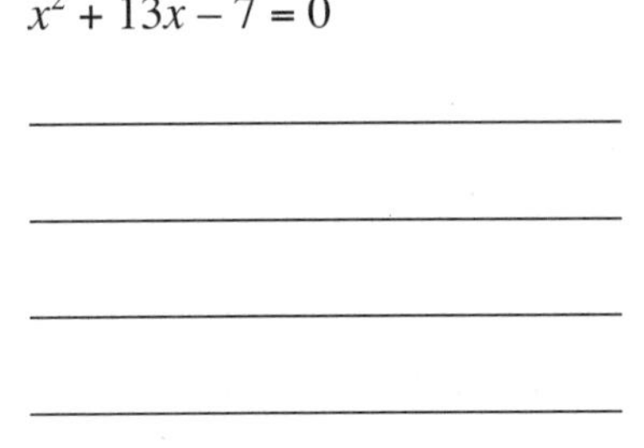

h $4x^2 + 8x - 9 = 0$

i $3x^2 - 5x - 7 = 0$

UNIT 12: Mixed quadratic equations

QUESTION 1 Solve these equations by any suitable method. Leave answers in simplest surd form where necessary.

a $m^2 - 5m - 24 = 0$

b $(3x - 2)(x + 5) = 0$

c $x^2 - 13x + 36 = 0$

d $(x - 2)^2 = 9$

e $y^2 - 7y = 0$

f $8n^2 - 72n = 0$

g $5a^2 + 3a - 1 = 0$

h $x^2 - 7x = 3$

i $3x^2 + 7x - 1 = 0$

j $x^2 = 8 - 7x$

k $x(x + 2) = 8$

l $3x + 1 = x^2 + x$

QUESTION 2 Solve the following quadratic equations. If necessary, give answers correct to two decimal places.

a $6x^2 - 3x - 1 = 0$

b $x^2 - 64 = 0$

c $x^2 - 5x - 5 = 0$

d $4x^2 - 8x + 3 = 0$

e $x^2 - 8x - 4 = 0$

f $4x^2 - 36 = 0$

g $x^2 + 8x = 20$

h $2x^2 - 11x + 3 = 0$

i $2t^2 - 3t - 5 = 0$

UNIT 13: The number of solutions of quadratic equations

QUESTION 1 Solve.

a $x^2 - 14x + 48 = 0$

b $x^2 - 12x + 36 = 0$

QUESTION 2 Use the quadratic formula to show that these equations have no real solution.

a $x^2 + 2x + 7 = 0$

b $3x^2 - 5x + 4 = 0$

QUESTION 3 Complete.

a A quadratic equation can have __________, __________ or __________ solutions.

b If $a \neq 0$ and $b^2 - 4ac > 0$ the equation $ax^2 + bx + c = 0$ will have __________ solutions.

If $a \neq 0$ and $b^2 - 4ac < 0$ the equation $ax^2 + bx + c = 0$ will have __________ solution(s).

If $a \neq 0$ and $b^2 - 4ac = 0$ the equation $ax^2 + bx + c = 0$ will have __________ solution(s).

QUESTION 4 Without solving, determine the number of solutions each quadratic equation below will have.

a $x^2 + 9x - 5 = 0$

b $x^2 - 3x + 6 = 0$

c $x^2 - 28x + 196 = 0$

d $2x^2 + 3x - 7 = 0$

e $3x^2 + 5x + 6 = 0$

f $5x^2 - 4x + 1 = 0$

g $4x^2 + 28x + 49 = 0$

h $7x^2 + 11x + 2 = 0$

i $9x^2 + 8x + 5 = 0$

j $13 + 8x - 2x^2 = 0$

k $x^2 - 12x = 5$

l $6x^2 = 14x + 9$

Quadratic expressions and equations

Excel Advanced-level Mathematics Study Guide Years 9–10
Pages 49–68

UNIT 14: Solving problems (1)

QUESTION 1 The height of an object thrown into the air from the top of a building is given by $h = 16 + 6t - t^2$ where h is the height in metres and t is the time in seconds.

a Find the values of t for which $h = 0$

b Only one of these values of t is a valid one. Which one is it? Briefly explain why the other value is not.

QUESTION 2

a Solve $x^2 + 12x - 64 = 0$

b A rectangle has one side 12 cm longer than the other. The area of the rectangle is 64 cm^2. Explain why the length of the shorter side can be found by solving the equation $x^2 + 12x - 64 = 0$.

c Briefly explain why only one solution is valid.

d What are the dimensions of the rectangle?

QUESTION 3 A rectangular piece of metal is 100 cm long and 36 cm wide. The long sides of the metal are to be folded up to form a gutter as shown in the diagram. The volume of the open gutter must be 16 000 cm^3.

a Show that the value of x can be found by solving the equation $x^2 - 18x + 80 = 0$.

b Find the possible dimensions of the gutter.

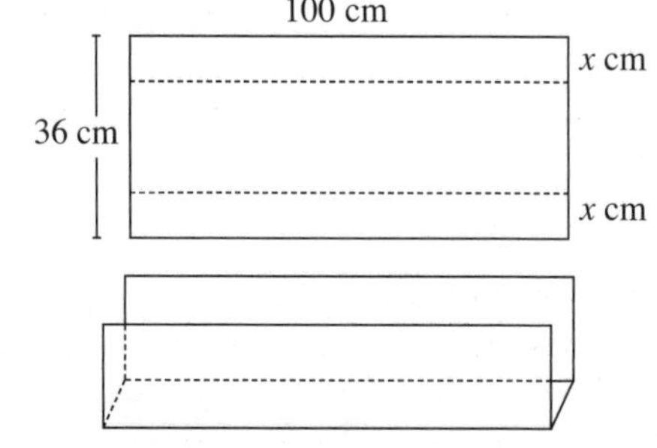

QUESTION 4 The sides of a right-angled triangle (in cm) are x, $x + 3$ and $2x - 3$. Show that there are two possible triangles and find the lengths of each side.

Quadratic expressions and equations

UNIT 15: Solving problems (2)

Question 1 In each of the following diagrams, find x. All measurements are in centimetres.

a

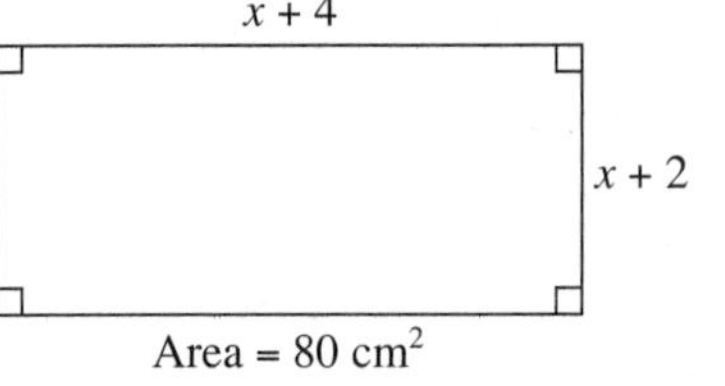

b

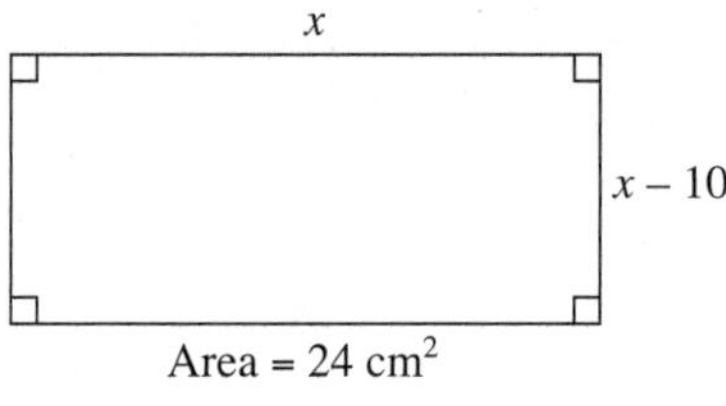

Question 2

a The sum of a positive number and its square is 110. Find the number.

b The sum of two numbers is 17 and their product is 72. Find the numbers.

Question 3

a What are the dimensions of a rectangle that has area 160 cm^2 and perimeter 56 cm?

b The product of two positive, consecutive, odd numbers is 143. What are the numbers?

UNIT 16: Simultaneous equations resulting in a quadratic

QUESTION **1** Solve the following simultaneous equations.

a $x + y = 3$
$xy = 2$

b $x + y = 5$
$xy = 4$

c $y = x^2 + 3x + 7$
$y = x + 10$

d $y = x^2 + 15x + 12$
$y = x - 1$

e $x + y = 7$
$x^2 + y^2 = 85$

f $x + y = 4$
$x^2 + y^2 = 10$

g $y = x$
$y = x^2$

h $y = x + 3$
$y = x^2 - x$

i $3x + y = 9$
$y = x^2 - x - 6$

j $y = 8 - 2x$
$y = 6 - x^2 + x$

k $y = 2x$
$y = x^2$

l $y = 2x - 1$
$y = x^2$

Quadratic expressions and equations

Excel Advanced-level Mathematics Study Guide Years 9–10
Pages 49–68

UNIT 17: Equations reducible to quadratics

QUESTION 1 By substituting u for x^2, solve these equations.

a $x^4 - 10x^2 + 9 = 0$

b $x^4 - 29x^2 + 100 = 0$

c $x^4 - 13x^2 + 36 = 0$

QUESTION 2 Find all possible solutions for these equations.

a $x^4 + 3x^2 - 28 = 0$

b $x^4 - 8x^2 + 15 = 0$

c $x^4 - 11x^2 + 18 = 0$

QUESTION 3 Use a suitable substitution to solve these equations.

a $x^6 + 7x^3 - 8 = 0$

b $x^6 - 16x^3 + 64 = 0$

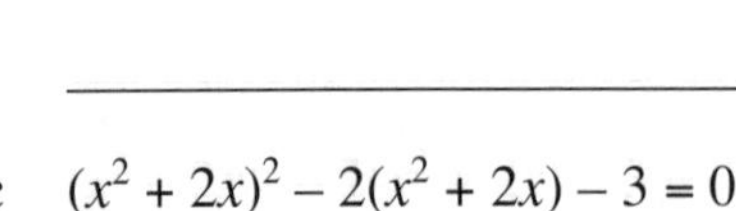

c $(x^2 + 2x)^2 - 2(x^2 + 2x) - 3 = 0$

d $(x^2 - 5x)^2 + 10(x^2 - 5x) + 24 = 0$

Quadratic expressions and equations

TOPIC TEST — PART A

Instructions
- This part consists of 10 multiple-choice questions.
- Fill in only ONE CIRCLE for each question.
- Each question is worth 1 mark.
- Calculators are NOT allowed.

Time allowed: 10 minutes — **Total marks: 10**

Marks

1 If $(x + 2)(x - 3) = 0$ then x is — 1

(A) 2 or –3 (B) –2 or –3 (C) 2 or 3 (D) –2 or 3

2 If $x(x - 2) = 0$ then x is — 1

(A) 2 (B) –2 (C) 0 or 2 (D) 0 or –2

3 If $x^2 - 9 = 0$ then x must be — 1

(A) 0 (B) 3 (C) ± 3 (D) ± 9

4 If $3x^2 - 48 = 0$ then x must be — 1

(A) 16 (B) ± 4 (C) 4 (D) 0

5 If $(x - 5)(4x - 3) = 0$ then x is — 1

(A) 5 or $-\frac{3}{4}$ (B) -5 or $\frac{3}{4}$ (C) 5 or $\frac{3}{4}$ (D) -5 or $-\frac{3}{4}$

6 If $x^2 - x - 5 = 0$ then x is — 1

(A) $\frac{1 \pm \sqrt{21}}{2}$ (B) $\frac{-1 \pm \sqrt{21}}{2}$ (C) $\frac{1 \pm \sqrt{19}}{2}$ (D) $\frac{-1 \pm \sqrt{19}}{2}$

7 Which is a factor of $2x^2 - x - 3$? — 1

(A) $2x - 3$ (B) $2x - 1$ (C) $2x + 1$ (D) $2x + 3$

8 If $4y^2 - 12y + P = (2y + Q)^2$ then — 1

(A) $P = 9, Q = -3$ (B) $P = -9, Q = -3$ (C) $P = 9, Q = 3$ (D) $P = -9, Q = 3$

9 Which one of the following is a perfect square for all values of x? — 1

(A) $x^2 + 49$ (B) $x^2 - 49$ (C) $x^2 - 14x + 49$ (D) $x^2 + 7x + 49$

10 $(x - 2)(x - 3)$ is the same as — 1

(A) $x^2 + 6$ (B) $x^2 + 5x - 6$ (C) $x^2 - 5x + 6$ (D) $x^2 - 5x - 6$

Total marks achieved for PART A

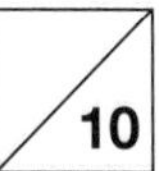

Quadratic expressions and equations

TOPIC TEST — PART B

Instructions
- This part consists of 6 questions.
- Each question part is worth 1 mark.
- Show all working.

Time allowed: 20 minutes — **Total marks: 15**

Marks

1 Factorise fully.

a $x^2 - 4x - 45$ **b** $3x^2 - 30x + 63$ **c** $6x^2 + 11x - 10$

3

2 Solve.

a $(2x - 1)^2 = 0$ **b** $x^2 - 16 = 0$ **c** $x^2 - 15x = 0$

3

3 Solve, by factorising.

a $7x^2 - 28 = 0$ **b** $x^2 - 12x + 27 = 0$ **c** $2x^2 + 9x - 5 = 0$

3

4 Solve.

a $x^2 + 4x - 12 = 0$ by completing the square

b $x^2 - 2x - 5 = 0$ giving the answers in simplest surd form

c $2x^2 + 7x - 8 = 0$ giving each answer correct to three decimal places.

5 **a** Solve $(x + 3)(x + 5) = 8$

b Find two numbers which, when added to their square, give 30

2

6 Solve these simultaneous equations. $x^2 + y^2 = 58$; $x + y = 4$

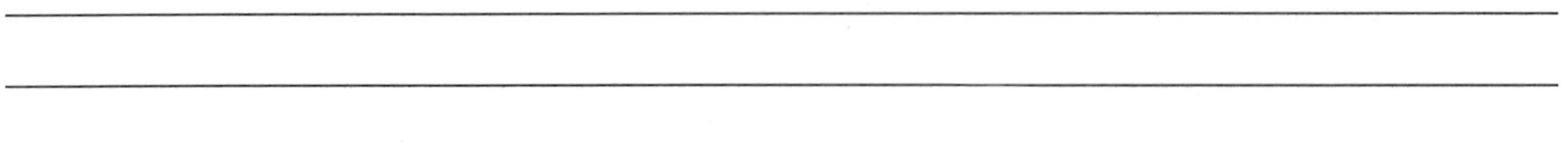

1

Total marks achieved for PART B

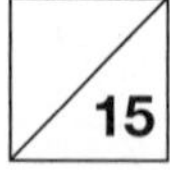

CHAPTER 3

Non-linear relationships

Excel Advanced-level Mathematics Study Guide Years 9–10
Pages 87–102

UNIT 1: Review of linear relationships (1)

QUESTION 1 On the same number plane draw the graphs of the following.

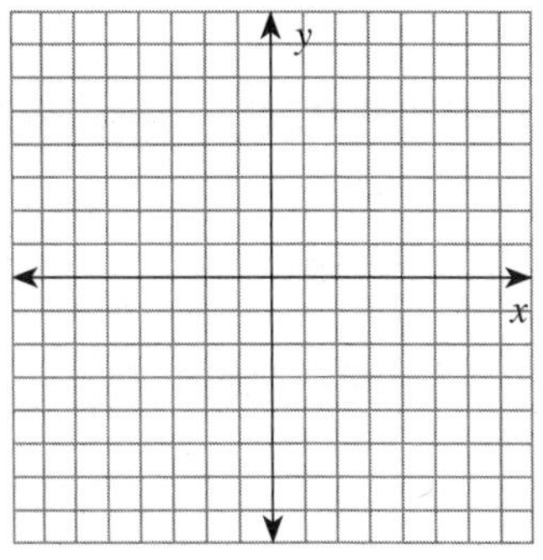

a $y = x$

b $y = -x$

c $y = \frac{1}{2}x$

d $y = 2x$

QUESTION 2 On the same number plane, draw the graphs of the following.

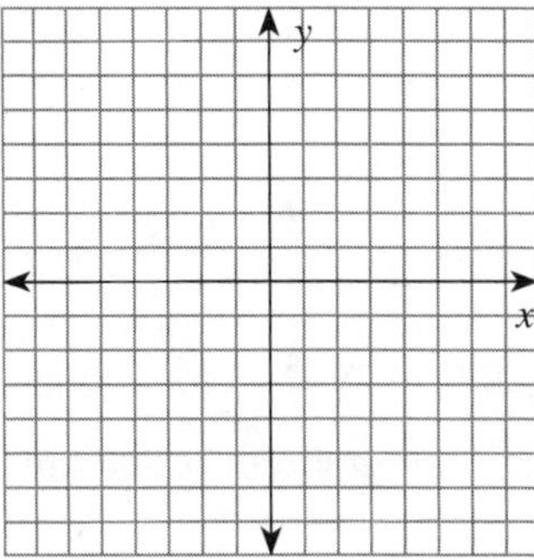

a $y = x$

b $y = x + 1$

c $y = x - 1$

d $y = x + 3$

QUESTION 3 On the same number plane, draw the graphs of the following.

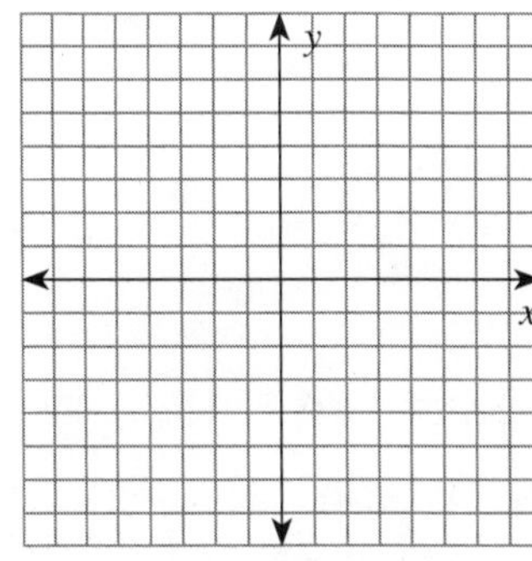

a $x = 0$

b $y = 0$

c $x = 2$

d $y = -3$

QUESTION 4 On the same number plane, draw the graphs of the following.

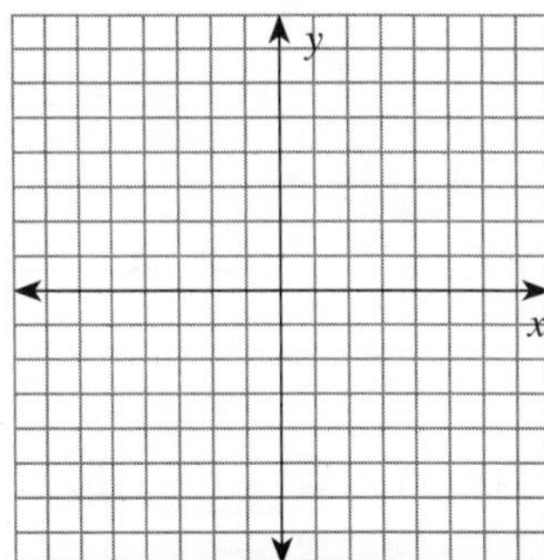

a $x + y = 2$

b $x - y = 3$

c $2x - y = 0$

d $x - 2y = 0$

QUESTION 5 On the same number plane, draw the graphs of the following.

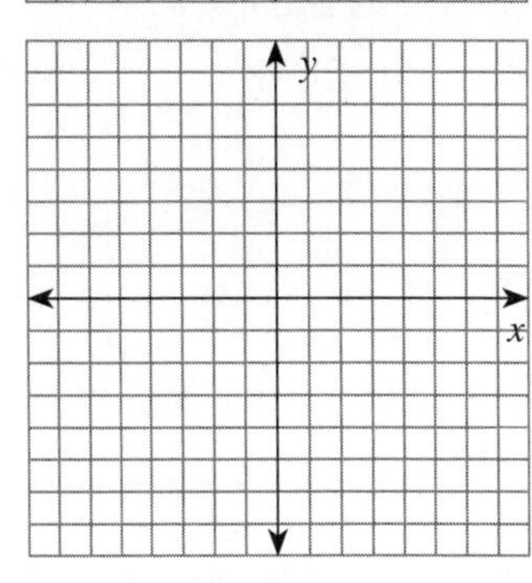

a $y = 2x + 1$

b $y = 2x - 1$

c $x + 2y = 4$

d $2x - 3y = 6$

Excel Advanced-level Mathematics Study Guide Years 9–10
Pages 87–102

UNIT 2: Review of linear relationships (2)

QUESTION 1 Line l has gradient $-\frac{2}{3}$ What is the gradient of any line:

a parallel to l? ____________________ **b** perpendicular to l? ____________

QUESTION 2 P is the point (5, –1) and Q is the point (–3, –5). Find:

a the coordinates of M, the midpoint of PQ

b the gradient of the line joining P and Q

c the distance from P to Q in simplest surd form

QUESTION 3 The equation of a line is $y = \frac{3}{4}x - 2$ What is:

a the gradient of the line?

b the y-intercept?

c the equation of the line in general form?

QUESTION 4 The equation of a line is $4x - 2y + 3 = 0$

a Write the equation in the form $y = mx + b$

b What is the gradient?

c What is the y-intercept?

QUESTION 5 Find the equation, in the form $y = mx + b$, of the line joining:

a (2, 3) and (7, –2)

b (–6, 1) and (4, 6)

Non-linear relationships

Excel Advanced-level Mathematics Study Guide Years 9–10
Pages 87–102

UNIT 3: Drawing quadratic relationships

QUESTION 1 Complete the table of values and then, on the same number plane, draw the graphs of the following.

a Complete

x	-3	-2	-1	0	1	2	3
$y = x^2$							
$y = 3x^2$							
$y = \frac{1}{3}x^2$							

b Explain the effect of the coefficient of x^2 on the graph.

____

____

____

QUESTION 2

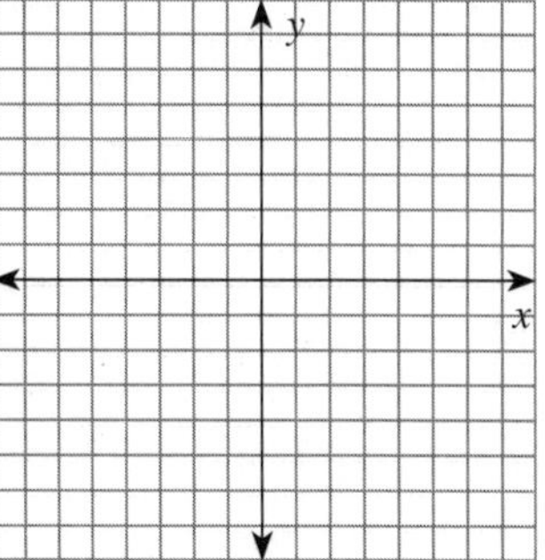

a Sketch the graphs of the following on the same number plane.

i $y = x^2$ **ii** $y = x^2 + 4$ **iii** $y = x^2 - 4$

b Explain how the graphs of $y = x^2 + 4$ and $y = x^2 - 4$ can be drawn using $y = x^2$

____

____

____

QUESTION 3

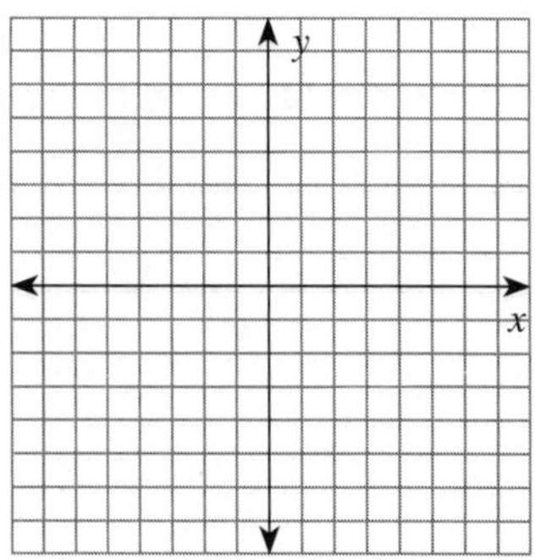

a Sketch the graphs of the following on the same number plane.

i $y = x^2$ **ii** $y = (x - 4)^2$ **iii** $y = (x + 4)^2$

b Explain the effect of the constant p in the equation $y = (x - p)^2$ on the graph.

____

____

____

QUESTION 4 Sketch these graphs on the same number plane.

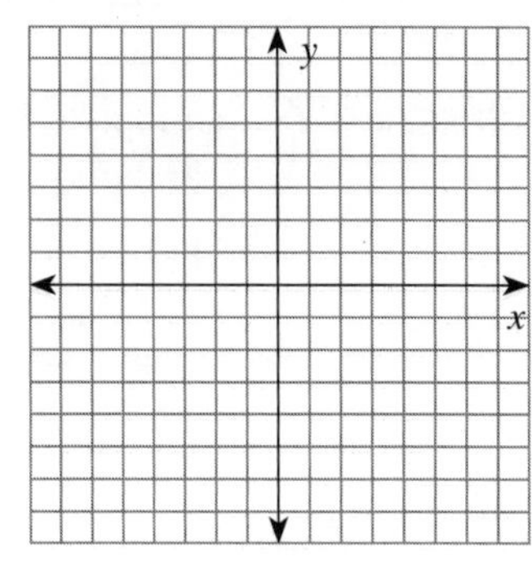

a $y = -x^2$

b $y = 5 - x^2$

c $y = -(5 - x)^2$

d $y = -(5 - x)^2 + 5$

Non-linear relationships

UNIT 4: Axis of symmetry and vertex of a parabola

QUESTION 1 Find where each parabola will cut the x-axis.

a $y = x^2 + 4x - 12$ **b** $y = x^2 - 12x + 27$ **c** $y = 2x^2 - 13x + 15$

QUESTION 2 Use the results of question 1 to find the equation of the axis of symmetry of each parabola.

a $y = x^2 + 4x - 12$ **b** $y = x^2 - 12x + 27$ **c** $y = 2x^2 - 13x + 15$

QUESTION 3 Use the results of question 2 to find the coordinates of the vertex of each parabola.

a $y = x^2 + 4x - 12$ **b** $y = x^2 - 12x + 27$ **c** $y = 2x^2 - 13x + 15$

QUESTION 4 Use the formula $x = -\dfrac{b}{2a}$ to find the equation of the axis of symmetry of each parabola.

a $y = 2x^2 + 8x - 1$ **b** $y = -x^2 + 6x + 9$ **c** $y = 3x^2 - 15x + 4$

QUESTION 5 Use the results of question 4 to find the coordinates of the vertex of each parabola.

a $y = 2x^2 + 8x - 1$ **b** $y = -x^2 + 6x + 9$ **c** $y = 3x^2 - 15x + 4$

QUESTION 6 Write down the coordinates of the vertex for each parabola.

a $y = (x - 1)^2 + 5$ **b** $y = (x + 3)^2 - 2$ **c** $y = (2x - 1)^2 - 3$

QUESTION 7 Complete the square on x for each equation and hence find the coordinates of the vertex of each parabola.

a $y = x^2 - 4x + 11$ **b** $y = 4x^2 + 12x + 5$ **c** $y = x^2 - 5x + 7$

Non-linear relationships

UNIT 5: Parabolas in the form $y = ax^2 + bx + c$

QUESTION 1 For each of the following parabolas:

i find the y-intercept. **ii** find the x-intercepts. **iii** find the axis of symmetry.

iv find the vertex. **v** sketch the graph.

a $y = x^2 - 6x + 8$

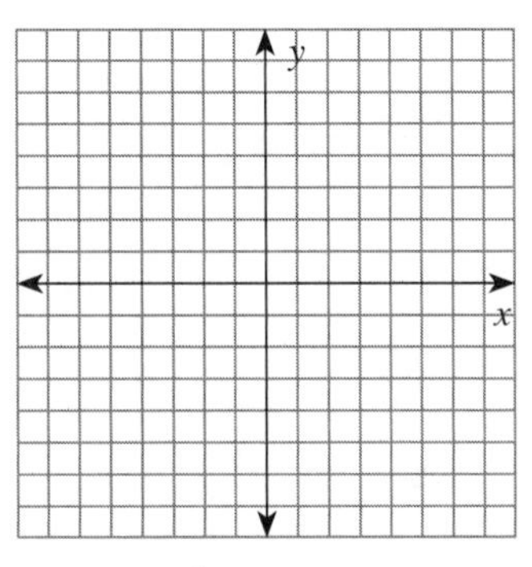

b $y = x^2 - 6x + 5$

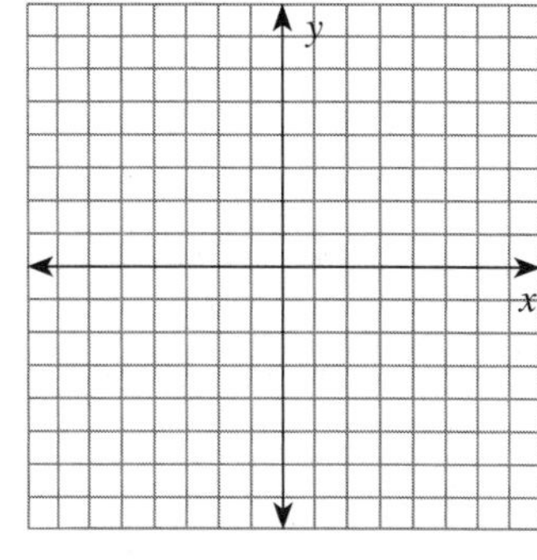

c $y = x^2 - 2x - 3$

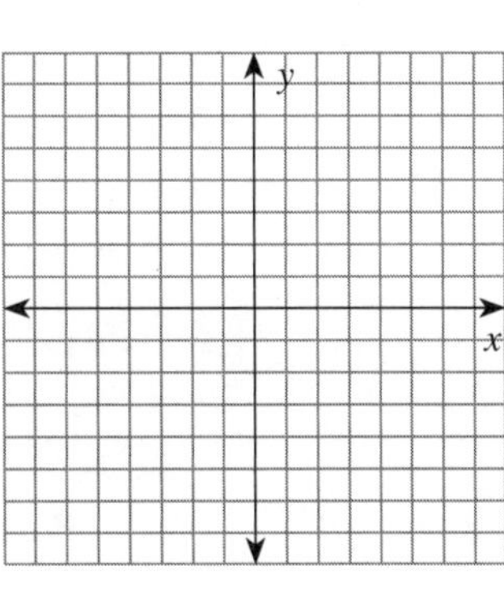

d $y = x^2 - 2x - 8$

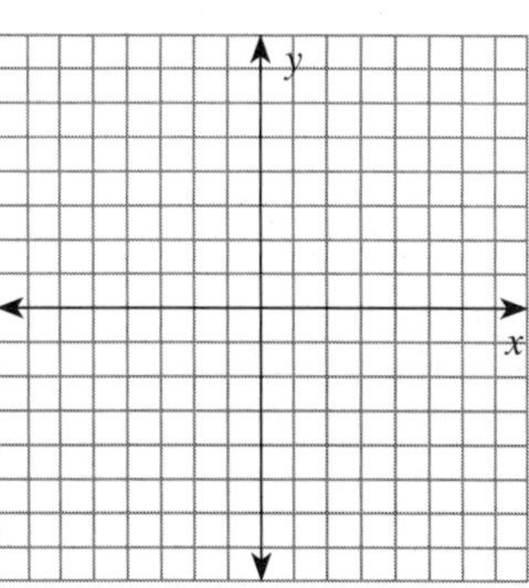

QUESTION 2 Sketch the following parabolas, showing x and y-intercepts, axes of symmetry and vertices.

a $y = x^2 - 4x + 3$

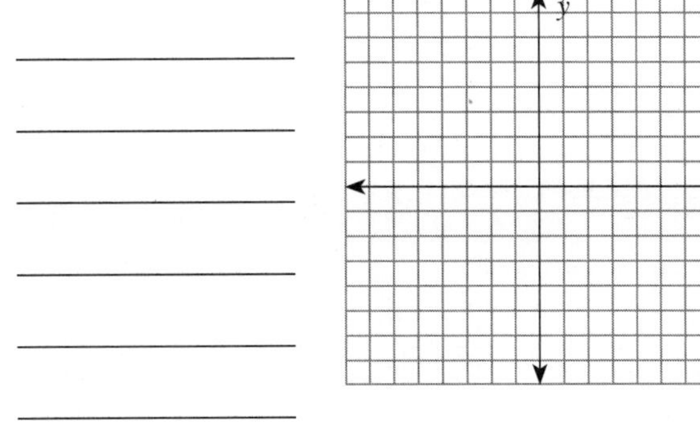

b $y = x^2 - 5x + 6$

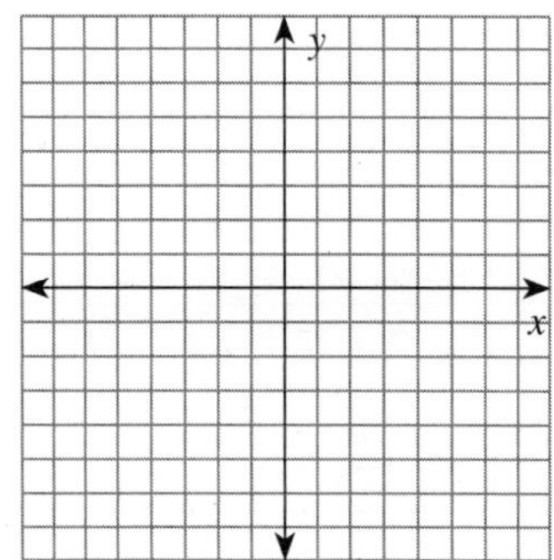

Non-linear relationships

Excel Advanced-level Mathematics Study Guide Years 9–10
Pages 87–102

UNIT 6: Further graphs of parabolas

QUESTION 1 Determine whether each parabola will be concave up or concave down.

a $y = x^2 - 5x + 8$

b $y = 4x^2 + 6x - 2$

c $y = -3x^2 + 8x + 9$

d $y = 12 - 8x - x^2$

e $y = -5x^2 - 8x - 6$

f $y = 17 - 6x + 2x^2$

QUESTION 2 Graph these parabolas, showing essential features.

a $y = 9 - x^2$

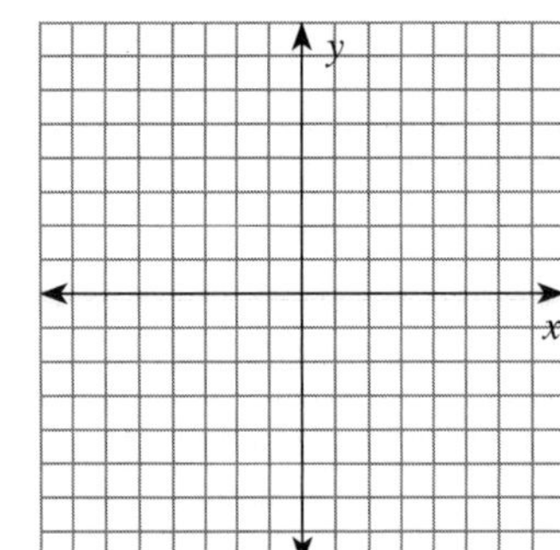

b $y = 4x - x^2$

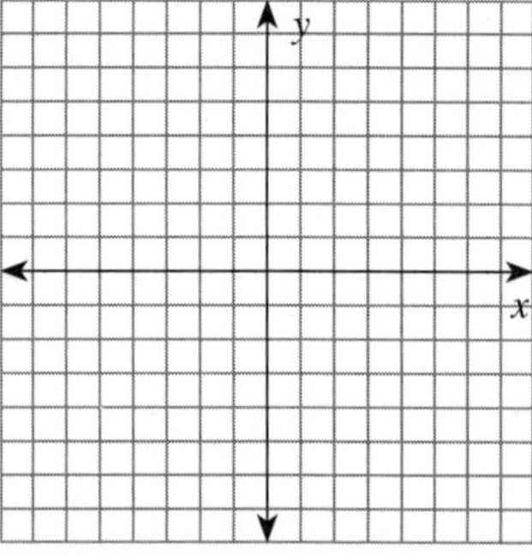

c $y = 2x^2 - 5x - 3$

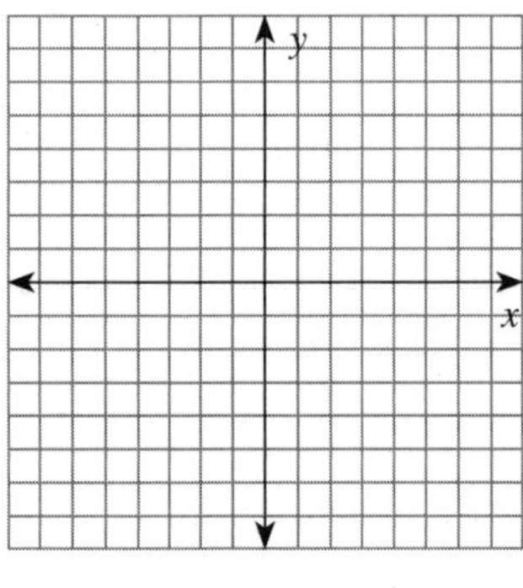

d $y = 4 - 11x - 3x^2$

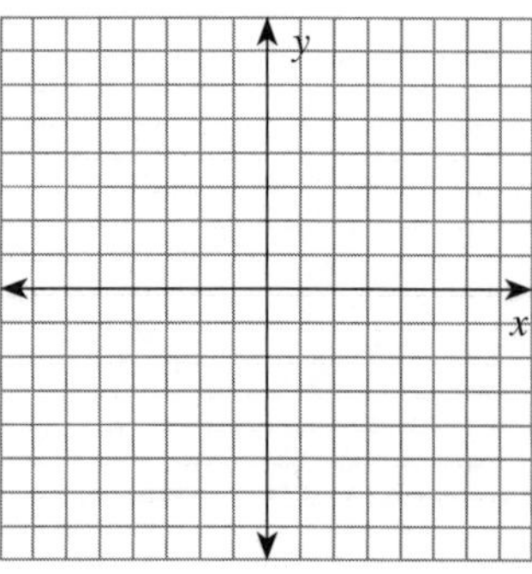

e $y = x^2 - 4x + 5$

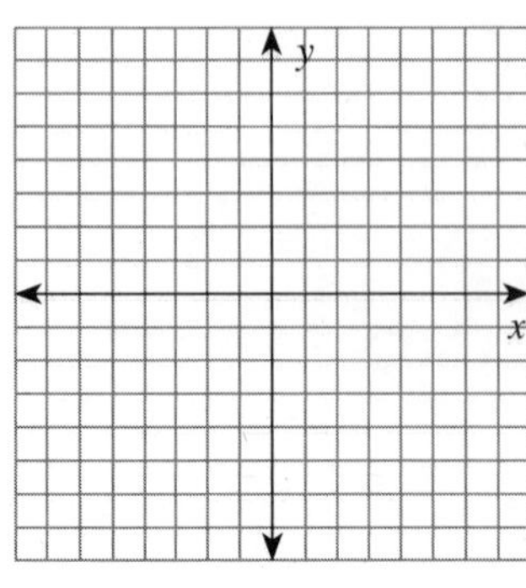

f $y = 4x^2 - 12x + 5$

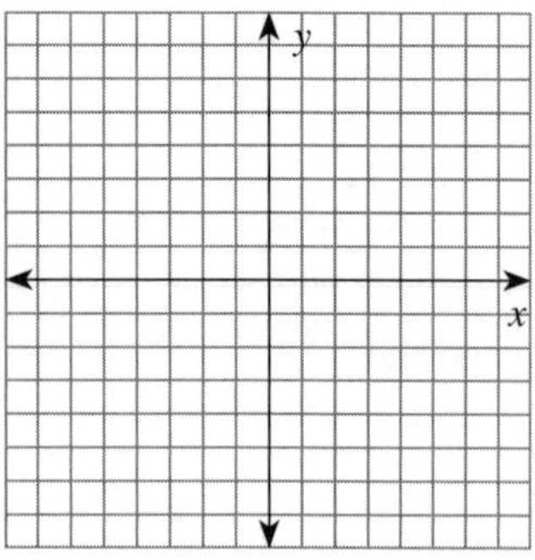

Excel Advanced-level Mathematics Study Guide Years 9–10
Pages 87–102

UNIT 7: Determining equations of parabolas

QUESTION 1 This parabola has equation of the form $y = ax^2 + c$.

a What is the value of c? ____________________

b Find the value of a.

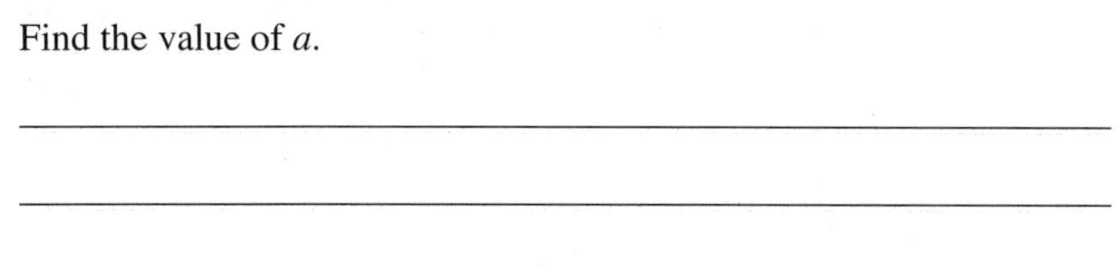

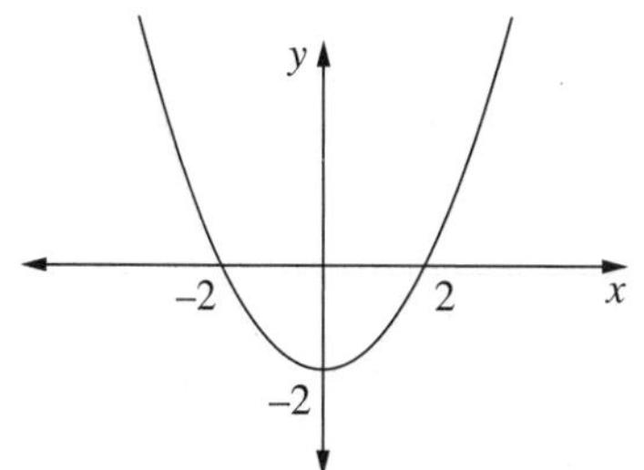

QUESTION 2 Find the equation of each of these parabolas given they are of the form $y = ax^2 + bx + c$.

a

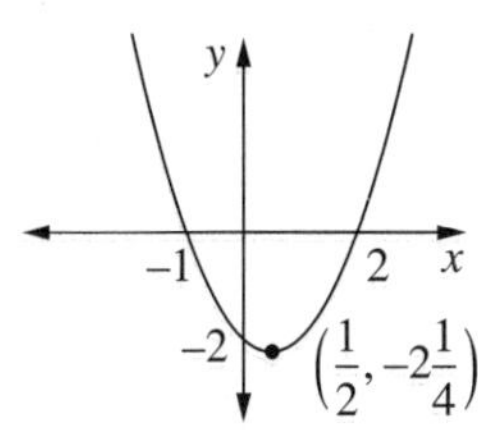

b

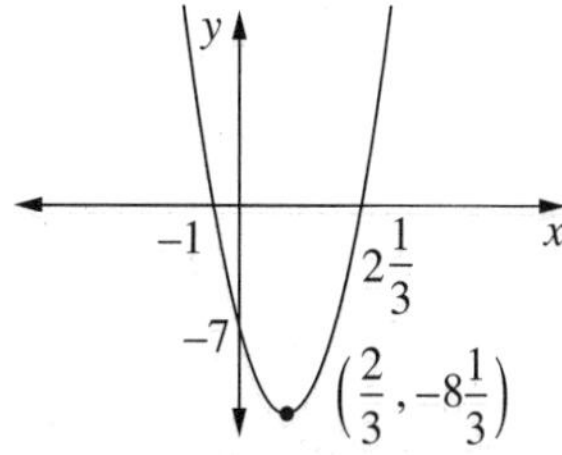

c

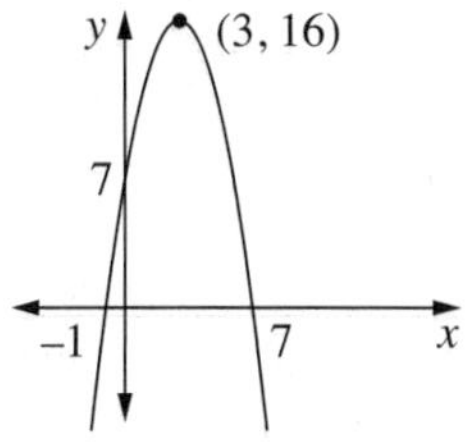

d

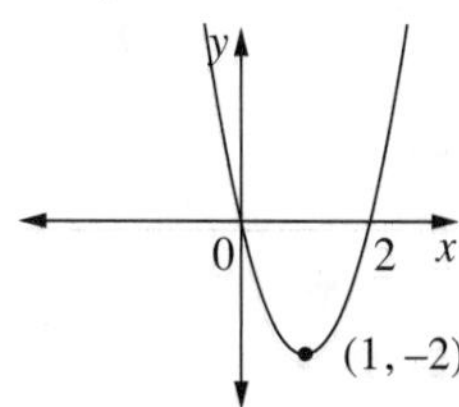

QUESTION 3 Find the equation that is satisfied by all the values for each table of values.

a

x	0	1	2	3	4	5
y	3	4	7	12	19	28

b

x	0	1	2	3	4	5
y	0	0	2	6	12	20

Non-linear relationships

UNIT 8: The hyperbola

Question 1

a Complete the table of values for $y = \frac{1}{x}$

x	-6	-5	-4	-3	-2	-1	$-\frac{1}{2}$	$-\frac{1}{3}$	$-\frac{1}{4}$	$-\frac{1}{5}$	$-\frac{1}{6}$	0	$\frac{1}{6}$	$\frac{1}{5}$	$\frac{1}{4}$	$\frac{1}{3}$	$\frac{1}{2}$	1	2	3	4	5	6
y												✗											

b Explain why there is no value of y when $x = 0$.

c What will happen to the graph when $x = 0$?

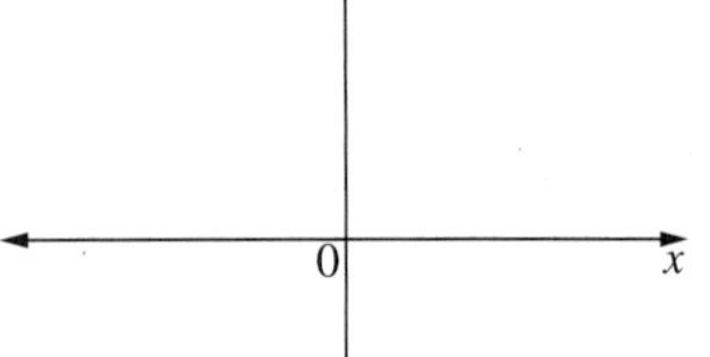

d Sketch the graph of $y = \frac{1}{x}$ on the axes provided.

e What happens to the y-values as the x-values get larger and larger?

f What happens to the y-values as the x-values get closer and closer to 0?

g What are the equations of the asymptotes of the curve $y = \frac{1}{x}$?

Question 2

Sketch the graphs of the following hyperbolas.

a $y = \frac{2}{x}$

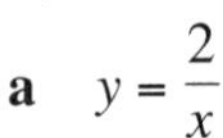

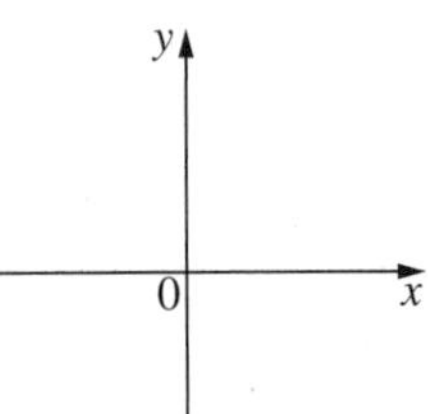

b $y = \frac{3}{x}$

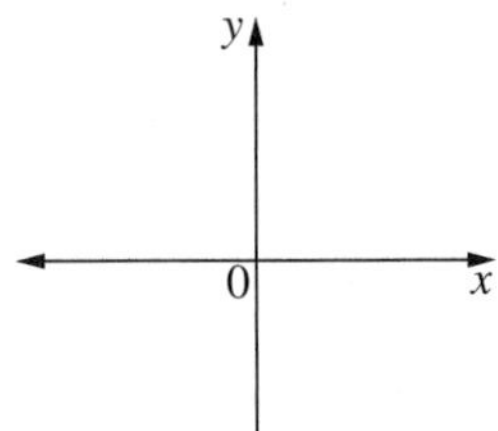

c $y = -\frac{1}{x}$

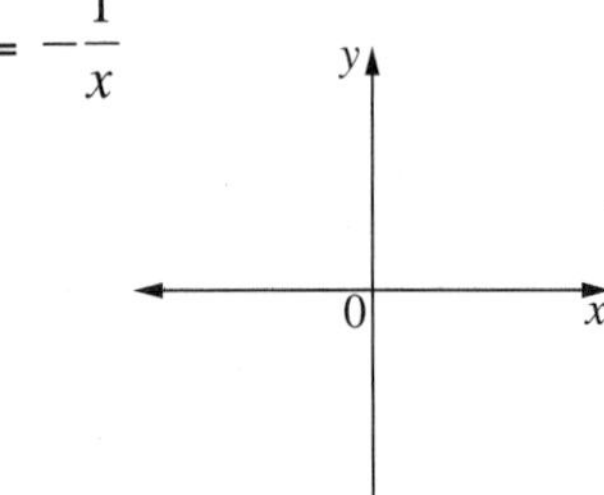

d $y = -\frac{3}{x}$

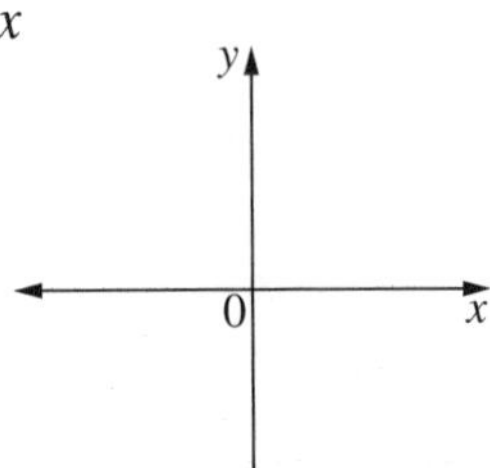

e $y = -\frac{4}{x}$

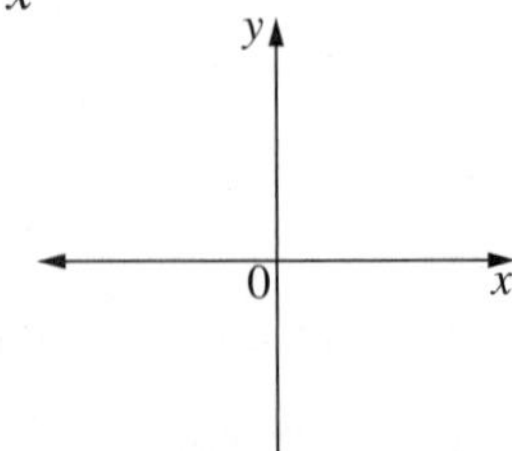

f $y = \frac{8}{x}$

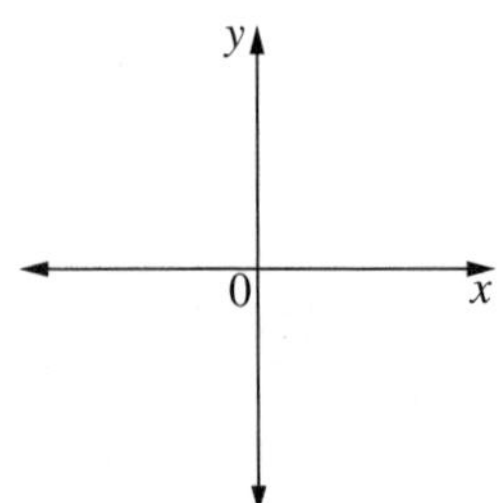

Question 3

Briefly describe the effect on the graph of $y = \frac{1}{x}$ of multiplying by:

a a positive constant

b a negative constant

Non-linear relationships

Excel Advanced-level Mathematics Study Guide Years 9–10
Pages 87–102

UNIT 9: Sketching hyperbolic curves

QUESTION 1 Consider the equations $y = \frac{6}{x}$, $y = \frac{6}{x} - 2$ and $y = \frac{6}{x} + 2$

a Sketch each hyperbola on separate axes.

i $y = \frac{6}{x}$

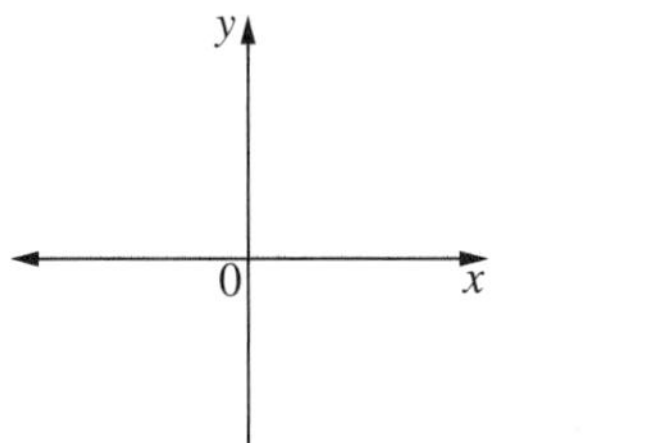

ii $y = \frac{6}{x} - 2$

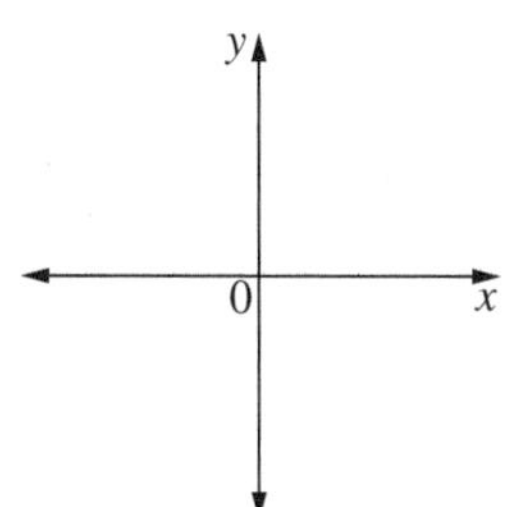

iii $y = \frac{6}{x} + 2$

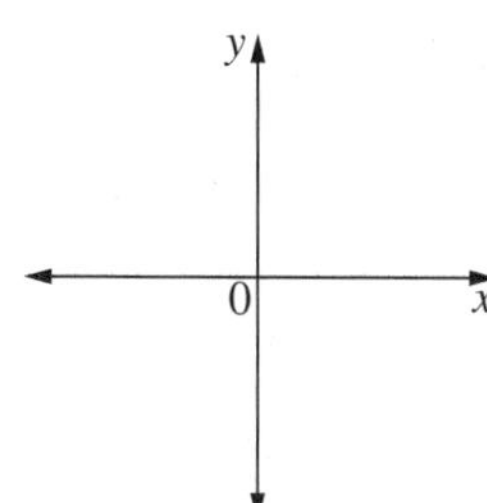

b Write down the equations of the asymptotes for each of the hyperbolas.

i ______________________ **ii** ______________________ **iii** ______________________

c Briefly explain the effect of adding or subtracting a constant on the graph of a hyperbola.

____

____

____

QUESTION 2 Consider the equations $y = \frac{2}{x}$, $y = \frac{2}{x+4}$ and $y = \frac{2}{x-4}$

a Sketch each hyperbola on separate axes.

i $y = \frac{2}{x}$

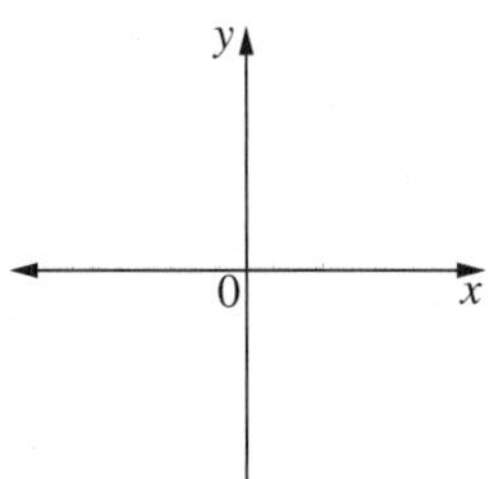

ii $y = \frac{2}{x+4}$

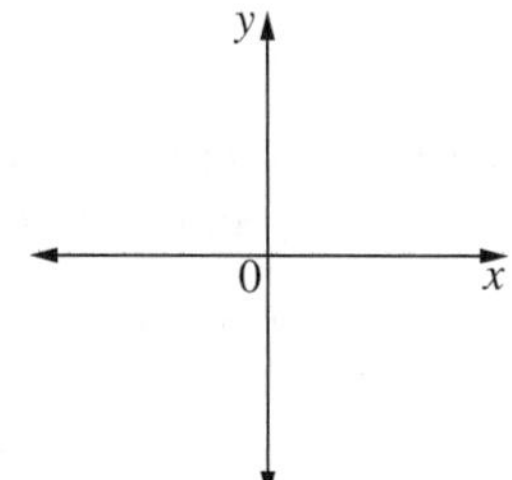

iii $y = \frac{2}{x-4}$

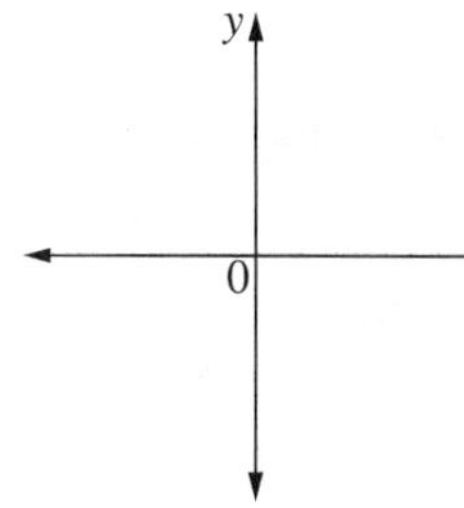

b Write down the equations of the asymptotes for each of the hyperbolas.

i ______________________ **ii** ______________________ **iii** ______________________

c Briefly explain the effect of adding or subtracting a constant to the denominator on the graph of a hyperbola.

____

____

____

Excel Advanced-level Mathematics Study Guide Years 9–10
Pages 87–102

UNIT 10: Further hyperbolas

QUESTION 1 For the hyperbola $y = \dfrac{5}{x - 1}$

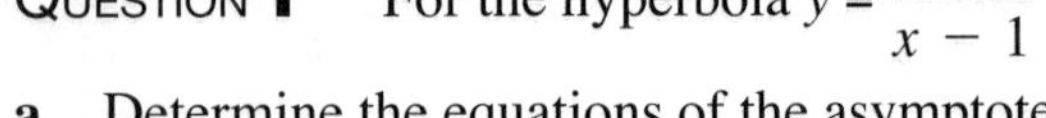

a Determine the equations of the asymptotes.

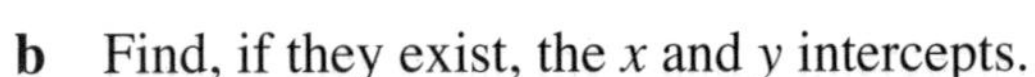

b Find, if they exist, the x and y intercepts.

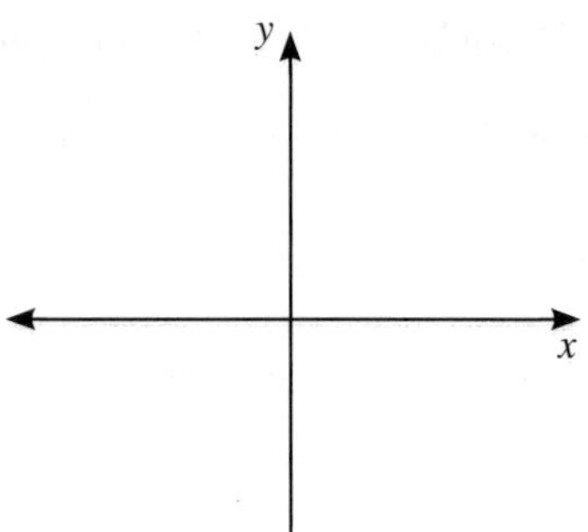

c Sketch the hyperbola on the axes provided showing essential features.

QUESTION 2 For the hyperbola $y = 3 - \dfrac{1}{x}$

a Determine the equations of the asymptotes.

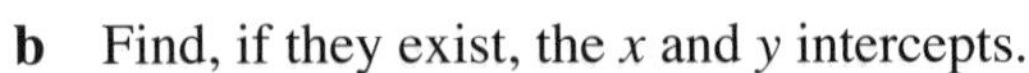

b Find, if they exist, the x and y intercepts.

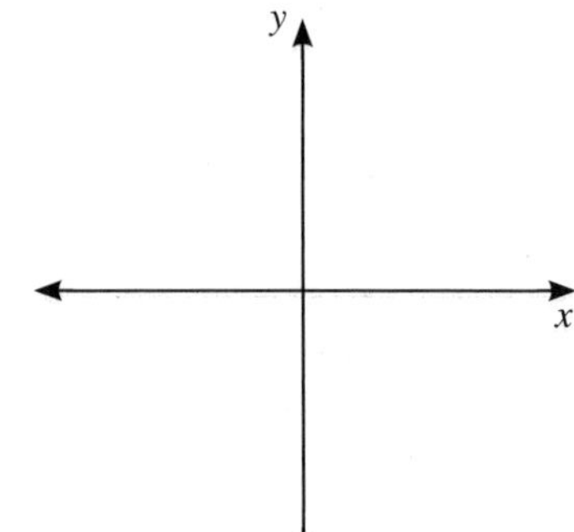

c Sketch the hyperbola on the axes provided showing essential features.

QUESTION 3 Sketch these hyperbolas.

a $y = -\dfrac{2}{x + 1}$ **b** $y = \dfrac{12}{x - 3}$ **c** $y = \dfrac{3}{x + 2} + 1$

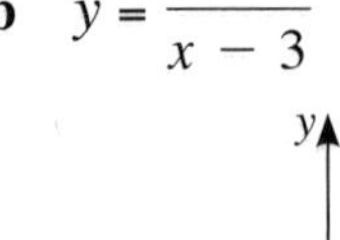

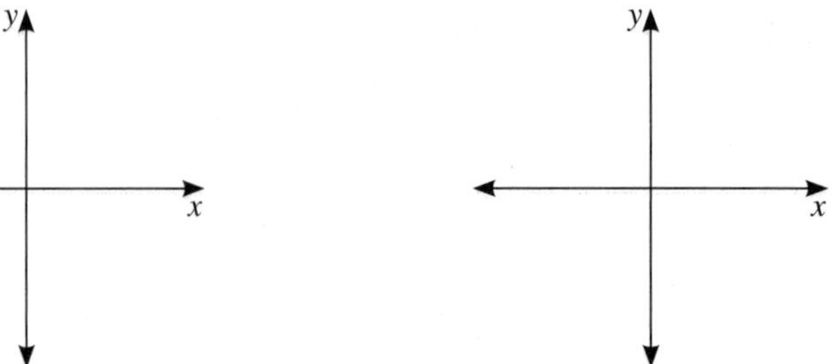

QUESTION 4 The graph below shows the width of a rectangle for given lengths if the area is always 600 units^2. All these questions refer to a rectangle of area 600 units^2.

a Briefly explain why only one branch of the hyperbola is used.

b How wide is a rectangle of length 30 cm?

c How long is a rectangle of width 15 cm?

d How much wider will a rectangle of length 50 m be than a rectangle of length 60 m?

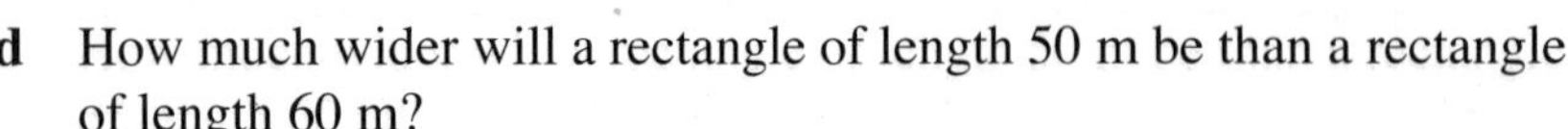

e What will be the perimeter of a rectangle of length 25 m?

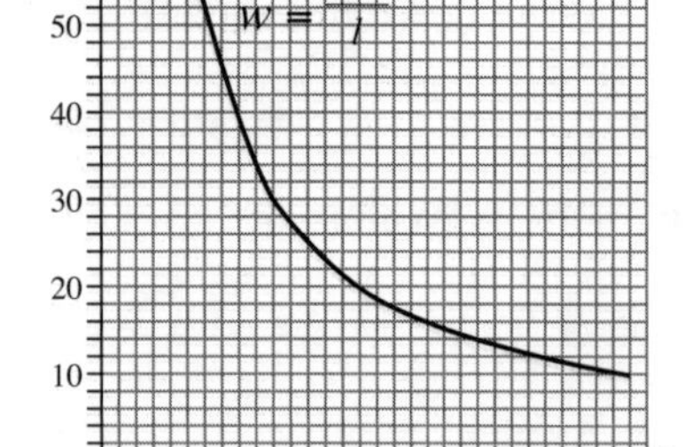

Non-linear relationships

UNIT 11: The circle $x^2 + y^2 = r^2$

QUESTION 1 Write the coordinates of the centre and the length of the radius for each of the following circles.

a $x^2 + y^2 = 400$ ________________________

b $x^2 + y^2 = 121$ ________________________

c $x^2 + y^2 = \frac{4}{9}$ ________________________

d $x^2 + y^2 = 81$ ________________________

QUESTION 2 Write the equation of each of the following circles, whose centre and radius are given.

a Centre (0, 0), radius = 12 units

____

____

b Centre (0, 0), radius = 7 units

____

____

c Centre (0, 0), radius = 2.5 units

____

____

d Centre (0, 0), radius = 10 units

____

____

QUESTION 3 Write the equation of each of the following circles.

a

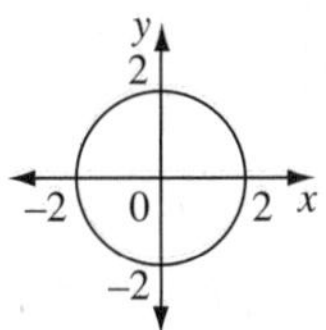

____

____

b

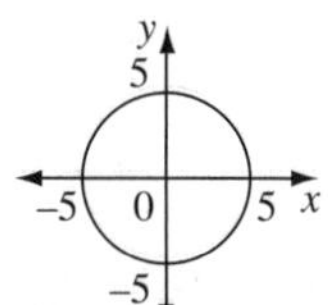

____

____

c

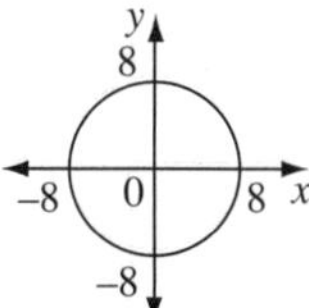

____

____

QUESTION 4 Graph each of the following circles, stating the radius and the centre.

a $x^2 + y^2 = 16$

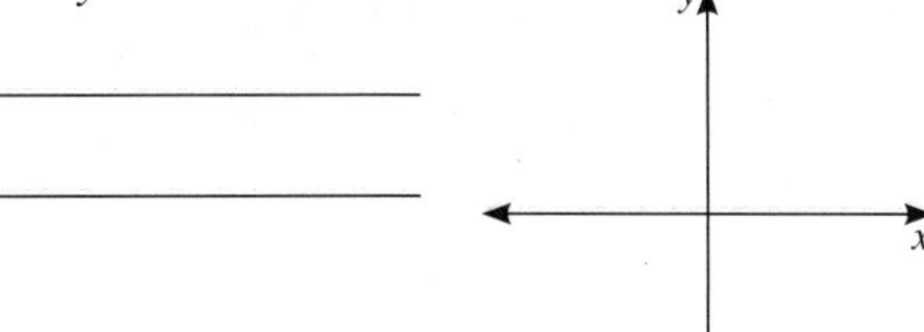

b $x^2 + y^2 = 1$

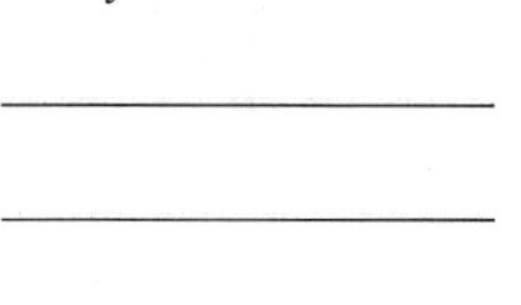

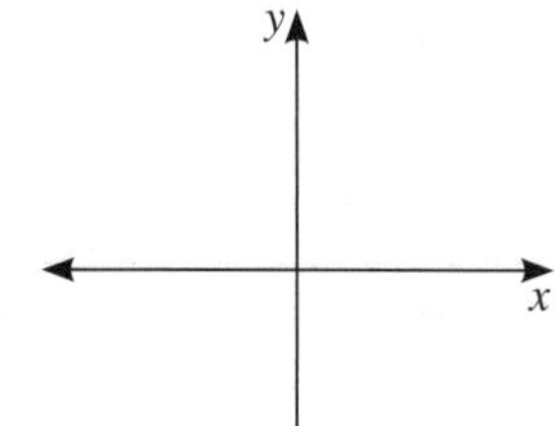

c $x^2 + y^2 = 9$

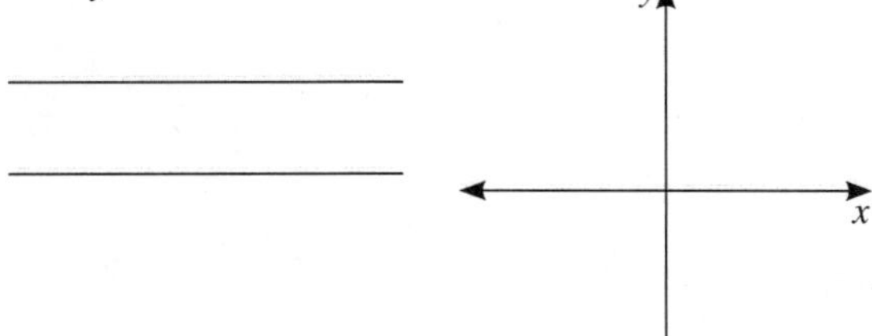

d $x^2 + y^2 = 36$

____

____

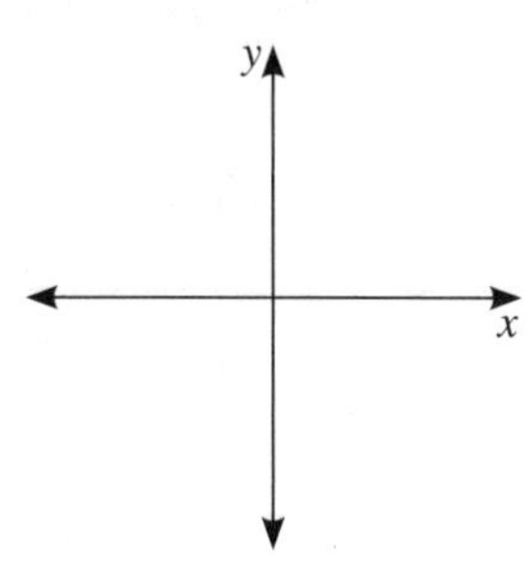

Non-linear relationships

UNIT 12: The circle $(x - h)^2 + (y - k)^2 = r^2$

Question 1 Write down the coordinates of the centre and the radius for each circle.

a $(x - 2)^2 + (y - 3)^2 = 16$ **b** $(x - 4)^2 + (y + 1)^2 = 25$ **c** $(x + 3)^2 + (y + 2)^2 = 49$

d $x^2 + (y - 4)^2 = 1$ **e** $(x + 5)^2 + y^2 = 81$ **f** $(x + 8)^2 + (y + 1)^2 = 64$

Question 2 Write down the equation, in the form $(x - h)^2 + (y - k)^2 = r^2$ of the circle with:

a centre (3, 5), radius 2 **b** centre (2, 6), radius 3 **c** centre (–3, 2), radius 10

d centre (4, –1), radius 6 **e** centre (–7, –2), radius 12 **f** centre (5, 0), radius 15

Question 3 Write down the equation, in the form $(x - h)^2 + (y - k)^2 = r^2$ of each of these circles.

a

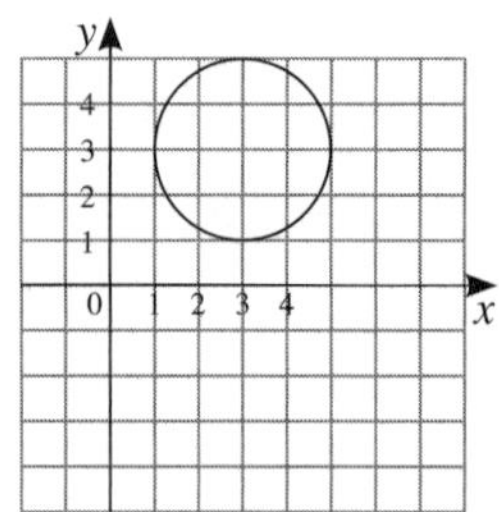

b

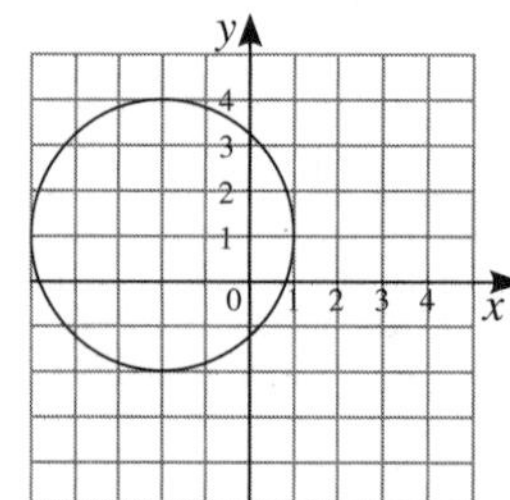

c

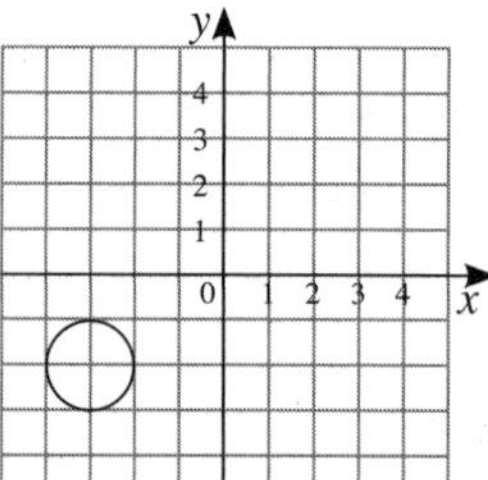

d

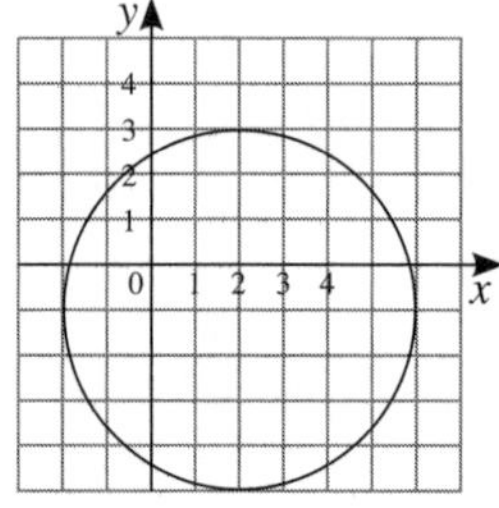

e

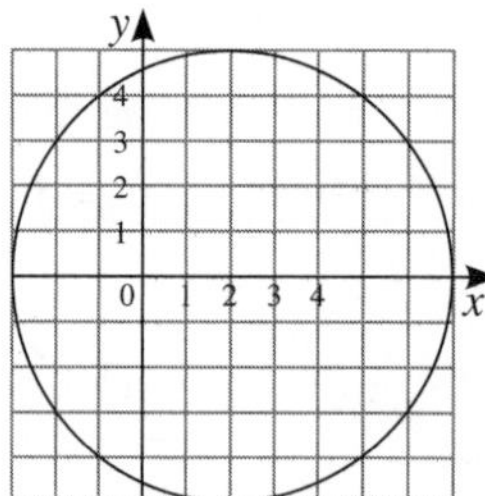

f

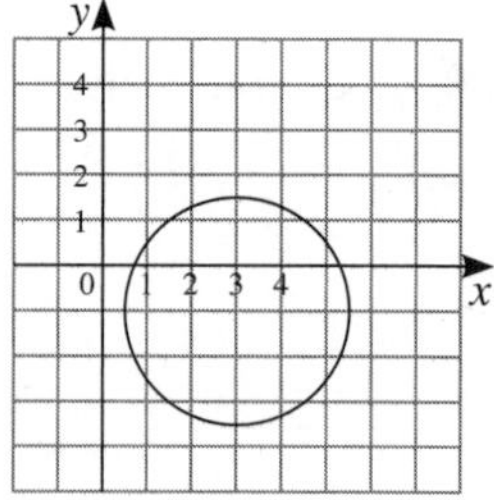

Question 4 Sketch each of these circles.

a $(x - 1)^2 + (y - 4)^2 = 9$ **b** $(x + 2)^2 + (y + 1)^2 = 16$

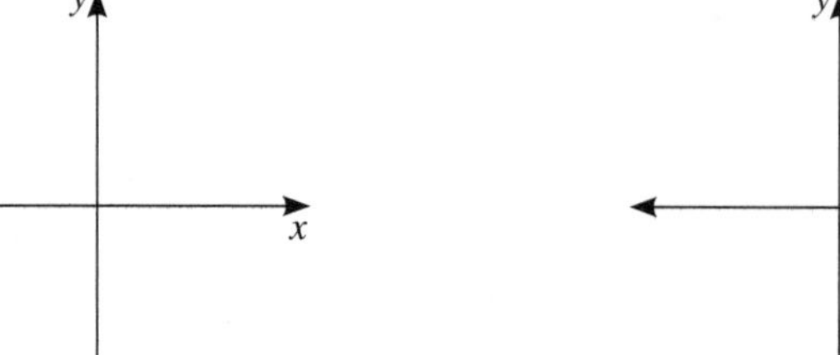

c $x^2 + (y + 3)^2 = 25$

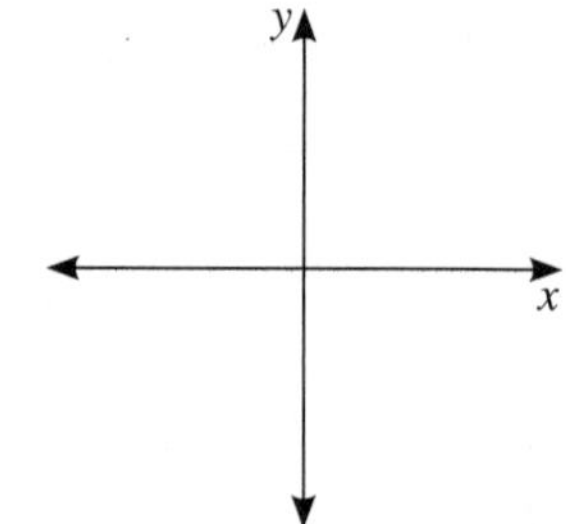

Excel Advanced-level Mathematics Study Guide Years 9–10
Pages 87–102

UNIT 13: Further circles

QUESTION 1 Find the equation of the circle with given centre that passes through the point P.

a centre (3, 2), $P(6, 6)$

b centre (–3, 1), $P(5, 7)$

c centre (–2, –1), $P(4, 2)$

d centre (5, –2), $P(3, 1)$

QUESTION 2 Determine whether the point P lies inside, on, or outside the circle.

a $x^2 + y^2 = 25$, $P(2, 4)$

b $(x - 3)^2 + y^2 = 49$, $P(-2, 5)$

c $(x + 2)^2 + (y - 2)^2 = 9$, $P(0, 0)$

d $(x + 1)^2 + (y + 6)^2 = 20$, $P(-3, -2)$

QUESTION 3 Find the centre and radius of these circles.

a $x^2 - 8x + y^2 + 8y = 4$

b $x^2 + y^2 + 12y = 28$

c $x^2 + 6x + y^2 - 2y = 6$

d $x^2 - 4x + y^2 - 10y + 20 = 0$

Non-linear relationships

Excel Advanced-level Mathematics Study Guide Years 9–10
Pages 87–102

UNIT 14: Points of intersection

QUESTION 1 Graph each curve and line on the same set of axes and then write down the coordinates of their points of intersection.

a $y = x^2 - 2x - 3$

$y = 2x - 3$

b $y = \frac{12}{x}$

$y = \frac{2}{3}x - 2$

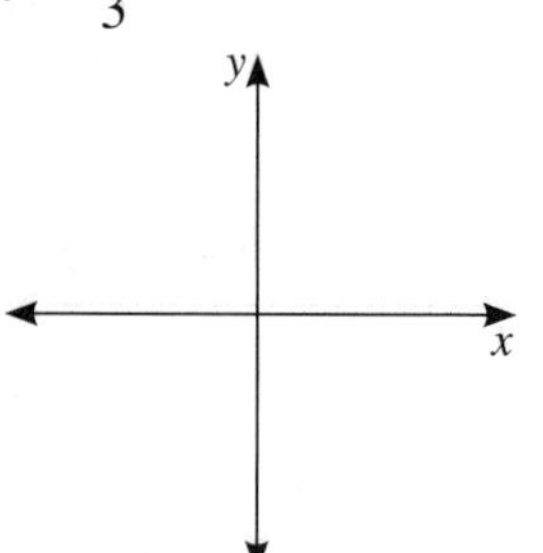

c $(x - 2)^2 + (y - 1)^2 = 25$

$x + 7y - 34 = 0$

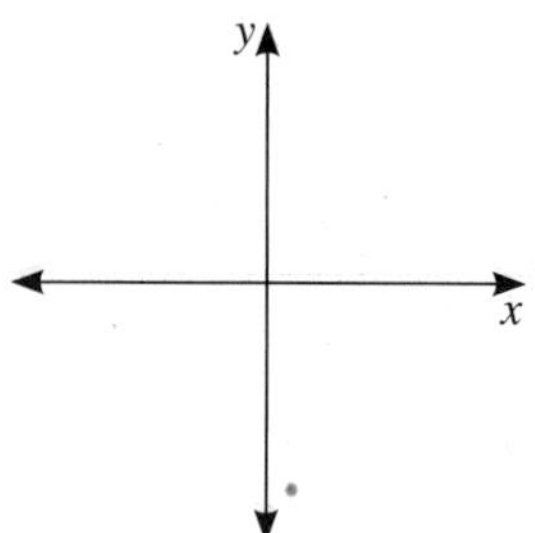

QUESTION 2 Find algebraically the coordinates of the points of intersection of these curves.

a $y = -x^2 + 8x - 7$

$y = 3x - 13$

b $(x + 3)^2 + (y - 4)^2 = 13$

$x + 5y - 4 = 0$

c $y = \frac{3}{x - 2}$

$y = 6 - x$

Non-linear relationships

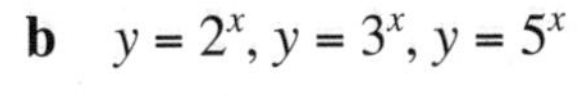

Excel Advanced-level Mathematics Study Guide Years 9–10
Pages 87–102

UNIT 15: Exponential Curves

QUESTION 1 Sketch the graphs of each group of exponentials on the same set of axes.

a $y = 2^x, y = 2^{-x}, y = -2^x, y = -2^{-x}$

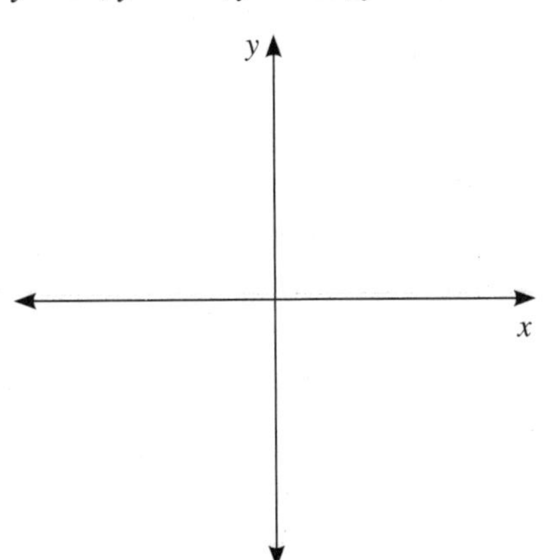

b $y = 2^x, y = 3^x, y = 5^x$

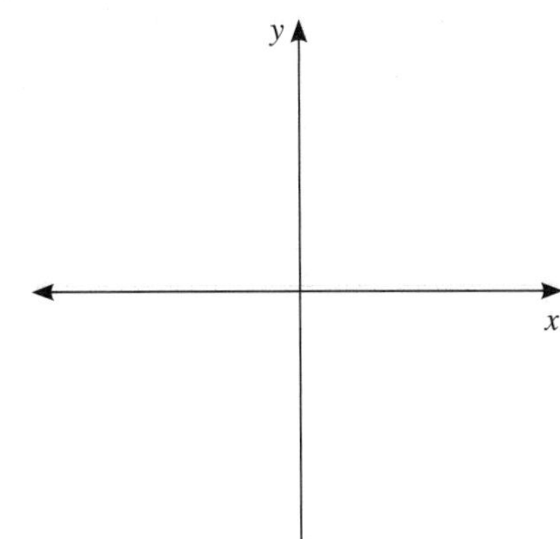

c $y = 4^x, y = 4^x + 1, y = 4^x - 2$

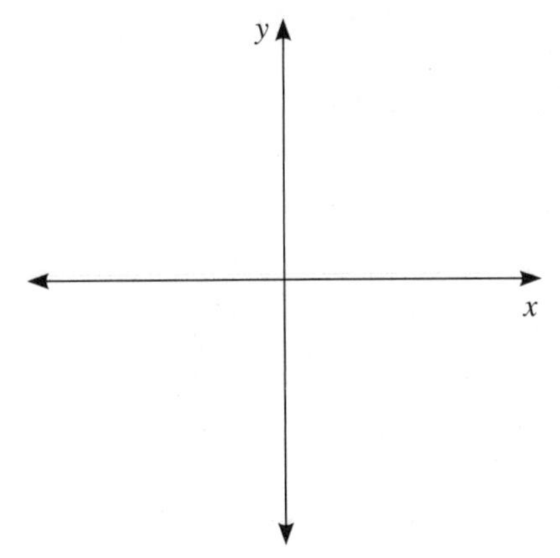

d $y = 6^x, y = \frac{1}{2}(6^x), y = 2(6^x)$

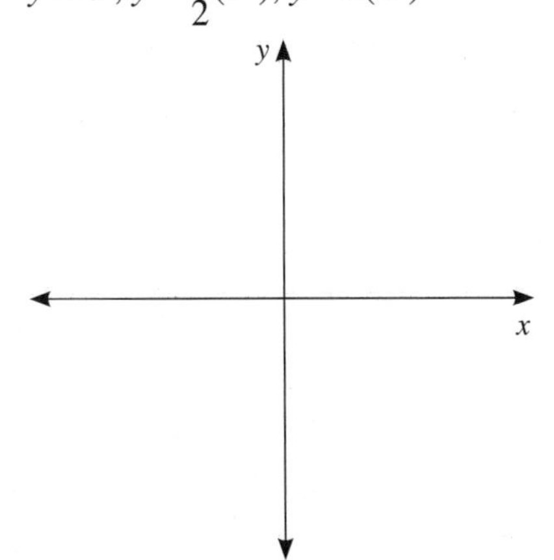

QUESTION 2 Complete the table of values and then draw the graph of $y = \dfrac{3^x + 3^{-x}}{2}$

x	−1	0	1	2	3
$y = 3^x$					
$y = 3^{-x}$					
$y = \dfrac{3^x + 3^{-x}}{2}$					

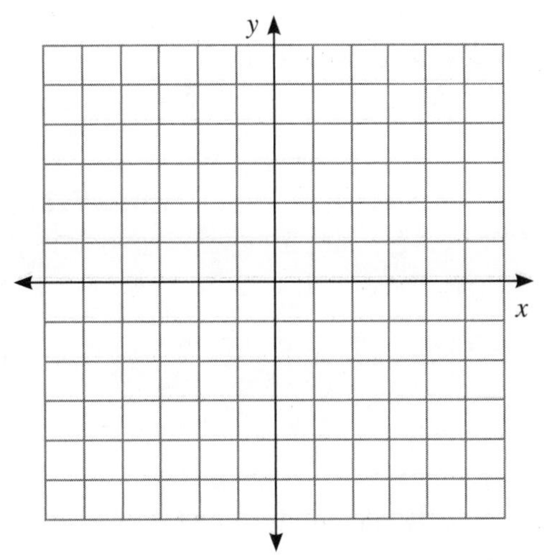

Non-linear relationships

UNIT 16: Cubic curves

QUESTION 1

a Complete the table of values for $y = x^3$

x	-2	$-1\frac{1}{2}$	-1	$-\frac{1}{2}$	0	$\frac{1}{2}$	1	$1\frac{1}{2}$	2
y									

b Graph $y = x^3$ on the axes provided.

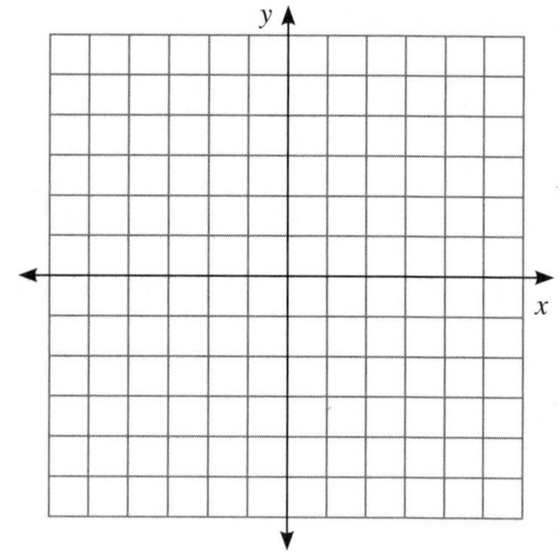

QUESTION 2 Sketch the graph of:

a $y = -x^3$

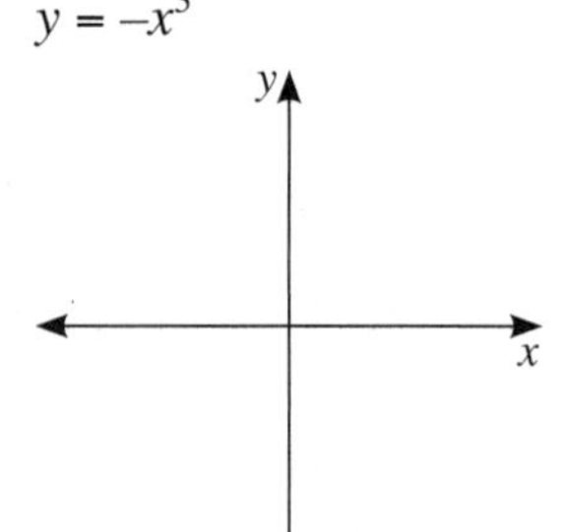

b $y = \frac{1}{2}x^3$

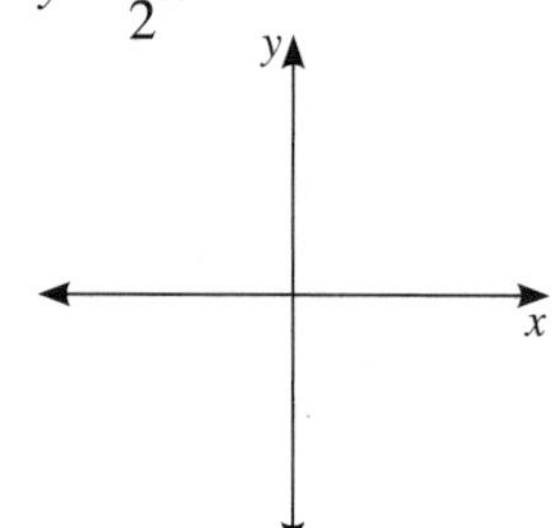

c $y = 2x^3$

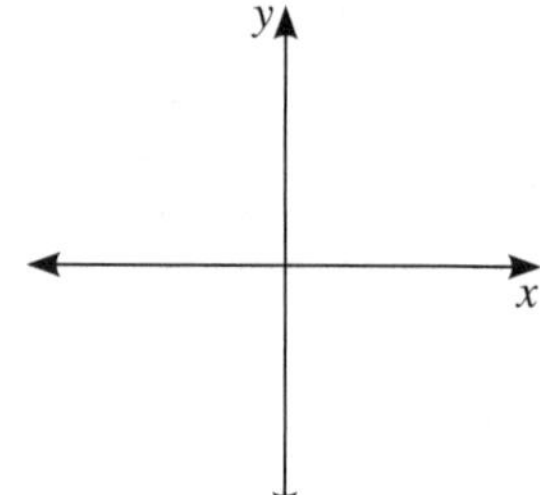

QUESTION 3 Briefly explain the effect of a in the graph of $y = ax^3$

QUESTION 4 Sketch the graph of:

a $y = x^3 + 2$

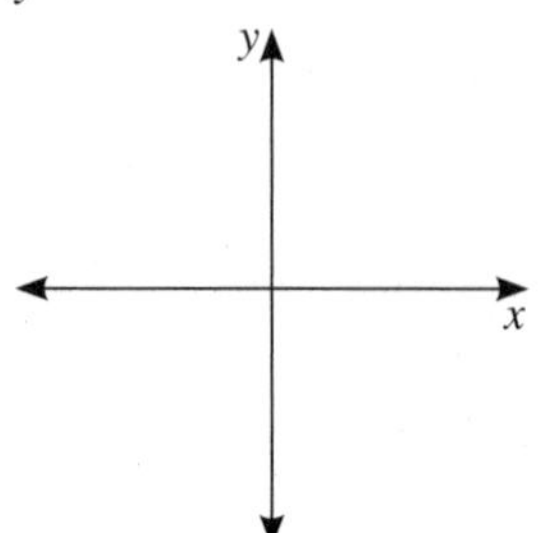

b $y = x^3 - 1$

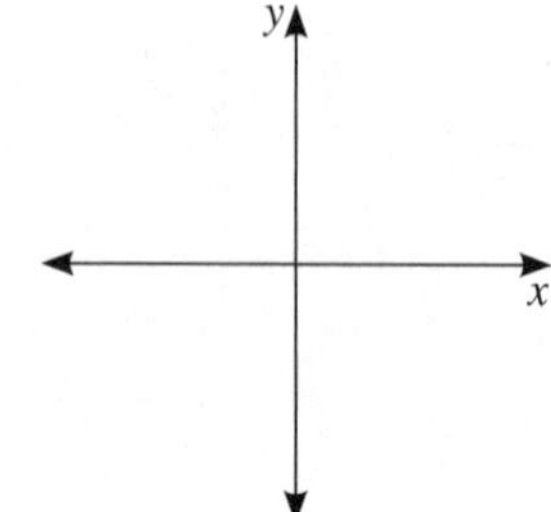

c $y = 3 - x^3$

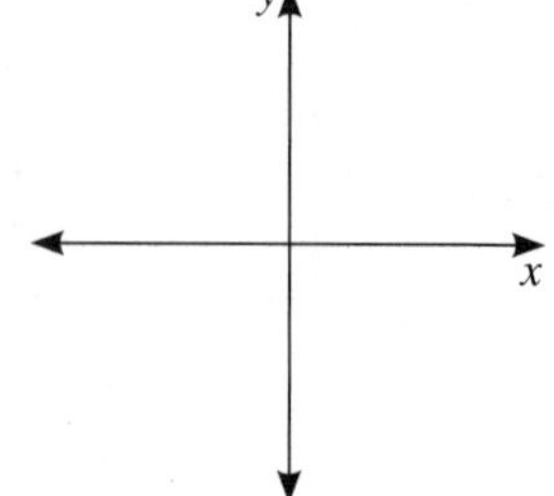

QUESTION 5 Briefly explain the effect of the constant k in the graph of $y = x^3 + k$

Non-linear relationships

UNIT 17: Further cubic and other curves

Question 1 Sketch the following curves.

a $y = (x-1)^3$

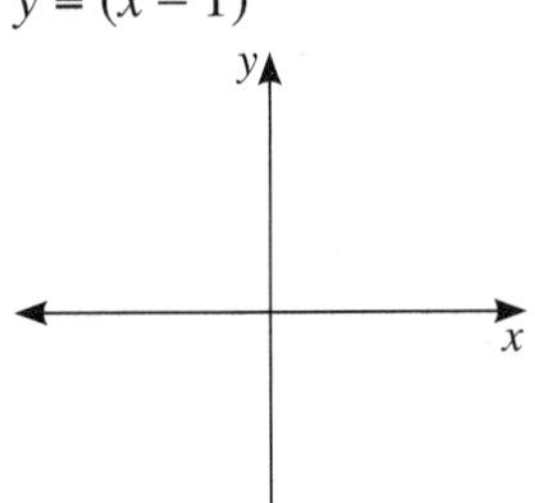

b $y = (x+2)^3$

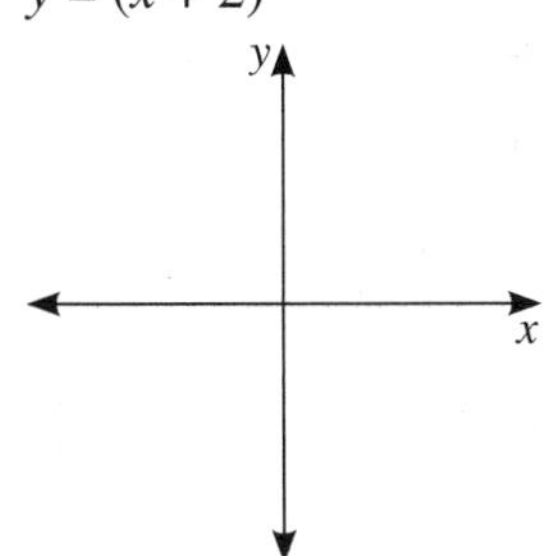

c $y = (x-3)^3$

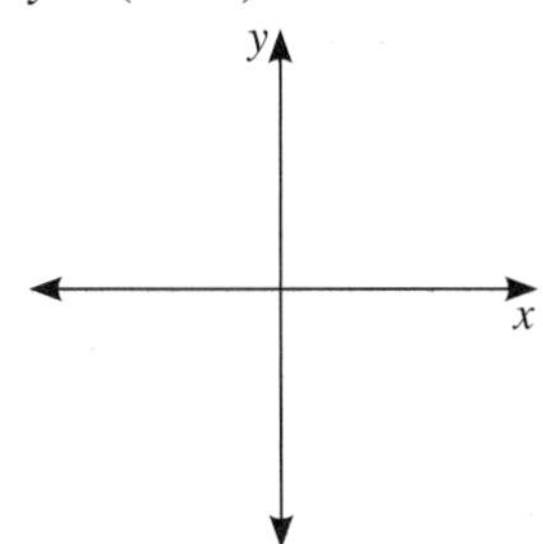

Question 2 Briefly explain the effect of the constant r in the graph of $y = (x-r)^3$

____

Question 3 Consider the equation $y = (x-1)(x+2)(x-3)$

a What is the value of y when $x = 0$?

____

b What happens to the y-value as the x-value gets very large?

____

c What happens to the y-value as the x-value gets very large negatively?

____

d For what values of x does $(x-1)(x+2)(x-3) = 0$?

____

____

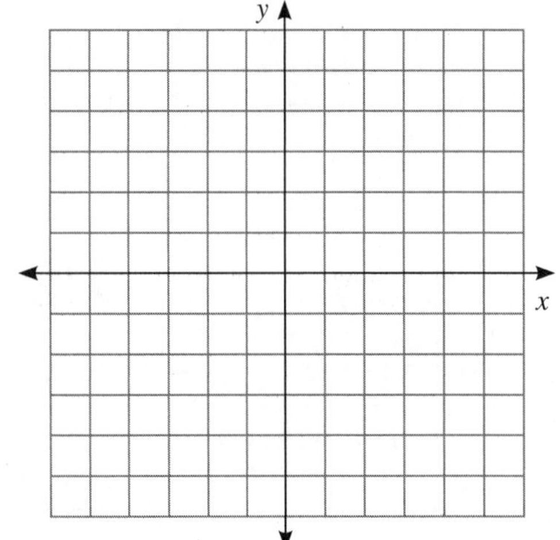

e Sketch the curve $y = (x-1)(x+2)(x-3)$.

Question 4 Each diagram shows the curve $y = x^4$. On the same diagram sketch:

a $y = x^4 + 2$

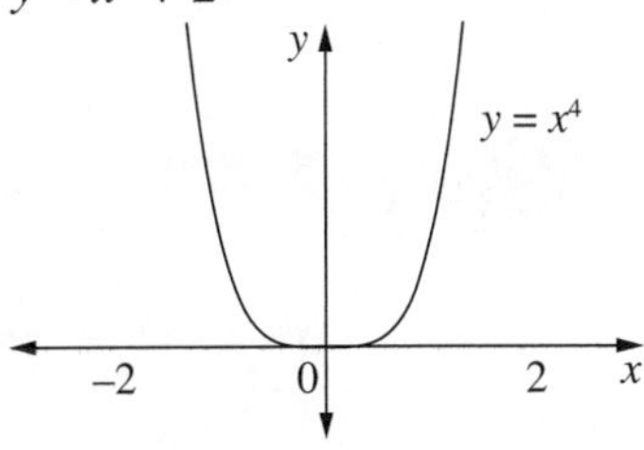

b $y = 2x^4$

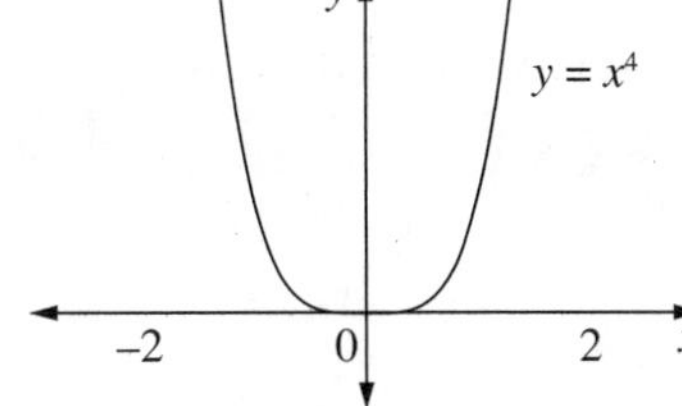

c $y = (x-2)^4$

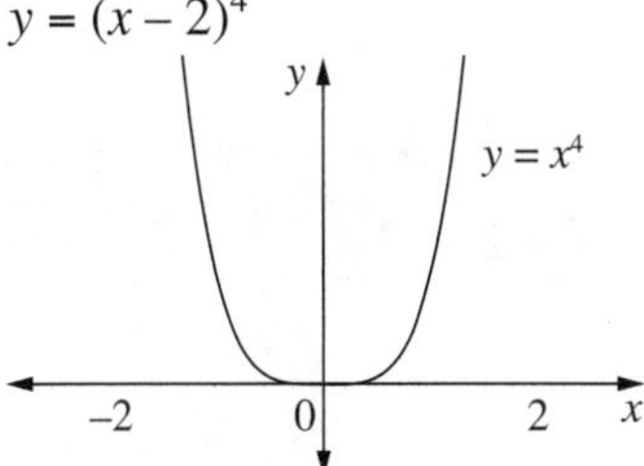

Question 5 Determine whether it is true or false that the y-axis will be an axis of symmetry for each of these curves.

a $y = x^2$ __________ b $y = x^3$ __________ c $y = x^4$ __________ d $y = x^5$ __________

e $y = x^6$ __________ f $y = x^{37}$ __________ g $y = x^{43}$ __________ h $y = x^{62}$ __________

Non-linear relationships

UNIT 18: Miscellaneous graphs

QUESTION 1 For the following equations, write whether the graphs are straight lines, parabolas, hyperbolas, circles, exponential functions, or none of these.

a $y = x$ ____________ **b** $xy = 6$ ____________ **c** $y = 0$ ____________

d $y = x^2 - 5x + 6$ ____________ **e** $y = x^2$ ____________ **f** $y = 3^{-x}$ ____________

g $y = x^2 - 1$ ____________ **h** $y = \frac{2}{x}$ ____________ **i** $y = x^3$ ____________

j $x^2 + y^2 = 16$ ____________ **k** $y = 2^x$ ____________ **l** $x^2 + y^2 = 64$ ____________

QUESTION 2 Match the equations with the graphs sketched below.

a $y = 2x + 1$ ________ **b** $y = \frac{4}{x}$ ________ **c** $y = 2^{-x}$ ________ **d** $x^2 + y^2 = 1$ ________

e $y = x^2 + 2$ ________ **f** $xy = -1$ ________ **g** $y = x$ ________ **h** $y = 2^x$ ________

i $y = x^2$ ________ **j** $y = x^2 - 4x + 3$ ________ **k** $y = -x$ ________ **l** $x^2 + y^2 = 25$ ________

A
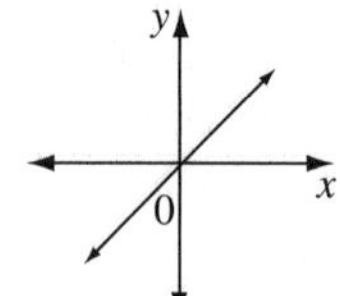

B
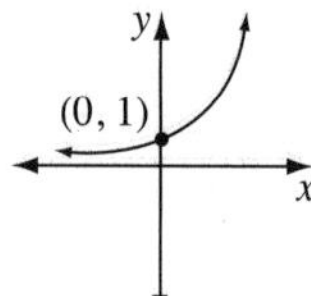

C
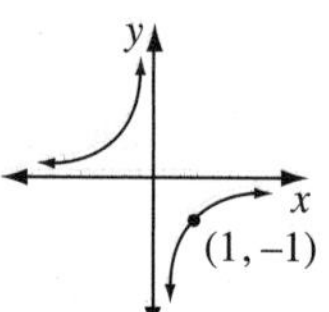

D
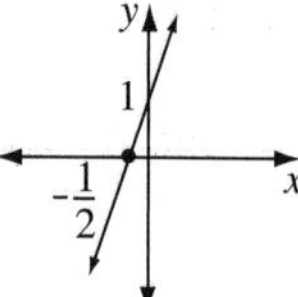

E
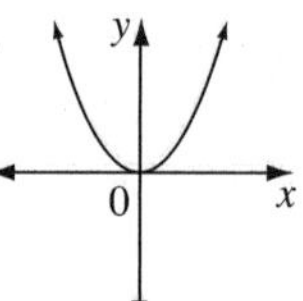

F
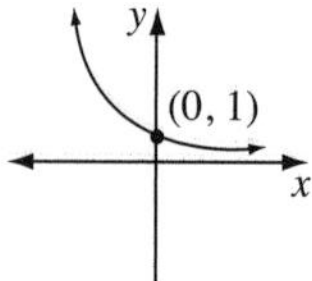

G
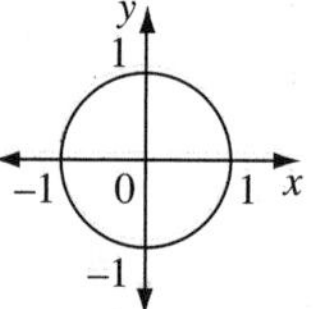

H
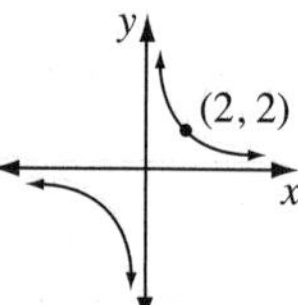

I
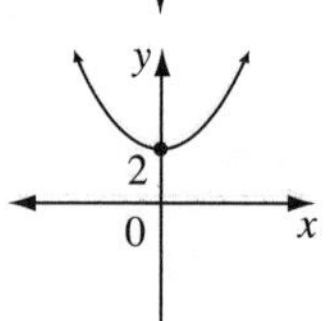

J
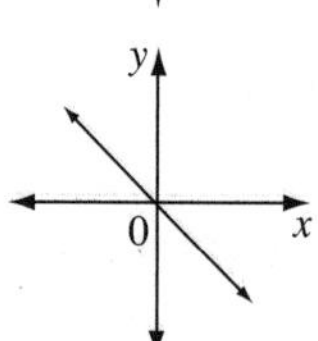

K
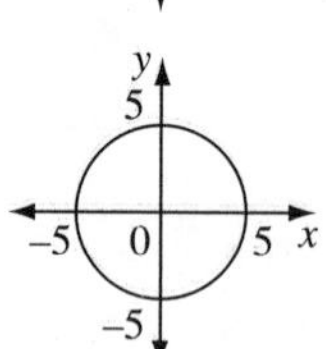

L
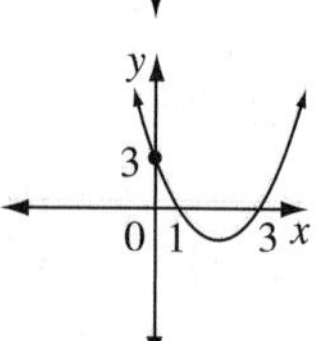

QUESTION 3 Draw a separate sketch of each of the following.

a $y = 2x + 3$

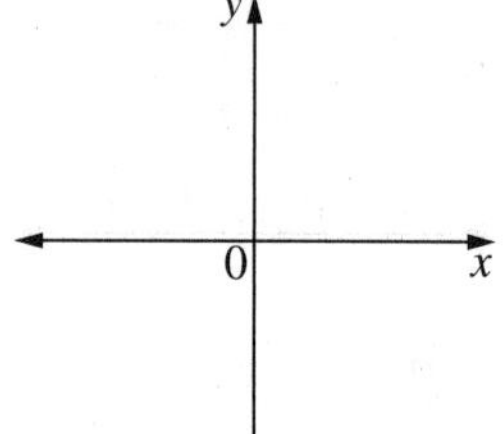

b $y = 2x^2$

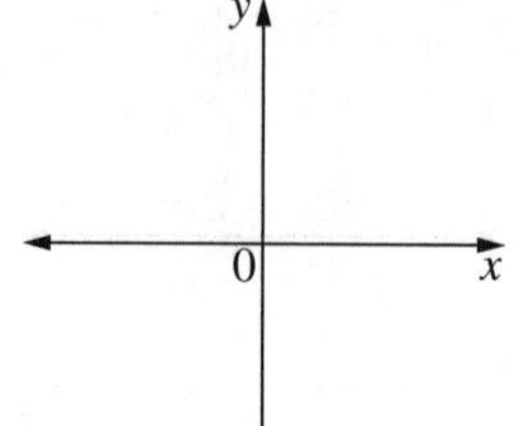

c $x^2 + y^2 = 9$

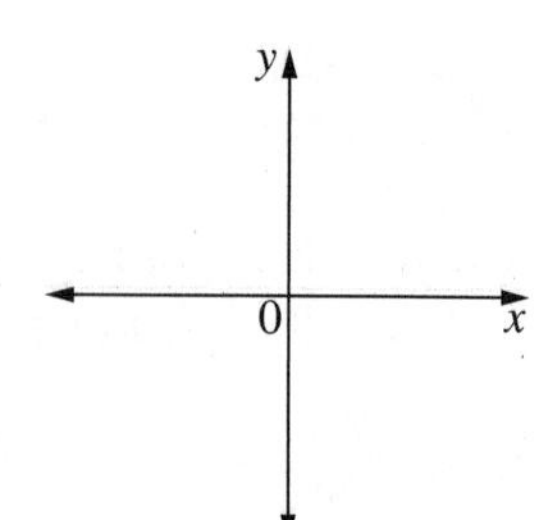

Non-linear relationships

UNIT 19: Mixed graphs

QUESTION 1 Name the type of graph (circle, parabola, hyperbola, etc.) represented by each of these equations.

a $y = x^2 + 5x - 2$ __________
b $y = \dfrac{x + 2}{3}$ __________
c $y = (x + 2)^2 + 3$ __________

d $(x + 2)^2 + y^2 = 3$ __________
e $y = 3^x + 2$ __________
f $y = \dfrac{3}{x - 3}$ __________

g $y = (x + 2)^3$ __________
h $(x - 3)^2 + (y - 2)^2 = 16$ __________
i $y = \dfrac{3}{x} + 2$ __________

j $2x + 3y - 5 = 0$ __________
k $y = x^3 - 4$ __________
l $y = (x - 4)^2$ __________

QUESTION 2 Write down the letter beside the graph that matches each equation.

a $y = 4x - x^2$ __________
b $y = 4 - x^2$ __________
c $y = 4^x$ __________

d $x^2 + (y - 2)^2 = 4$ __________
e $y = \dfrac{4}{x - 2}$ __________
f $y = \dfrac{x - 2}{4}$ __________

g $y = \dfrac{x}{4} - 2$ __________
h $y = \dfrac{4}{x} - 2$ __________
i $(x - 2)^2 + y^2 = 4$ __________

A

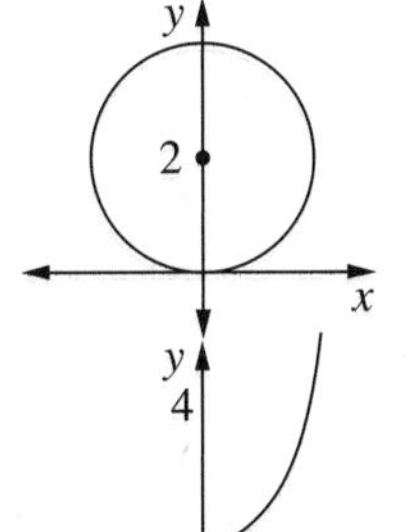

B

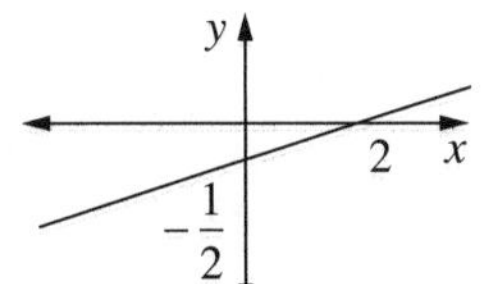

C

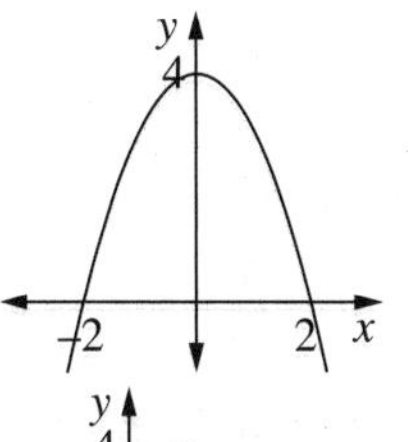

D

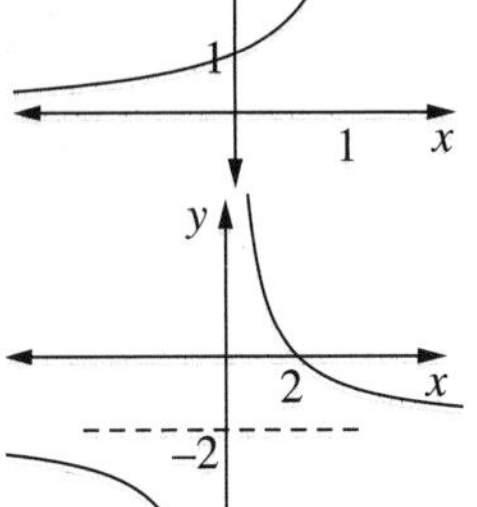

E

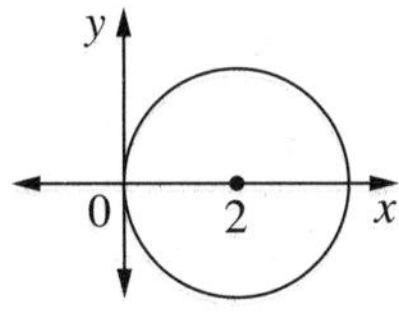

F

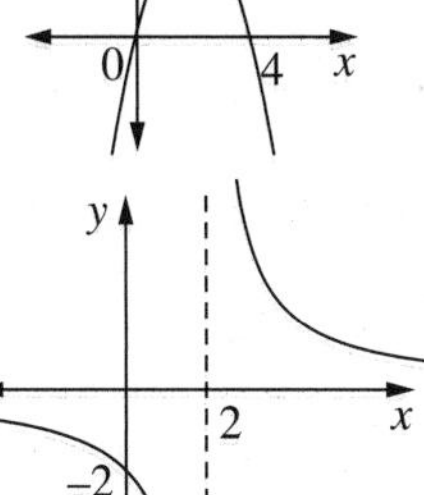

G

H

I

QUESTION 3 Determine the equation of each graph.

a

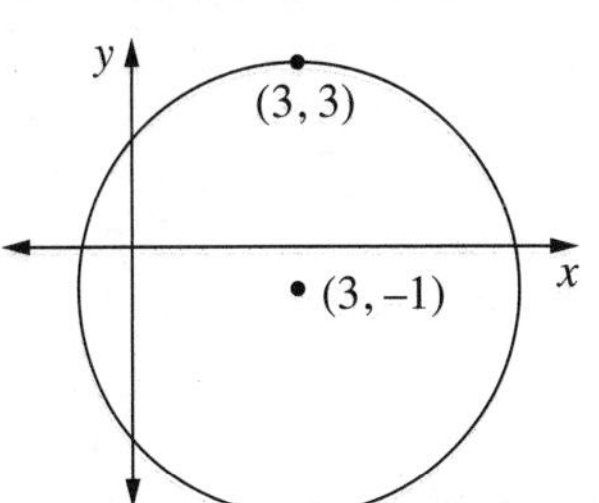

____

b

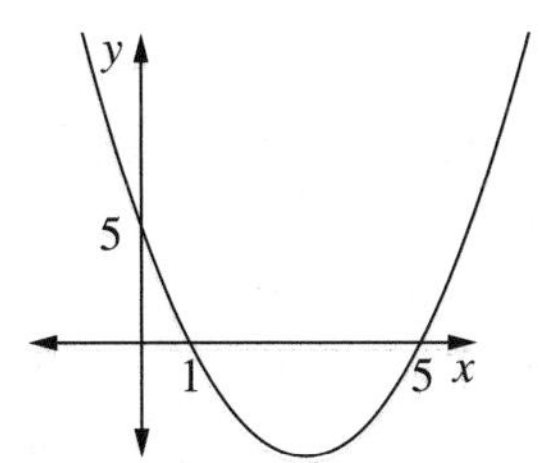

____

c

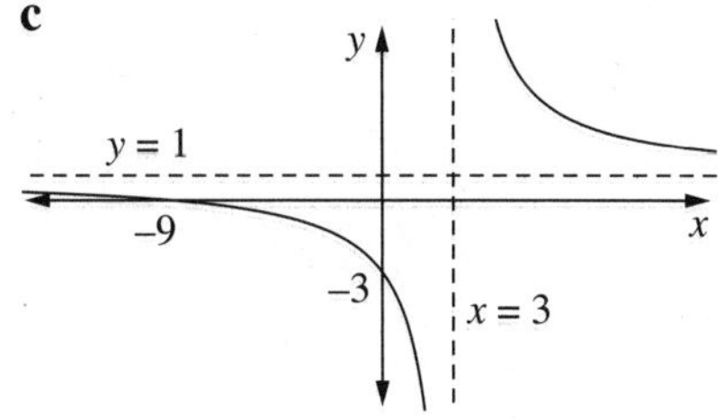

____

Non-linear relationships

TOPIC TEST — PART A

Instructions
- This part consists of 10 multiple-choice questions.
- Fill in only ONE CIRCLE for each question.
- Each question is worth 1 mark.
- Calculators are NOT allowed.

Time allowed: 10 minutes — **Total marks: 10**

Marks

1 The equation $y = \frac{3}{x}$ represents

Ⓐ a line. Ⓑ a parabola. Ⓒ a hyperbola. Ⓓ a circle. — 1

2 The radius of the circle $x^2 + y^2 = 4$ is equal to

Ⓐ 2 units. Ⓑ 4 units. Ⓒ 16 units. Ⓓ none of these. — 1

3 Which graph best represents $y = x^2$? — 1

Ⓐ 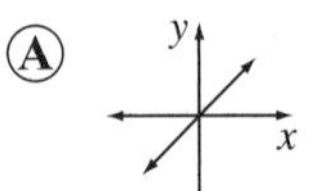

Ⓑ 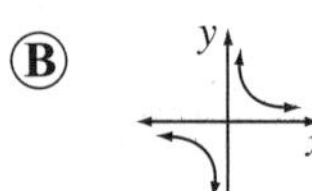

Ⓒ 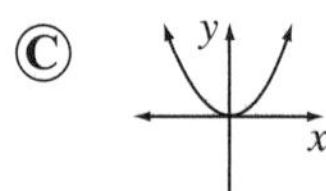

Ⓓ 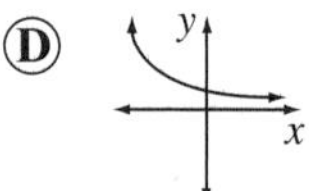

4 The graph shown could be part of the graph with equation

Ⓐ $y = 2^{-x}$ Ⓑ $y = 2^x$ Ⓒ $y = -\frac{1}{x}$ Ⓓ $y = \frac{1}{x}$ — 1

5 Which of the following could be the equation of the graph?

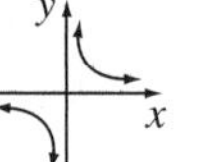

Ⓐ $y = x^2$ Ⓑ $y = 2^x$ Ⓒ $y^2 = 4x$ Ⓓ $xy = 8$ — 1

6 Which of these is an asymptote of the curve $y = \frac{1}{x + 2}$?

Ⓐ $x = 2$ Ⓑ $x = -2$ Ⓒ $y = 2$ Ⓓ $y = -2$ — 1

7 A circle has its centre at $(2, -1)$ and radius of length 3 units. What is its equation?

Ⓐ $(x - 2)^2 + (y + 1)^2 = 3$ Ⓑ $(x - 2)^2 + (y + 1)^2 = 9$

Ⓒ $(x + 2)^2 + (y - 1)^2 = 3$ Ⓓ $(x + 2)^2 + (y - 1)^2 = 9$ — 1

8 What are the coordinates of the vertex of the parabola $y = (x - 1)^2 + 4$?

Ⓐ $(1, 4)$ Ⓑ $(-1, 4)$ Ⓒ $(1, 5)$ Ⓓ $(-1, 8)$ — 1

9 A circle has equation $x^2 + y^2 = 16$. P is the point $(-3, 5)$. Where does P lie?

Ⓐ inside the circle Ⓑ outside the circle Ⓒ on the circle Ⓓ not enough information — 1

10 Which could be the graph of $y = (2 - x)(4 - x)$? — 1

Ⓐ y, 2, 4, x

Ⓑ 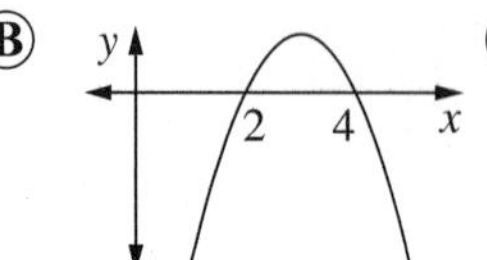

Ⓒ 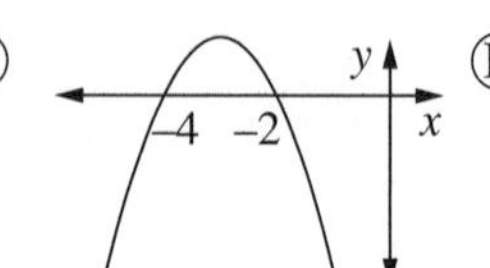

Ⓓ 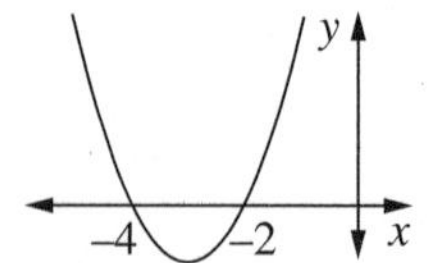

Total marks achieved for PART A ___ / 10

Non-linear relationships

TOPIC TEST — PART B

Instructions
- This part consists of 4 questions.
- Each question part is worth 1 mark, except question 3c.
- Show all working.

Time allowed: 20 minutes **Total marks: 15**

Marks

1 Consider the parabola $y = x^2 - 8x + 7$

a What is the y-intercept? ________________

b Find the x-intercepts. ________________

c What is the equation of the axis of symmetry? __________

d What are the coordinates of the vertex? __________

e Sketch the graph of the parabola.

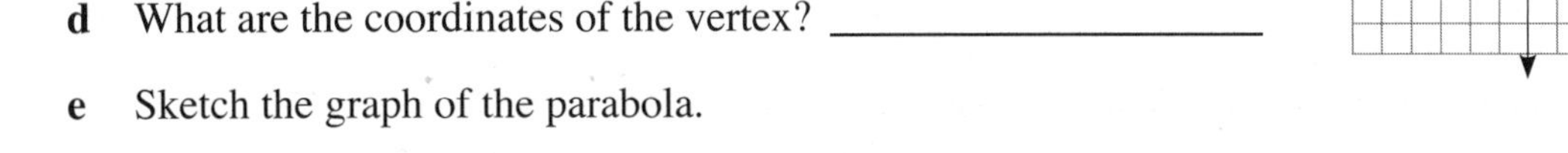

5

2 Consider the circle given by the equation $x^2 - 6x + y^2 + 4y = 36$

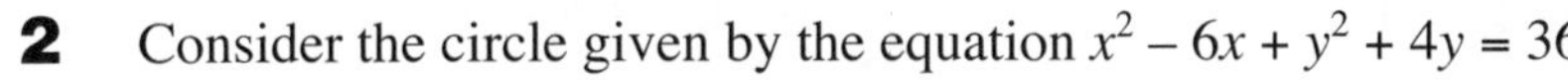

a Find the coordinates of the centre of the circle.

b What is the radius of the circle?

c Find the points of intersection of the circle with the line $y = -x - 6$

3

3 Consider the hyperbola $y = \dfrac{6}{x + 3} - 1$

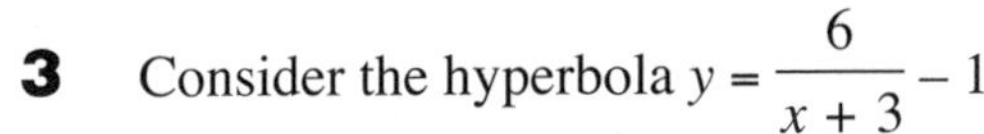

a What is the y-intercept?

b What is the x-intercept?

2

c Write down the equations of the two asymptotes.

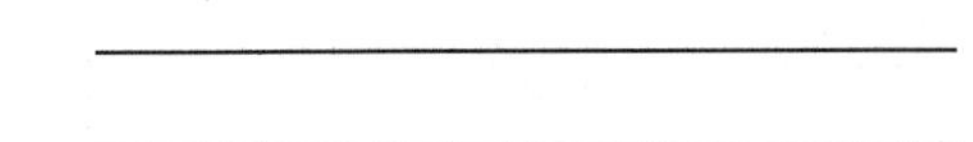

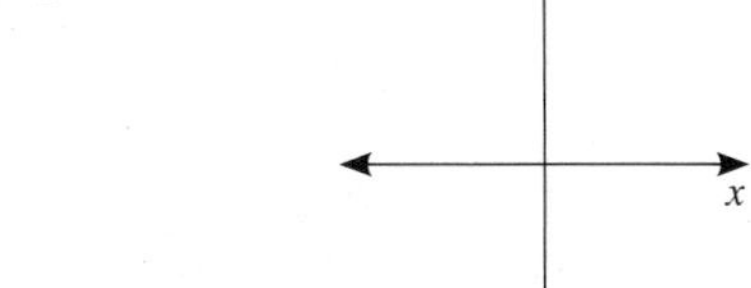

2

d Sketch the graph of the hyperbola.

1

4 Sketch graphs of:

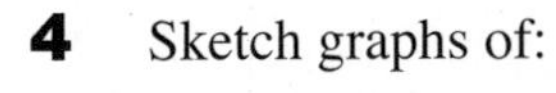

a $y = 4^{-x} + 1$

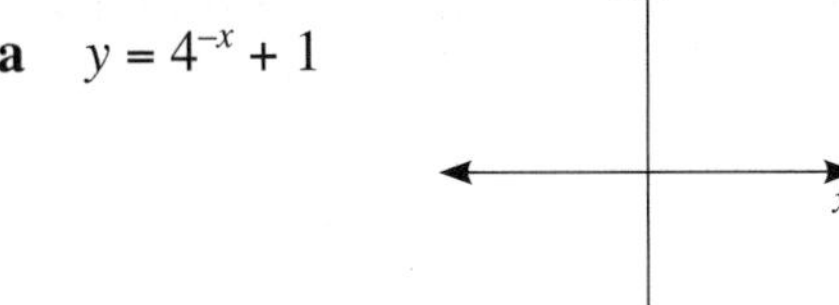

b $y = (x - 2)^3$

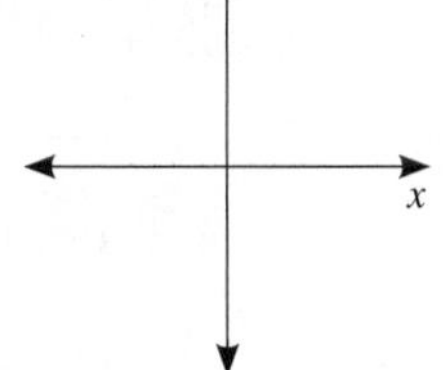

2

Total marks achieved for PART B /15

CHAPTER 4

Logarithms, exponentials and functions

Excel Advanced-level Mathematics Study Guide Years 9–10
Pages 242–262

UNIT 1: Logarithms

QUESTION 1 Write each of the following in logarithmic form.

a $3^2 = 9$ **b** $64 = 4^3$ **c** $125 = 5^3$

d $2^5 = 32$ **e** $3^4 = 81$ **f** $2^{-3} = \frac{1}{8}$

g $64 = 2^6$ **h** $343 = 7^3$ **i** $\frac{1}{9} = 3^{-2}$

j $3^{-1} = \frac{1}{3}$ **k** $25^{-\frac{1}{2}} = \frac{1}{5}$ **l** $4 = 8^{\frac{2}{3}}$

QUESTION 2 Write each of the following in index form.

a $\log_2 2 = 1$ **b** $\log_3 27 = 3$ **c** $\log_5 \sqrt{5} = \frac{1}{2}$

d $\log_3 9 = 2$ **e** $\log_8 2 = \frac{1}{3}$ **f** $\log_{27} 9 = \frac{2}{3}$

g $\log_2 32 = 5$ **h** $\log_2 16 = 4$ **i** $\log_3 1 = 0$

j $\log_4 64 = 3$ **k** $\log_2 \sqrt{2} = \frac{1}{2}$ **l** $\log_2 128 = 7$

QUESTION 3 Evaluate the following.

a $\log_2 4 =$ ________ **b** $\log_2 8 =$ ________ **c** $\log_2 16 =$ ________

d $\log_5 25 =$ ________ **e** $\log_2 64 =$ ________ **f** $\log_7 49 =$ ________

g $\log_3 9 =$ ________ **h** $\log_3 27 =$ ________ **i** $\log_3 81 =$ ________

j $\log_5 125 =$ ________ **k** $\log_6 216 =$ ________ **l** $\log_7 343 =$ ________

QUESTION 4 Write down the value of x for which:

a $\log_2 x = 3$ **b** $\log_3 x = 4$ **c** $\log_5 x = 2$ **d** $\log_6 x = -1$

e $\log_3 27 = x$ **f** $\log_5 125 = x$ **g** $\log_7 7 = x$ **h** $\log_9 3 = x$

i $\log_x 27 = 3$ **j** $\log_x 64 = 2$ **k** $\log_x 81 = 4$ **l** $\log_x \sqrt{5} = \frac{1}{2}$

UNIT 2: The log laws (1)

QUESTION 1 Complete.

a $\log_a x + \log_a y = \log_a$ ________

b $\log_a x - \log_a y = \log_a$ ________

c $\log_a x^n =$ ________ $\log_a x$

d $\log_a a^x =$ ________

e $\log_a a =$ ________

f $\log_a 1 =$ ________

g $\log_a\left(\frac{1}{x}\right) =$ ________ $\log_a x$

QUESTION 2 Complete.

a $\log_2 1 =$ ________

b $\log_7 1 =$ ________

c $\log_m 1 =$ ________

d $\log_3 3 =$ ________

e $\log_5 5 =$ ________

f $\log_2 2 =$ ________

g $\log_7 7^3 =$ ________

h $\log_2 2^4 =$ ________

i $\log_3 3^5 =$ ________

j $3 \log_a 2 = \log_a$ ________

k $2 \log_a 5 = \log_a$ ________

l $4 \log_a 3 = \log_a$ ________

m $\log_a 49 =$ ________ $\log_a 7$

n $\log_a 1000 =$ ________ $\log_a 10$

o $\log_a 3^2 =$ ________ $\log_a 3$

QUESTION 3 Express as a single log.

a $\log_a 3 + \log_a 5 =$ ________

b $\log_n 12 - \log_n 3 =$ ________

c $\log_m 2 + \log_m 3 + \log_m 4 =$ ________

d $\log_x 18 - \log_x 2 - \log_x 3 =$ ________

e $2 \log_e 5 + \log_e 2 =$ ________

f $4 \log_b 6 - 3\log_b 2 =$ ________

QUESTION 4 Evaluate.

a $\log_{10} 2 + \log_{10} 50$

b $\log_3 24 - \log_3 8$

c $2\log_5 10 - \log_5 4$

d $\dfrac{\log_a 32}{\log_a 8}$

e $\log_2 72 - 2 \log_2 3$

f $2\log_m 8 - 3 \log_m 4$

g $\log_6 9 + \log_6 12 - \log_6 3$

h $\log_2 20 - \log_3 7 - \log_2 5 + \log_3 63$

UNIT 3: The log laws (2)

QUESTION 1 Simplify the following.

a $\log_2 32 + \log_2 4$ **b** $\log_{10} 20 + \log_{10} 5$ **c** $\log_3 81 - \log_3 9$

d $\log_5 125 - \log_5 25$ **e** $\log_a a^3 - \log_a a^2$ **f** $\dfrac{\log_a x^3}{\log_a x}$

QUESTION 2 Use the logarithm laws to expand the following.

a $\log_a (xy)$ **b** $\log_a \left(\dfrac{xy^2}{z}\right)$

c $\log_a \left(\dfrac{2x}{x-1}\right)$ **d** $\log_a (x^3\sqrt{y})$

QUESTION 3 Use the logarithm laws to simplify the following.

a $\log_a x + \log_a y - \log_a z^2$ **b** $3\log_a x - 2\log_a y$

c $\dfrac{1}{2}\log_a x + 2\log_a y$ **d** $2\log_a x - 3\log_a y + \dfrac{1}{2}\log_a z$

QUESTION 4 If $\log_a 2 = 0.7285$ and $\log_a 3 = 1.1546$, evaluate:

a $\log_a 16$ **b** $\log_a 27$ **c** $\log_a 6$ **d** $\log_a 36$

e $\log_a 12$ **f** $\log_a \left(\dfrac{1}{2}\right)$ **g** $\log_a(\sqrt{3})$ **h** $\log_a 3a$

UNIT 4: Logarithmic graphs

QUESTION 1

a Complete the table of values for $y = \log_2 x$.

x	$\frac{1}{8}$	$\frac{1}{4}$	$\frac{1}{2}$	1	2	4	8
y							

b Sketch the graph of $y = \log_2 x$ on the axes provided.

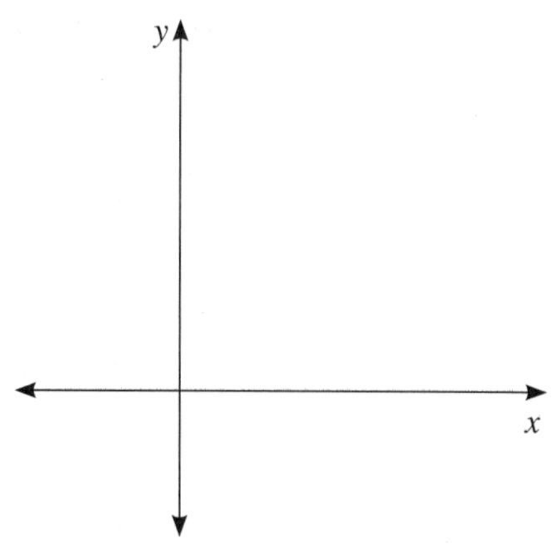

QUESTION 2

a Show the line $y = x$ on the axes at right.

b Sketch the graphs of $y = 3^x$ and $y = \log_3 x$ on the axes provided.

c Briefly explain the role of the line $y = x$ in the graphs of $y = 3^x$ and $y = \log_3 x$.

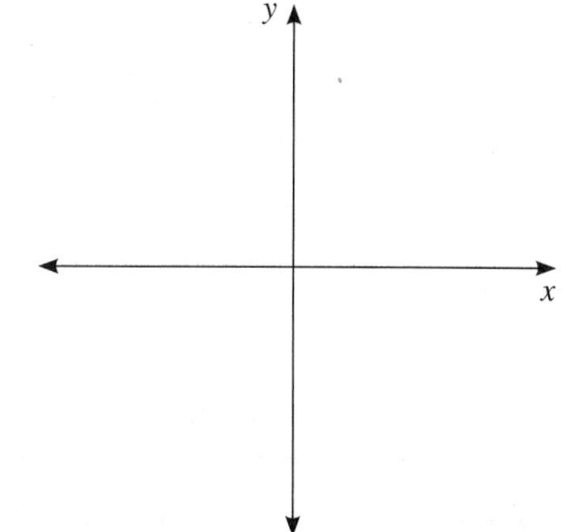

QUESTION 3

a Sketch the graphs of $y = \log_5 x$, $y = \log_{10} x$, $y = 5^x$ and $y = 10^x$ on the same set of axes.

b Briefly comment on any similarities and differences between the graphs.

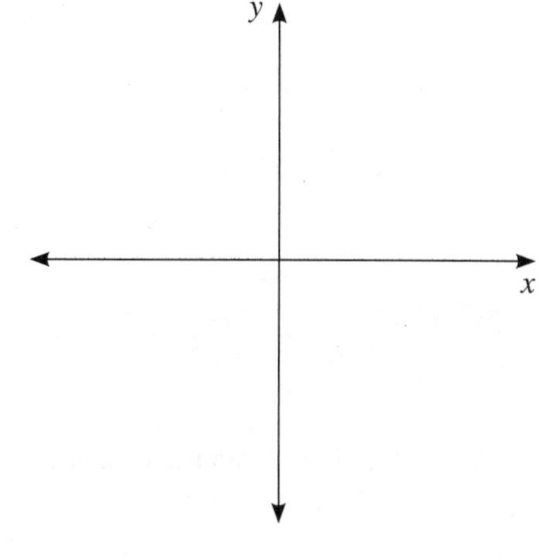

UNIT 5: Simple exponential equations

Question 1 Solve the following exponential equations.

a $2^x = 32$ **b** $2^x = 64$ **c** $2^x = 512$

d $3^x = 9$ **e** $3^x = 81$ **f** $3^x = 243$

g $4^x = 256$ **h** $7^x = 2401$ **i** $9^x = 729$

j $10^x = 10\,000$ **k** $5^x = 625$ **l** $6^x = 216$

Question 2 Solve the following equations.

a $3^x = \frac{1}{9}$ **b** $4^x = \frac{1}{64}$ **c** $2^x = \frac{1}{16}$ **d** $25^x = 5$

e $49^x = 7$ **f** $128^x = 2$ **g** $27^x = \frac{1}{3}$ **h** $8^x = \frac{1}{64}$

i $81^x = \frac{1}{27}$ **j** $3^x = \sqrt{3}$ **k** $9^x = \frac{1}{3}$ **l** $25^x = \frac{1}{125}$

m $2^x = \frac{1}{\sqrt{2}}$ **n** $8^x = 4$ **o** $16^x = 8$ **p** $5^x = 5\sqrt{5}$

q $9^x = 27$ **r** $4^x = \sqrt{2}$ **s** $100^x = 1000$ **t** $9^x = \frac{1}{\sqrt{3}}$

Question 3 Use a calculator and the 'guess and check' method to find the value of x to one decimal place.

a $2^x = 7$ **b** $3^x = 5$ **c** $7^x = 3.2$ **d** $0.8^x = 2.3$

Excel Advanced-level Mathematics Study Guide Years 9–10
Pages 242–262

UNIT 6: Logarithmic and exponential equations

QUESTION 1 Solve these equations.

a $\log_2 x = 9$

b $\log_{15} x = 1$

c $\log_{13} x = 0$

d $\log_5 a = -1$

e $\log_4 x = \frac{1}{2}$

f $\log_9 m = -2$

g $\log_3 81 = x$

h $\log_2 64 = x$

i $x = \log_{10} 100\,000$

j $\log_8 2 = x$

k $\log_9 3 = n$

l $\log_7 \sqrt{7} = x$

m $\log_2 0.5 = a$

n $\log_2 2\sqrt{2} = x$

o $\log_8 x = -\frac{2}{3}$

QUESTION 2 Solve the following equations.

a $2^{x+1} = 16$

b $2^{2x-1} = 128$

c $2^{3-x} = 512$

d $3^{x-1} = 243$

e $3^{2x+1} = \frac{1}{243}$

f $3^{3x-1} = 9$

g $\left(\frac{1}{4}\right)^x = 64$

h $\left(\frac{1}{2}\right)^{x-1} = 32$

i $\left(\frac{1}{3}\right)^{2x+1} = 27^2$

j $8^{2x-1} = 256$

k $9^{1-x} = 729$

l $4^{x-2} = 64$

m $3^{3x-2} = 2187$

n $4^{2x} = 8$

o $9^{3-2x} = 27^{x-1}$

p $36^{n+1} = 6\sqrt{6}$

q $8^m = \frac{1}{2\sqrt{2}}$

r $27^{2a+1} = \frac{1}{9\sqrt{3}}$

Excel Advanced-level Mathematics Study Guide Years 9–10
Pages 242–262

UNIT 7: Functions

Question 1 State whether the following sets of ordered pairs represent a function or not.

a $\{(1, 3), (2, 5), (3, 7), (4, 9), (5, 11)\}$ ______

b $\{(0, 0), (1, 1), (-1, 1), (2, 4), (-2, 4)\}$ ______

c $\{(0, 1), (1, 2), (1, 4), (3, 6)\}$ ______

d $\{(1, 1), (2, 2), (3, 3), (4, 4), (5, 5)\}$ ______

Question 2 If $f(x) = 2x + 3$, find:

a $f(0) =$ ______ **b** $f(4) =$ ______ **c** $f(-1) =$ ______

d $f(1) =$ ______ **e** $f(2) =$ ______ **f** $f(-2) =$ ______

Question 3 If $f(x) = 2^x + 5$, find:

a $f(0) =$ ______ **b** $f(2) =$ ______ **c** $f(4) =$ ______

d $f(-1) =$ ______ **e** $f(-2) =$ ______ **f** $f(3) =$ ______

Question 4 If $g(x) = 5x - 3$, find x if:

a $g(x) = 17$

______

b $g(x) = 32$

______

Question 5 If $f(x) = 5 - 3x$, find:

a $f(3a) =$

______

b $f\left(\frac{1}{2a}\right) =$

______

Question 6

a If $f(x) = x^2$, show that $f(-a) = f(a)$

______

b If $f(x) = x^3$, show that $f(-a) = -f(a)$

______

Question 7 Which of the graphs given below represent functions? ______

A

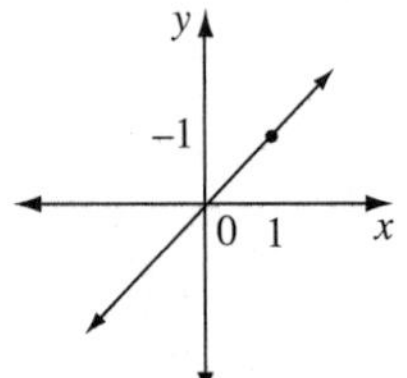

B

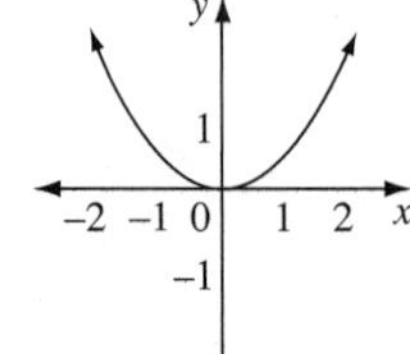

C

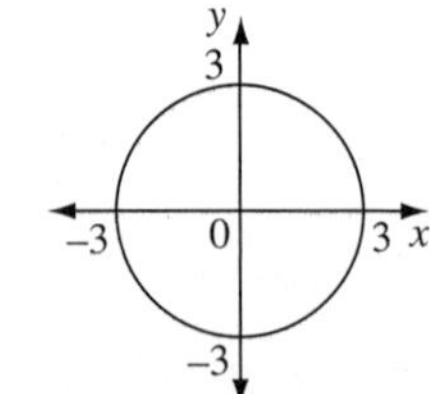

Excel Advanced-level Mathematics Study Guide Years 9–10
Pages 242–262

UNIT 8: Further work with functions

QUESTION 1 Write down the permissible x-values (domain) and permissible y-values (range) for each function.

a $y = x + 3$ **b** $y = x^2 + 2$ **c** $y = \frac{4}{x}$

d $y = 9 - x^2$ **e** $y = 5 - 3x$ **f** $y = \log_{10} x$

QUESTION 2 Consider the function $y = 2x + 4$

a What are the permissible x-values?

b What are the permissible y-values?

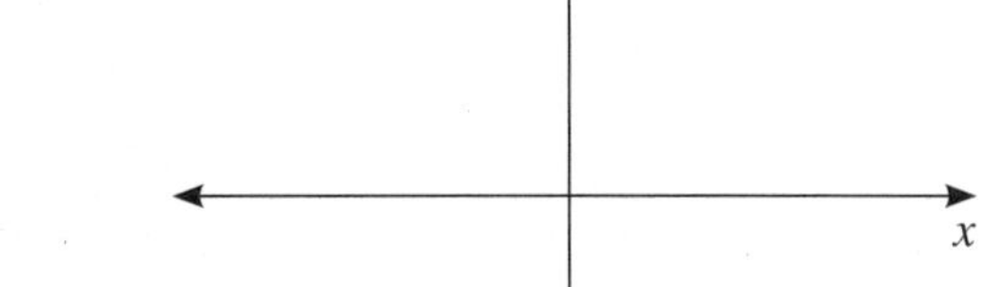

c Determine the inverse function.

d Sketch the graphs of the function and inverse function on the same set of axes.

QUESTION 3 Consider the function $f(x) = x^2, x \geq 0$

a Briefly describe why the domain needed to be restricted if the function is to have an inverse.

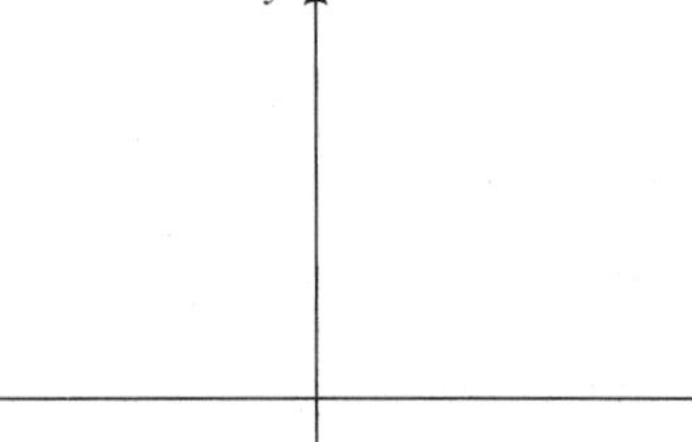

b Determine the inverse function.

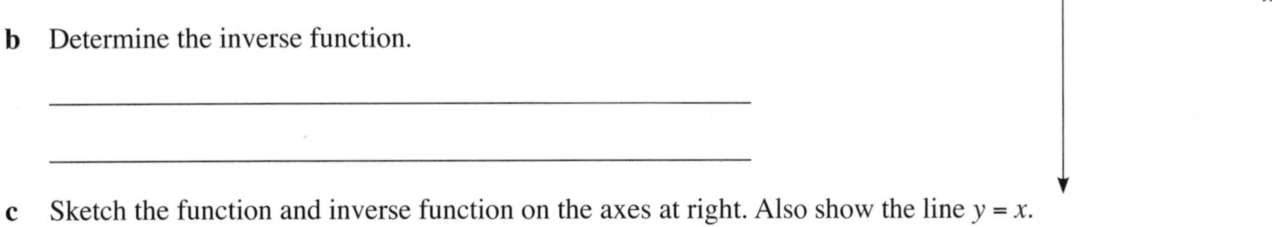

c Sketch the function and inverse function on the axes at right. Also show the line $y = x$.

d Briefly explain the role of the line $y = x$ in the diagram.

QUESTION 4 Determine the inverse function.

a $y = \frac{3}{x}$ **b** $y = x^3 + 1$ **c** $y = \log_3 x$

UNIT 9: Miscellaneous questions

QUESTION **1** If $f(x) = 3x - 4$, find:

a $f(1)$ ______ **b** $f(5)$ ______ **c** $f(2)$ ______

d $f(4)$ ______ **e** $f(-2)$ ______ **f** $f(3)$ ______

g $f(a)$ ______ **h** $f(-a)$ ______ **i** $f\left(\frac{1}{a}\right)$ ______

QUESTION **2** For the following exponential functions, complete the tables and, on the same axes, draw the graphs.

a $y = 2^x$ for $-2 \le x \le 2$

x	-2	-1	0	1	2
y					

b $y = 2^{2x}$ for $-2 \le x \le 2$

x	-2	-1	0	1	2
y					

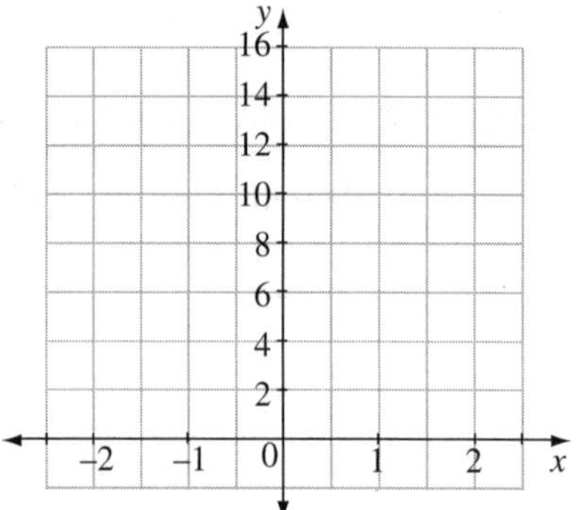

QUESTION **3** Write each of the following in logarithmic form.

a $2^5 = 32$ ______ **b** $3^3 = 27$ ______ **c** $4^5 = 1024$ ______

d $5^4 = 625$ ______ **e** $6^3 = 216$ ______ **f** $7^3 = 343$ ______

g $9^3 = 729$ ______ **h** $10^4 = 10\,000$ ______ **i** $3^7 = 2187$ ______

QUESTION **4** Write each of the following in index form.

a $\log_3 27 = 3$ ______ **b** $\log_2 128 = 7$ ______ **c** $\log_3 81 = 4$ ______

d $\log_5 625 = 4$ ______ **e** $\log_2 32 = 5$ ______ **f** $\log_3 243 = 5$ ______

g $\log_3 2187 = 7$ ______ **h** $\log_6 216 = 3$ ______ **i** $\log_{\sqrt{x}} 32 = 10$ ______

QUESTION **5** Use logarithm laws to simplify the following.

a $\log_4 64 + \log_4 16$ ______ **b** $\log_5 125 - \log_5 25$ ______ **c** $\log_a x^3 - \log_a x^2$ ______

d $\log_7 49 - \log_7 343$ ______ **e** $\log_6 4 + \log_6 9$ ______ **f** $\log_8 64 - \log_8 8$ ______

QUESTION **6** Solve the following equations.

a $3^x = 729$ ______ **b** $5^x = \frac{1}{625}$ ______ **c** $2^{3x-1} = 32$ ______

Logarithms, exponentials and functions

TOPIC TEST **PART A**

Instructions
- This part consists of 10 multiple-choice questions.
- Fill in only ONE CIRCLE for each question.
- Each question is worth 1 mark.
- Calculators are NOT allowed.

Time allowed: 10 minutes **Total marks: 10**

Marks

1 The graph of $y = 2^x$ passes through the point

Ⓐ $(0, 1)$ Ⓑ $(1, 0)$ Ⓒ $(0, -1)$ Ⓓ $(-1, 0)$ **1**

2 If $2^x = 32$ then x equals

Ⓐ 2 Ⓑ 32 Ⓒ 5 Ⓓ $\frac{1}{5}$ **1**

3 $\log_2 8$ equals

Ⓐ 2 Ⓑ 3 Ⓒ 4 Ⓓ 8 **1**

4 $\frac{1}{3}\log_2 64$ equals

Ⓐ 2 Ⓑ 3 Ⓒ 4 Ⓓ 6 **1**

5 Simplify $\log_a a^2$.

Ⓐ a Ⓑ 2 Ⓒ a^2 Ⓓ 2a **1**

6 Which statement is equivalent to $y = a^x$?

Ⓐ $y = \log_a x$ Ⓑ $y = \log_x a$ Ⓒ $x = \log_a y$ Ⓓ $x = \log_y a$ **1**

7 $\log_a x + \log_a y$ equals

Ⓐ $\log_a\left(\frac{x}{y}\right)$ Ⓑ $\log_a(xy)$ Ⓒ $\log_a x^y$ Ⓓ $\log_a y^x$ **1**

8 If $f(x) = 2x - 5$ then $f(-2)$ equals

Ⓐ 9 Ⓑ −9 Ⓒ 1 Ⓓ −1 **1**

9 $\log_2 100 - 2\log_2 5 =$

Ⓐ 10 Ⓑ 4 Ⓒ 2 Ⓓ 0.6 **1**

10 $\log_m m^6 + \log_m m^3 =$

Ⓐ 2 Ⓑ 3 Ⓒ 9 Ⓓ 18 **1**

Total marks achieved for PART A /10

Logarithms, exponentials and functions

TOPIC TEST **PART B**

Instructions
- This part consists of 5 questions.
- Each question part is worth 1 mark.
- Show all working.

Time allowed: 20 minutes **Total marks: 15**

Marks

1 Evaluate.

a $\log_3 27$ **b** $\log_2\left(\frac{1}{2}\right)$ 2

2 Simplify.

a $3\log_2 20 - 3\log_2 10$. **b** $\log mn - \log np + \log p$ 2

3 Rewrite the equation $x = a^b$ with b as the subject. 1

4 Solve.

a $3^x = 81$ **b** $2^{5x-1} = 16$ **c** $\log_3 x = 2$

d $\log_x 125 = 3$ **e** $4^x = 128$ **f** $7^{1-x} = 343$ 6

5 Consider the function $y = \log_4 x$

a What are the permissible x-values?

b What are the permissible y-values?

c What is the inverse function?

d Sketch $y = \log_4 x$ and its inverse function on the same set of axes. 4

Total marks achieved for PART B

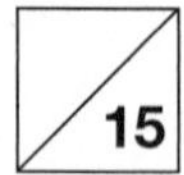

Chapter 5
Polynomials and curve sketching

Excel Advanced-level Mathematics Study Guide Years 9–10
Pages 229–241

UNIT 1: Polynomials in general

Question 1 Which of the following algebraic expressions are not polynomials? ____________

A $x^2 + 3x - 1$ B $2x^3 + 4x^2 - 8x - 3$ C $3^x + 9$

D $5x^3 - 8x^2 + 7x$ E $2x^2 - 3x^{\frac{1}{3}} + 6$ F $x^3 - \frac{1}{x^2} + 3x - 1$

Question 2 For the following polynomials, write the degree, the leading term, the leading coefficient and the constant term.

a $6x^3 + 3x^2 - 6x + 7$ b $4x^3 - 3x^2 - 6x - 2$ c $8x^5 + 3x^4 + 5x^3 - 6x^2$

d $5x^8 - 6x^3 + 7x^2 - 8x - 3$ e $5 - x + 3x^2 + 9x^3$ f $x^4 - 2x^3 + 9x - 5$

g $9x^2 + 11x - 8$ h $x^3 + 2x^2 - 7x + 2$ i $6x^5 - 4x^4 - x + 9$

Question 3 Which of the following are monic polynomials? ____________

A $3x^2 + 5x + 7$ B $x^5 - 4x^3 + 3x^2 - x + 1$ C $x^6 + 19$

D $8x^2 - 9x$ E $2x^3 + 3x^2 - 9x + 3$ F $x^4 + 3x^2 - 5x + 7$

G $x^3 - 3x^2 + 7x - 8$ H $x^3 - x^2 + 7$ I $8x^2 + 6x - 9$

Question 4 For the following polynomials, find the values indicated.

a $P(x) = x^2 + x - 1$ i $P(1) =$ ______ ii $P(3) =$ ______ iii $P(2) =$ ______

b $P(x) = x^3 + 3$ i $P(0) =$ ______ ii $P(1) =$ ______ iii $P(-1) =$ ______

c $P(x) = 8 - 2x$ i $P(1) =$ ______ ii $P(2) =$ ______ iii $P(-5) =$ ______

d $P(x) = x^3 + x^2$ i $P(0) =$ ______ ii $P(1) =$ ______ iii $P(2) =$ ______

Question 5 What is the degree of each polynomial?

a $5x^3(8x^2 + 7x - 6)$ ____________ b $(6x^5 + 9x^3 - 6x^2) + (7x - 3)$ ____________

Polynomials and curve sketching

Excel Advanced-level Mathematics Study Guide Years 9–10
Pages 229–241

UNIT 2: Addition of polynomials

QUESTION 1 Add the following polynomials.

a $(x^3+2x+1)+(4x^3+3x^2-9)=$ ____________

b $(3x^3-2x^2+5x-2)+(x^3+4x^2+9x-7)=$ ____________

c $(6x^5+3x^4-8x^2)+(9x^4+9x^2+9x)=$ ____________

d $(3x^6+8x^2+5x)+(x^2+6x-3)=$ ____________

e $(8x^2+7x-6)+(x^5-3)=$ ____________

f $(9x^3-8x^2+9)+(x^4-2x+7)=$ ____________

QUESTION 2 If $P(x)=3x^2+2x-5$, $Q(x)=x^3+8x^2-9x+7$, $A(x)=2x^3+6x^2+x$ and $B(x)=3x^5+x^4-x^3+11$, find the following.

a $P(x)+Q(x)=$ ____________

b $A(x)+B(x)=$ ____________

c $P(x)+A(x)=$ ____________

d $Q(x)+B(x)=$ ____________

e $P(x)+B(x)=$ ____________

f $Q(x)+A(x)=$ ____________

QUESTION 3 Add the following polynomials and write the degree of $P(x)+Q(x)$.

a $P(x)=x^3+2x^2-9,\ Q(x)=2x^4+9x-6$ ____________

b $P(x)=x^4+9x^2-5,\ Q(x)=x^3+x^2+x+1$ ____________

c $P(x)=5x^3-1,\ Q(x)=2x^2-9$ ____________

d $P(x)=6x^3+4x^2+5x,\ Q(x)=3x^2-6x-9$ ____________

e $P(x)=x^4-3x^2+8x,\ Q(x)=x^3-3x^2+8x$ ____________

f $P(x)=5x^3+6x^2-5x+7,\ Q(x)=8x^2+4x-9$ ____________

g $P(x)=7x^4+3x-6,\ Q(x)=x^4+3x^3+2x^2-x+12$ ____________

Polynomials and curve sketching

UNIT 3: Subtraction of polynomials

QUESTION 1 Subtract the following polynomials.

a $(x^3 + 2x^2 - x + 1) - (5x^2 + 7x - 9) =$ ____________________

b $(4x^3 - 3x^2 + 7x - 3) - (x^3 - x^2 + x + 1) =$ ____________________

c $(6x^5 + 3x^4 - 9x^2) - (8x^4 + 6x^2 - 3x) =$ ____________________

d $(2x^6 + 7x^3 - 8x) - (5x^4 + 9x^3 - 7x) =$ ____________________

e $(9x^2 + 5x - 7) - (x^2 - 3x - 11) =$ ____________________

f $(5x^3 - 7x^2 + 7) - (x^4 - 3x - 6) =$ ____________________

QUESTION 2 If $P(x) = 5x^2 + 4x - 7$, $Q(x) = 2x^3 + 6x^2 - 7x + 9$, $A(x) = 3x^3 + 7x^2 - x$ and $B(x) = 7x^5 + x^4 - x^3 + 2$, find the following.

a $P(x) - Q(x) =$ ____________________

b $A(x) - B(x) =$ ____________________

c $P(x) - A(x) =$ ____________________

d $Q(x) - B(x) =$ ____________________

e $P(x) - B(x) =$ ____________________

f $Q(x) - A(x) =$ ____________________

QUESTION 3 For the following polynomials find $P(x) - Q(x)$ and state its degree.

a $P(x) = x^3 + 3x - 7$, $Q(x) = 3x^4 + 7x - 6$ ____________________

b $P(x) = x^5 + 7x^2 - 9$, $Q(x) = 3x^2 - 2x + 1$ ____________________

c $P(x) = 8x^3 - 2x^2 + 5$, $Q(x) = 9x^2 - 7x + 3$ ____________________

d $P(x) = x^4 - 3x^3 + 7x^2$, $Q(x) = x^3 - 4x^2 + 7x$ ____________________

e $P(x) = x^5 - 3x^3 + 7x$, $Q(x) = 9x^2 - 6x + 3$ ____________________

f $P(x) = 6x^3 + 7x^2 - 9x + 2$, $Q(x) = 9x^2 - 6x + 3$ ____________________

g $P(x) = 8x^3 + 5x - 4$, $Q(x) = x^2 - 3x + 7$ ____________________

UNIT 4: Multiplication of polynomials

QUESTION 1 Expand and simplify the following.

a $(x+1)(x^2+5x+2)$

b $(x+2)(x^2-3x+4)$

c $(x-2)(x^2-3x-5)$

d $(x+3)(x^2+7x-8)$

e $(2x+1)(x^2-5x-3)$

f $(3x-2)(x^3+x^2-x-2)$

QUESTION 2 Find the following products.

a $(x^2-3)(x^2+7x-9)$

b $(x^2+2x)(x^2+2x+1)$

c $(6x^2-7)(x^3+2x^2-3x+4)$

d $(4x^3-2x)(8x^2-3x+5)$

e $(2x^3-3)(3x^2-8x+1)$

f $(3x^2+2x+1)^2$

QUESTION 3 Find the product of the following polynomials and state the degree of the product.

a $5x^2+3x-9\ \times$
$4x+1$

b $2x^3+3x^2-x\ \times$
x^2-2

c $3x^4-5x^2+3\ \times$
$2x+1$

Polynomials and curve sketching

UNIT 5: Division of polynomials

QUESTION 1 Complete these divisions.

a $x + 1\overline{)x^2 + 3x + 5}$

b $x - 1\overline{)x^3 + 3x^2 + 2x - 1}$

c $3x - 2\overline{)6x^3 + 11x^2 - 19x + 2}$

QUESTION 2 Divide $P(x)$ by $A(x)$ and express the result in the form $P(x) = A(x)Q(x) + R$

a $P(x) = 5x^2 + 6x + 3 \quad A(x) = x + 2$

b $P(x) = x^2 - 5x + 16 \quad A(x) = x - 1$

c $P(x) = 6x^3 + 4x^2 - 7x + 5 \quad A(x) = x + 1$

d $P(x) = x^4 - 3x^3 - 5x^2 + 7x - 9 \quad A(x) = x - 4$

e $P(x) = x^3 + 2x^2 - 8 \quad A(x) = x + 5$

f $P(x) = x^4 + 3x^2 + 2x \quad A(x) = x + 1$

UNIT 6: The remainder theorem

QUESTION 1 Use the remainder theorem to find the remainder for the following divisions.

a $(2x^2 - 3x + 6) \div (x + 1)$

b $(x^2 + 5x + 6) \div (x + 1)$

c $(3x^2 + 4x - 5) \div (x + 2)$

d $(2x^2 + 3x - 14) \div (x - 1)$

e $(2x^3 + 5x^2 - 7x + 6) \div (x - 3)$

f $(3x^4 - 2x^3 + 9x - 5) \div (x - 1)$

QUESTION 2 Find the remainder when $P(x)$ is divided by $A(x)$.

a $P(x) = 3x^2 + 9x - 6,\ A(x) = x + 2$

b $P(x) = 2x^2 + 7x - 8,\ A(x) = x + 1$

c $P(x) = x^3 + 5x^2 - 10x - 6,\ A(x) = x + 4$

d $P(x) = x^4 - 7x^3 + 6x^2,\ A(x) = x + 1$

e $P(x) = 2x^3 + 2x^2 - 8x + 3,\ A(x) = x + 1$

f $P(x) = 5x^3 + 7x^2 - 8x + 1,\ A(x) = x - 2$

QUESTION 3

a When $px^3 + x^2 - 3x - 2$ is divided by $(x + 2)$, the remainder is zero. Find the value of p.

b When $x^3 + x^2 - mx + 4$ is divided by $(x - 2)$, the remainder is 8. Find the value of m.

Polynomials and curve sketching

UNIT 7: The factor theorem

QUESTION 1 In each of the following, show that the first polynomial is a factor of $P(x)$ and hence find all the factors of $P(x)$.

a $(x+1),\ P(x)=5x^2+8x+3$

b $(x-1),\ P(x)=x^3+x^2-5x+3$

c $(2x-1),\ P(x)=2x^3-11x^2+17x-6$

d $(x+3),\ P(x)=x^3+12x^2+47x+60$

QUESTION 2 Use the factor theorem to show that $A(x)$ is a factor of $P(x)$ and hence find all the factors of $P(x)$.

a $P(x)=x^3+6x^2+11x+6,\quad A(x)=x+2$

b $P(x)=2x^3+5x^2-23x+10,\quad A(x)=x-2$

c $P(x)=2x^3+5x^2+7x+6,\quad A(x)=2x+3$

d $P(x)=2x^3-3x^2-2x+3,\quad A(x)=2x-3$

QUESTION 3 Use the factor theorem to factorise the following.

a $4x^3-17x^2-16x+5$

b $3x^3+4x^2-5x-2$

c $2x^3+7x^2+2x-3$

d $4x^3-27x^2+33x+10$

QUESTION 4 If x^3-x^2+px+2 is divisible by $x-1$, find the value of p.

Polynomials and curve sketching

Excel Advanced-level Mathematics Study Guide Years 9–10
Pages 229–241

UNIT 8: Zeros of a polynomial

QUESTION 1 Consider the polynomial $P(x) = x^3 - 3x^2 - 6x + 8$

a Show that $(x - 1)$ is a factor of $P(x)$

b Divide $P(x)$ by $(x - 1)$

c Write $P(x)$ in fully factorised form.

d How many zeros does $P(x)$ have?

QUESTION 2 Consider the polynomial $P(x) = x^4 + 2x^3 - 13x^2 - 14x + 24$

a Show that $(x - 3)$ is a factor of $P(x)$.

b $P(x) = (x - 3)Q(x)$. Find $Q(x)$.

c Show that $(x + 2)$ is a factor of $Q(x)$.

d Divide $Q(x)$ by $(x + 2)$

e Write $P(x)$ in fully factorised form.

f How many zeros does $P(x)$ have? ____________

QUESTION 3 Complete: a polynomial of degree n has at most ____________ zeros.

QUESTION 4 Find all zeros of these polynomials.

a $x^3 + 4x^2 + x - 6$

b $x^3 + 8x^2 + 19x + 12$

c $x^3 - 3x^2 - 16x - 12$

d $x^4 + 3x^3 - 4x^2 - 12x$

Polynomials and curve sketching

UNIT 9: Solving equations involving polynomials

Question 1 For what values of x does $P(x) = 0$?

a $P(x) = (x - 1)(x + 3)(x + 7)$

b $P(x) = x(x + 1)(x + 2)(x - 8)$

Question 2

a Show that $(x - 3)$ is a factor of $P(x) = x^3 - 7x - 6$

b Hence solve $x^3 - 7x - 6 = 0$

Question 3 Given that $x = \frac{3}{4}$ is a root of the equation $4x^3 - 23x^2 + 39x - 18 = 0$, find the other roots.

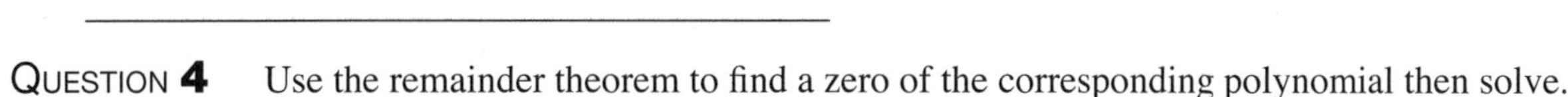

Question 4 Use the remainder theorem to find a zero of the corresponding polynomial then solve.

a $x^3 + 6x^2 + 3x - 10 = 0$

b $x^3 - 6x^2 - x + 30 = 0$

c $2x^3 + x^2 - 12x + 9 = 0$

d $x^4 - x^3 - 10x^2 - 8x = 0$

Polynomials and curve sketching

UNIT 10: Curve sketching

QUESTION 1 Sketch the following curves.

a $y = x$

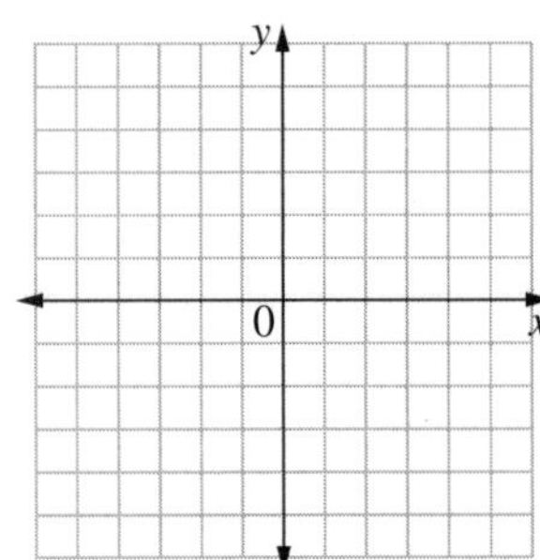

b $y = x^2$

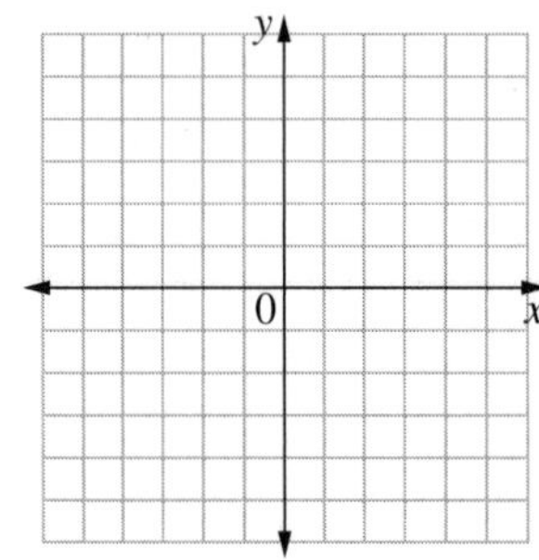

c $y = x^3$

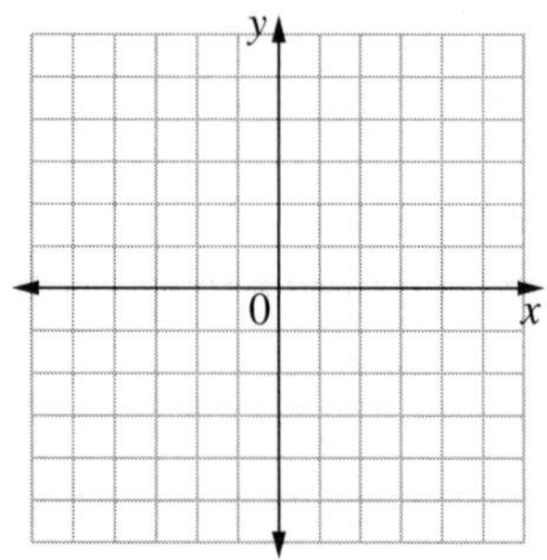

d $y = x^4$

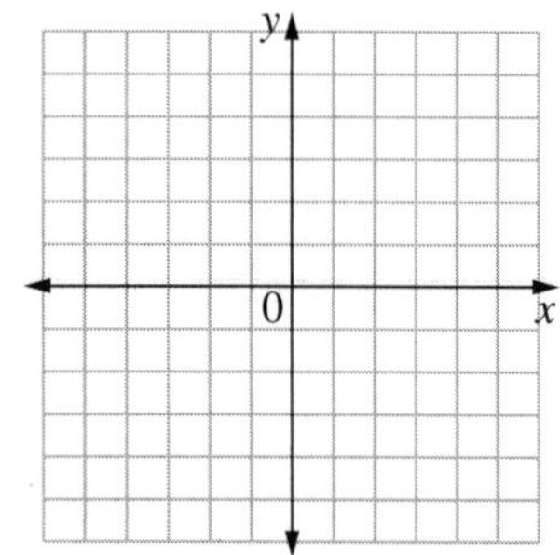

QUESTION 2 Make sketches of the following curves.

a $y = 3x + 5$

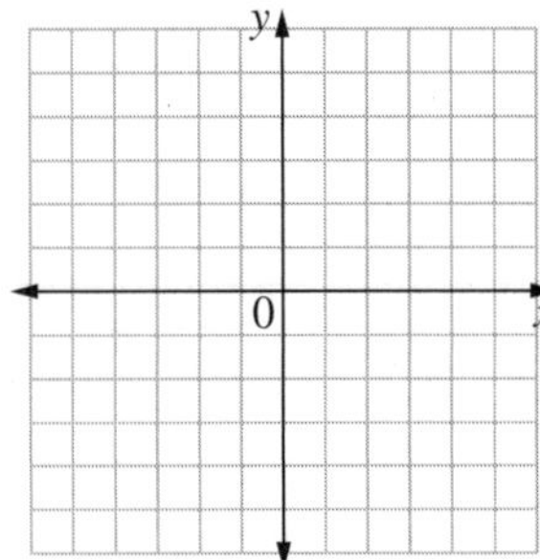

b $y = x^2 + 3$

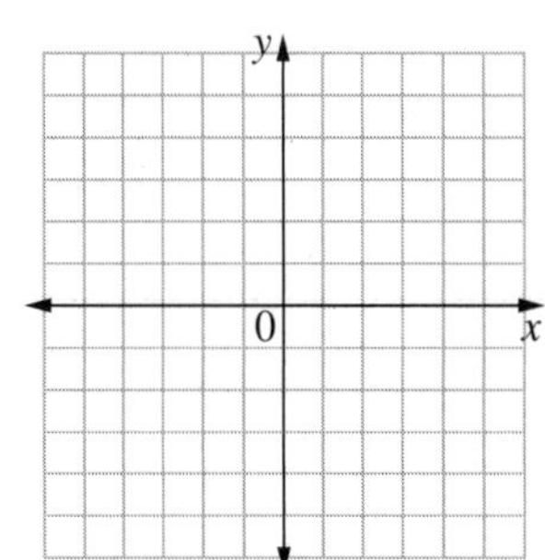

c $y = 2x^3$

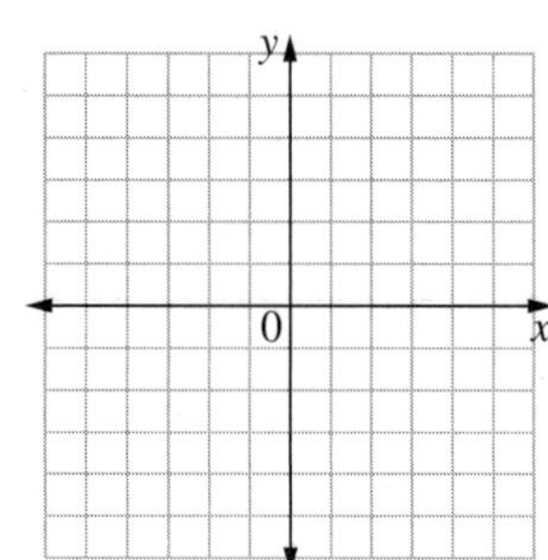

d $y = -x^3 + 1$

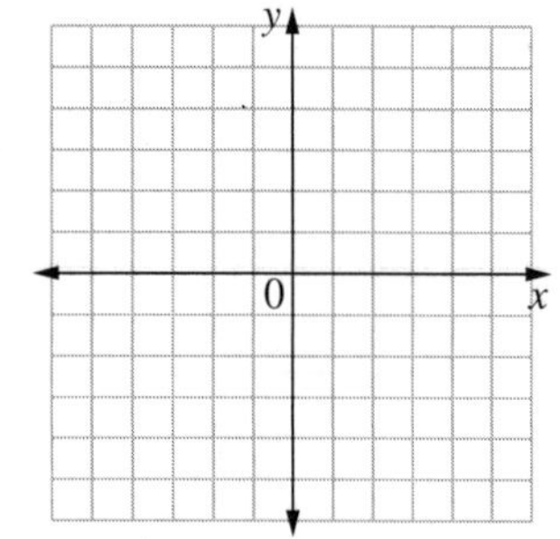

QUESTION 3 Sketch each of the following curves.

a $y = (x - 1)^2$

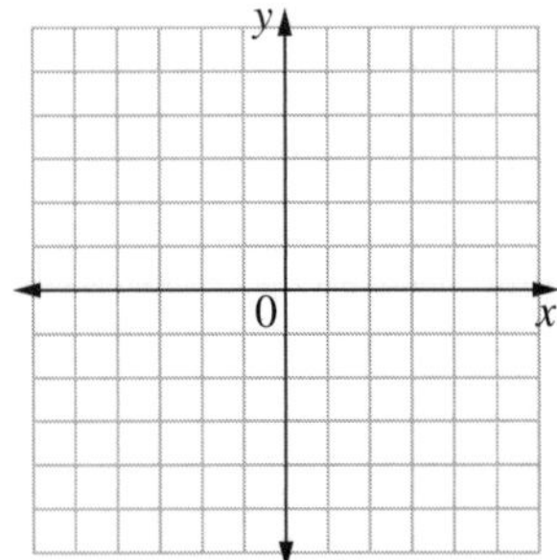

b $y = 3(x + 2)^2$

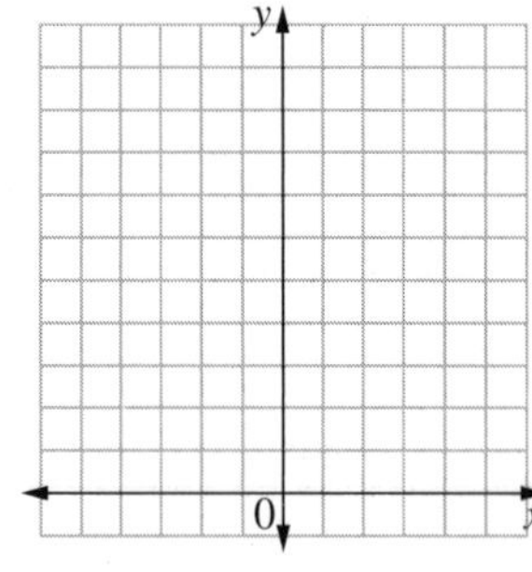

c $y = (x - 3)^3$

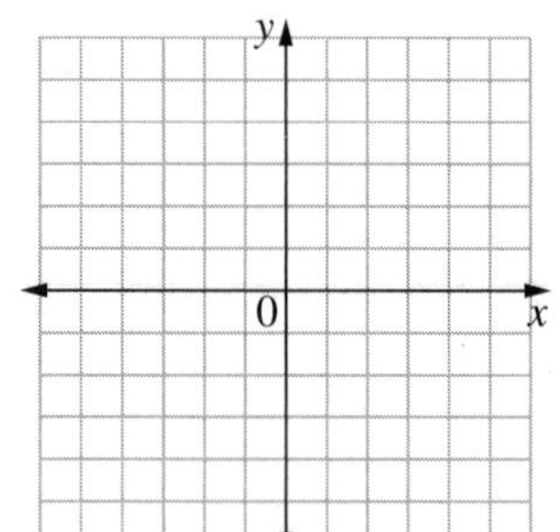

d 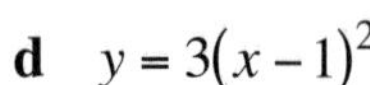 $y = 3(x - 1)^2$

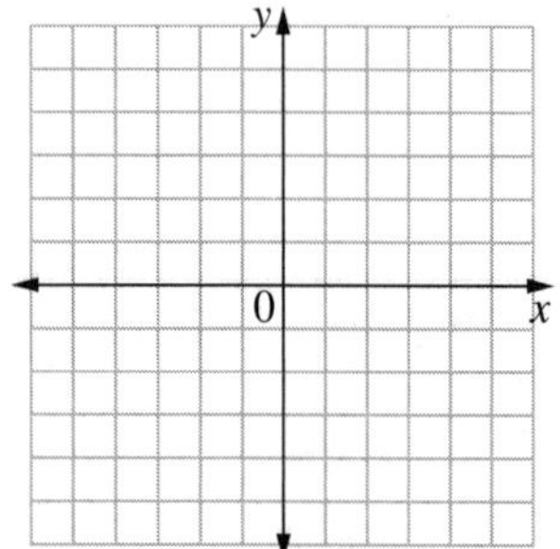

QUESTION 4 Sketch each of the following curves.

a $y = (x - 1)(x + 2)(x - 3)$

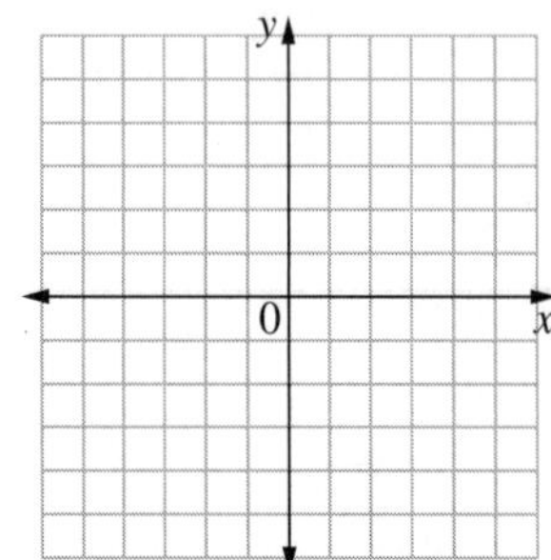

b $y = 3(x - 1)(x - 4)(x + 2)$

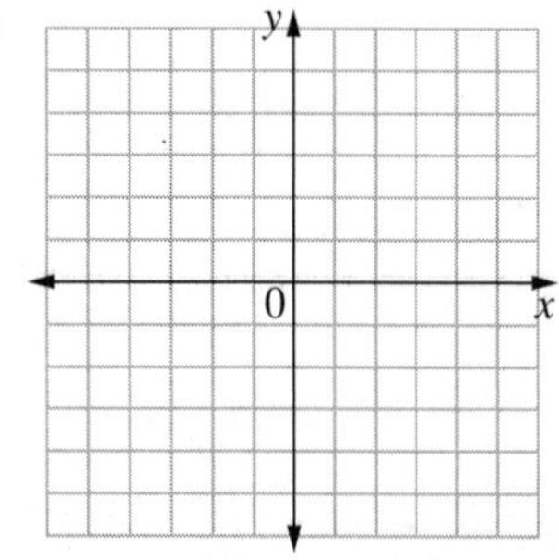

Polynomials and curve sketching

Excel Advanced-level Mathematics Study Guide Years 9–10
Pages 229–241

UNIT 11: Sketching polynomials (1)

QUESTION 1 $P(x) = x^3 + 3x^2 - 6x - 8 = (x + 4)(x + 1)(x - 2)$. Consider the curve $y = P(x)$.

a What is the y-intercept?

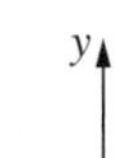

____

b What are the x-intercepts?

____

c What happens to y when x gets very large positively?

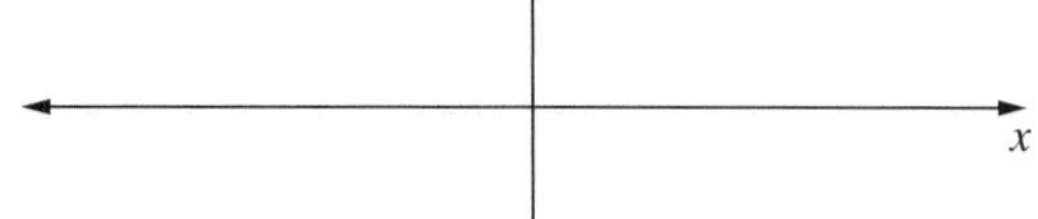

____

d What happens to y when x gets very large negatively?

____

e Sketch the curve $y = P(x)$ on the axes provided.

QUESTION 2 Sketch the graph of $y = P(x)$ given that:

a $P(x) = -x^3 + 3x^2 + 28x - 60$
$= -(x + 5)(x - 2)(x - 6)$

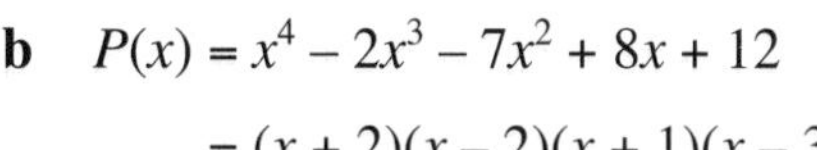

b $P(x) = x^4 - 2x^3 - 7x^2 + 8x + 12$
$= (x + 2)(x - 2)(x + 1)(x - 3)$

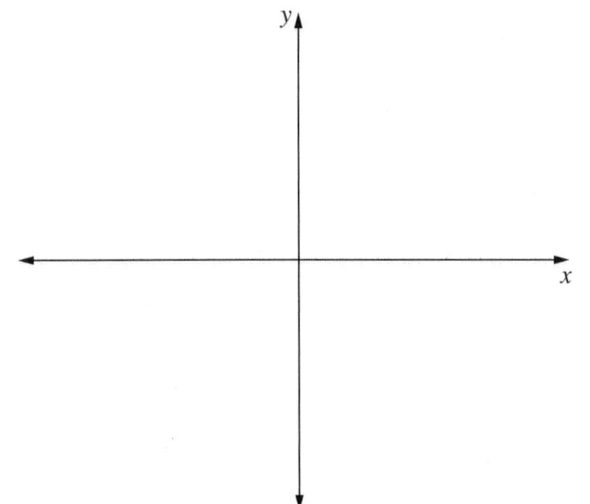

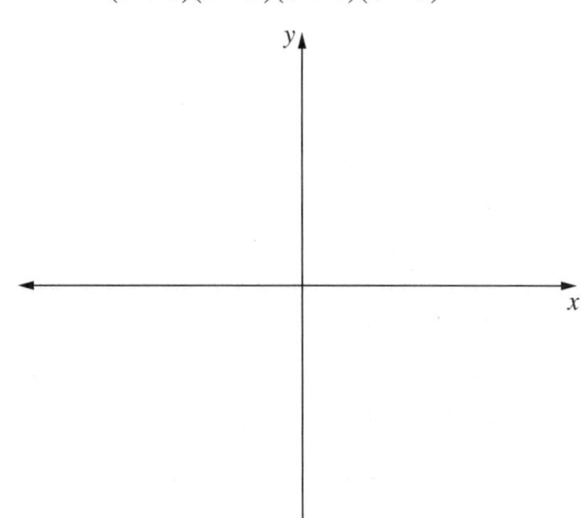

QUESTION 3

a Factorise $x^3 + 7x^2 + 4x - 12$

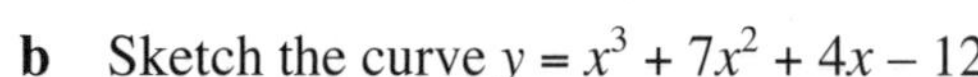

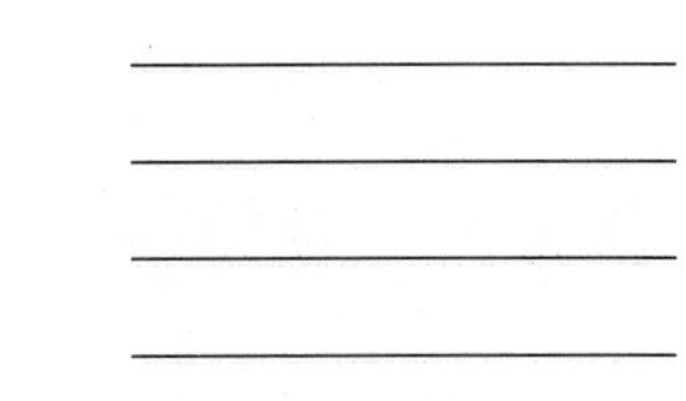

b Sketch the curve $y = x^3 + 7x^2 + 4x - 12$

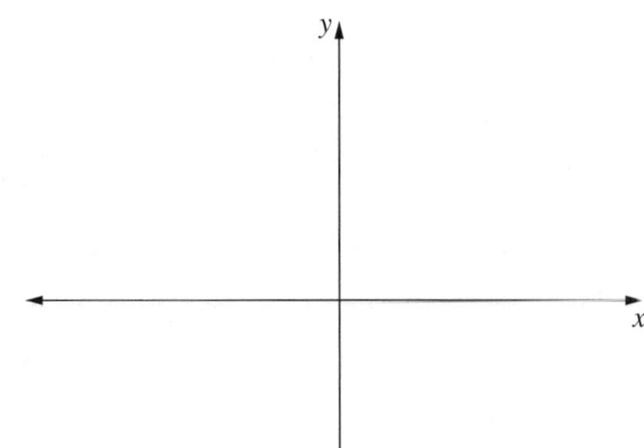

Polynomials and curve sketching

UNIT 12: Sketching polynomials (2)

QUESTION 1 Sketch the graph of $y = 2x^4 - 8x^3 + 2x^2 + 12x$

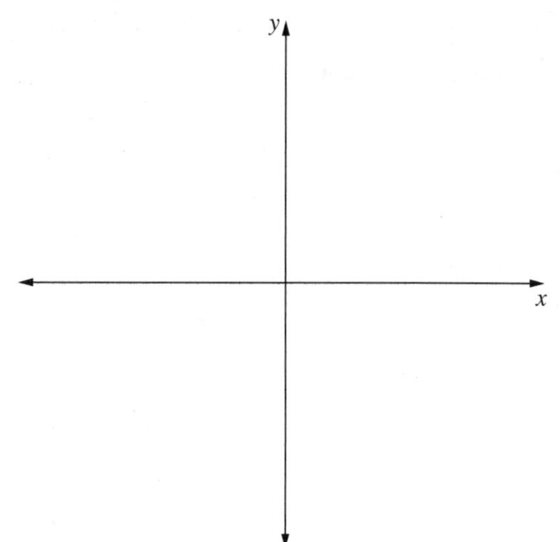

QUESTION 2 Consider the polynomial $P(x) = x^3 + x^2 - 5x + 3$

a Factorise $P(x)$.

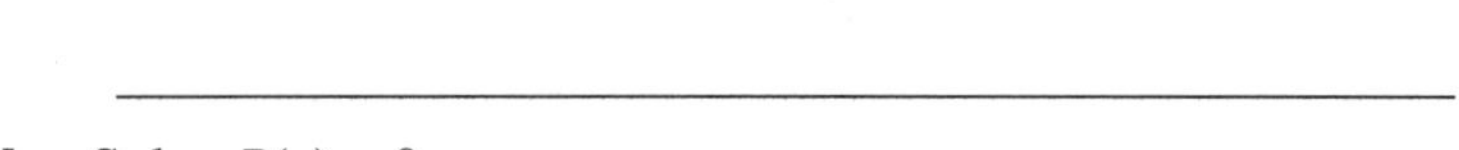

b Solve $P(x) = 0$

c Which root is a double root of $P(x) = 0$? ________

d Find $P(0)$ ________

e Find $P(2)$ ________

f At how many points will the graph **cut** the x-axis? ________

g Sketch $y = P(x)$.

QUESTION 3 Consider $P(x) = x^4 - 6x^2 + 8x - 3 = (x + 3)(x - 1)^3$

a Which is a triple root of $P(x) = 0$?

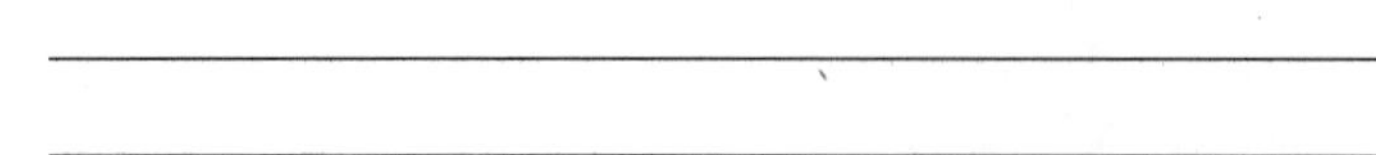

b Find $P(0)$ ________

c Find $P(2)$ ________

d Does the graph cut the x-axis at $x = 1$? ________

e Sketch the curve $y = P(x)$.

Excel Advanced-level Mathematics Study Guide Years 9–10
Pages 229–241

UNIT 13: Sketching polynomials (3)

QUESTION 1 The diagram shows a sketch of $y = P(x)$. On the same diagram sketch:

a $y = P(x) + 3$

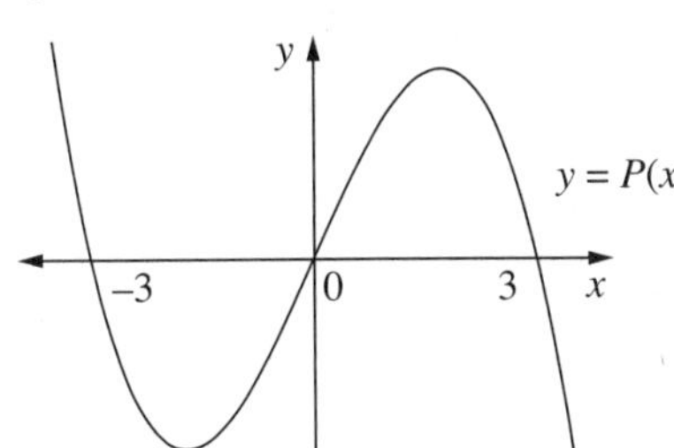

b $y = 2P(x)$

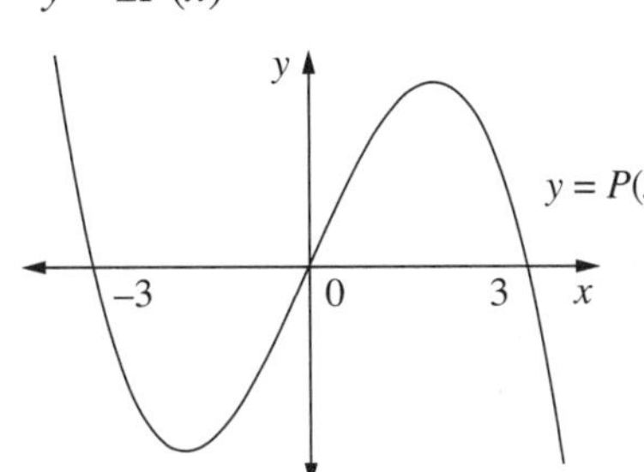

c $y = -P(x)$

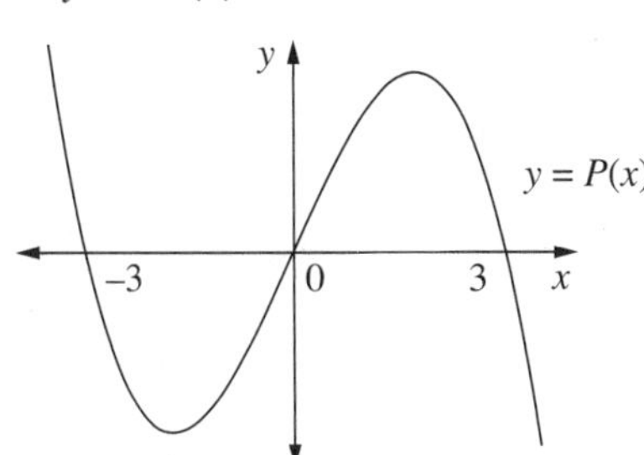

QUESTION 2 Sketch $y = P(-x)$ on the same axes.

a

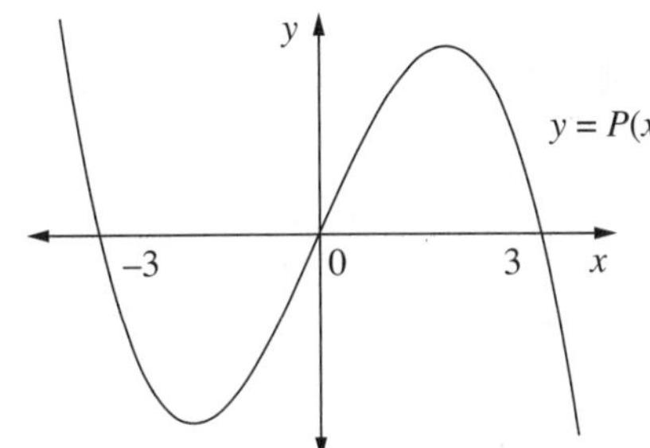

b

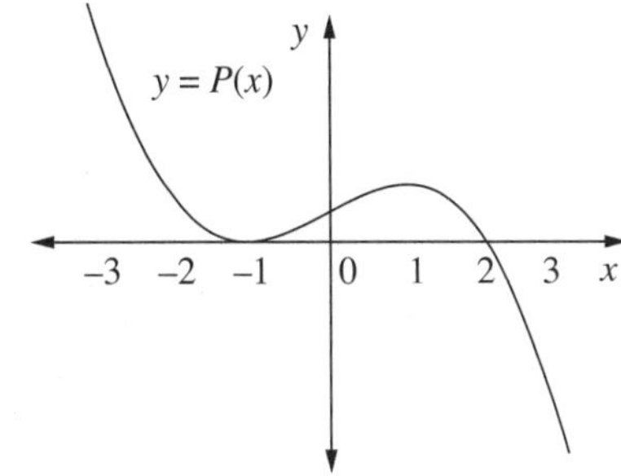

c

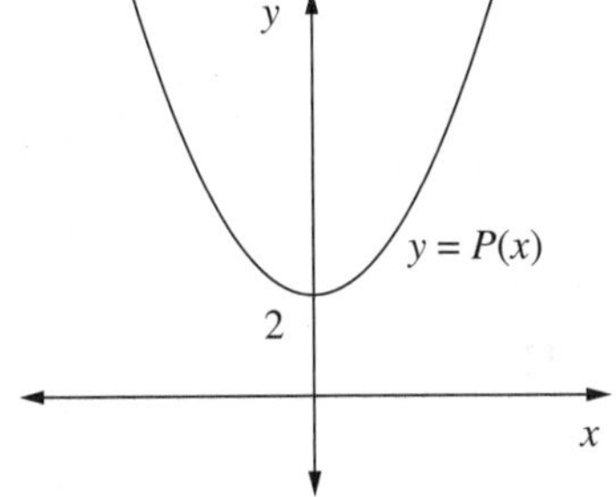

d

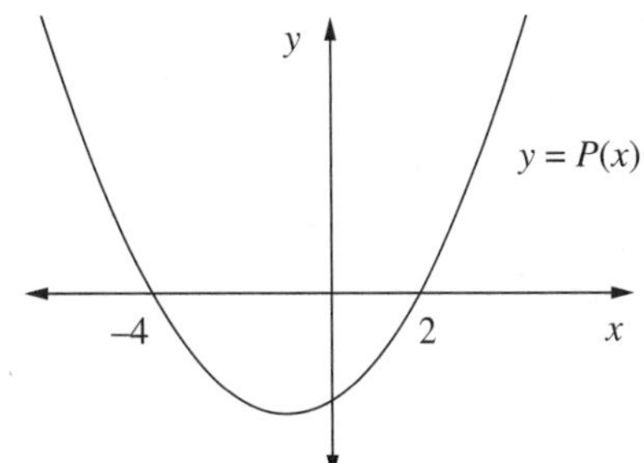

QUESTION 3 Briefly describe the effect on the graph of $y = P(x)$ of:

a adding or subtracting a constant ______________________________

____

b multiplying or dividing by a positive constant ______________________________

____

c multiplying by –1 ______________________________

____

d replacing x with $-x$ ______________________________

____

Polynomials and curve sketching

TOPIC TEST — PART A

Instructions
- This part consists of 10 multiple-choice questions.
- Fill in only ONE CIRCLE for each question.
- Each question is worth 1 mark.

Time allowed: 10 minutes — **Total marks: 10**

Marks

1 Which of these is not a polynomial? 

Ⓐ $3x + 1$ Ⓑ $x^2 + 2x - 3$ Ⓒ $7 + 5x - 6x^2 + x^8$ Ⓓ $3x^3 - 2x^{-2} + 5x + 1$

2 The degree of the polynomial $7 - 2x + x^4$ is — 1

Ⓐ 0 Ⓑ 1 Ⓒ 3 Ⓓ 4

3 In the polynomial $2x^3 + 8x^2 - 9x + 7$, the leading coefficient is — 1

Ⓐ 2 Ⓑ 8 Ⓒ 9 Ⓓ 7

4 In the polynomial $x^5 + 3x^2 + 7x + 5$, the constant term is — 1

Ⓐ x^5 Ⓑ $3x^2$ Ⓒ $7x$ Ⓓ 5

5 The polynomial is monic if the leading coefficient is — 1

Ⓐ 1 Ⓑ 2 Ⓒ 3 Ⓓ 4

6 The degree of the polynomial $6x^2\left(x^5 + 3x^2 - 1\right)$ is — 1

Ⓐ 2 Ⓑ 4 Ⓒ 5 Ⓓ 7

7 What is the remainder when $x^4 - 3x^3 + 5x^2 - 7x + 11$ is divided by $(x - 3)$? — 1

Ⓐ 2 Ⓑ 239 Ⓒ 35 Ⓓ 11

8 If $P(x) = (x - 2)(x^2 - 5x + 7) + 3$ which is $P(x)$?

Ⓐ $x^3 - 7x^2 + 17x - 17$ Ⓑ $x^3 - 7x^2 + 17x - 11$

Ⓒ $x^3 - 7x^2 - 3x - 17$ Ⓓ $x^3 - 7x^2 - 17x - 17$ 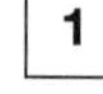

9 Which of these is a factor of $x^3 + 3x^2 - 13x - 15$? 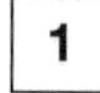

Ⓐ $x - 1$ Ⓑ $x + 1$ Ⓒ $x + 3$ Ⓓ $x - 5$

10 $(x + 2)$ is a factor of $x^3 + bx^2 + 7x - 6$. What is the value of b? — 1

 7 Ⓑ –2 Ⓒ 4 Ⓓ –4

Total marks achieved for PART A 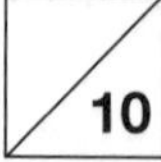 /10

Polynomials and curve sketching

TOPIC TEST — PART B

Instructions
- This part consists of 5 questions.
- Each question part is worth 1 mark.
- Show all working.

Time allowed: 20 minutes — **Total marks: 15**

	Marks
1 For the polynomial $x^3 + 4x^2 - 9$, write down the: **a** degree ________ **b** leading coefficient ________ **c** coefficient of x^2 ________ **d** constant term ________	4
2 Find in simplest form. **a** $\left(x^3 - x^2 + 2x - 8\right) + \left(5x^3 + 3x^2 - 9x + 11\right)$ ________ **b** $\left(4x^2 - 8x - 12\right) - \left(2x^2 + 7x + 9\right)$ **c** $(x+2)\left(x^2 - 5x + 6\right)$ ________ ________	3
3 Complete these long divisions and write the result in the form $P(x) = A(x)Q(x) + R$. **a** $x - 2\overline{)x^3 + 7x^2 - 2x - 4}$ **b** $x + 7\overline{)x^4 + 5x^3 - 9x^2 + 34x - 4}$ ________ ________	2
4 Find the remainder. **a** $x - 3\overline{)x^2 + 2x - 7}$ **b** $x + 1\overline{)2x^2 + 6x - 1}$ ________ ________	2
5 **a** Factorise $y = x^3 - 3x^2 - 10x$ fully 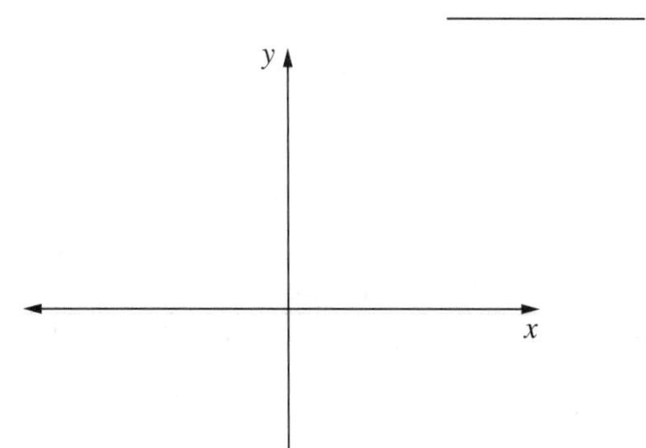 **b** What is the y-intercept? ________ **c** What are the x-intercepts? ________ **d** Sketch the curve $y = x^3 - 3x^2 - 10x$.	4

Total marks achieved for PART B ___ / 15

Chapter 6

Surface area and volume

Excel Advanced-level Mathematics Study Guide Years 9–10
Pages 144–158

UNIT 1: Surface area of different solids

Question 1 Find the surface area of the following rectangular prisms.

a

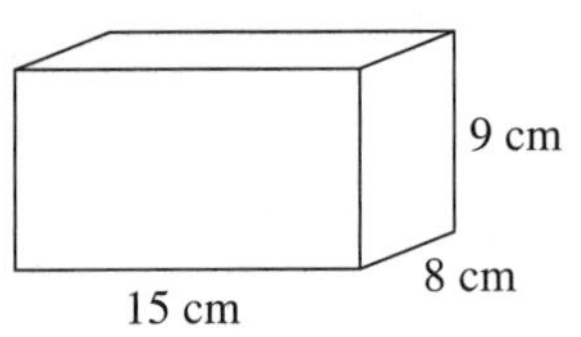

b

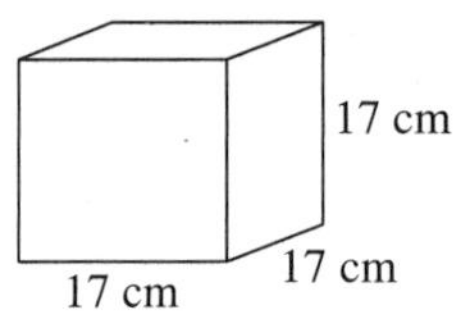

c

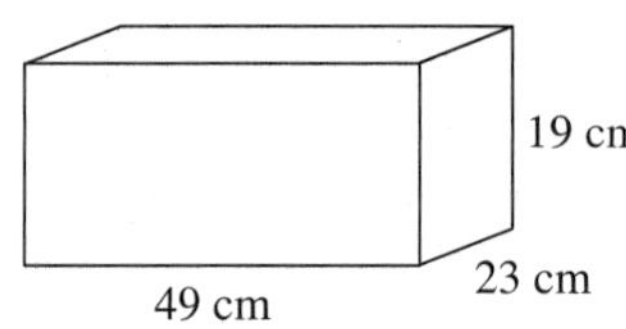

Question 2 Find the surface area of the following triangular prisms.

a

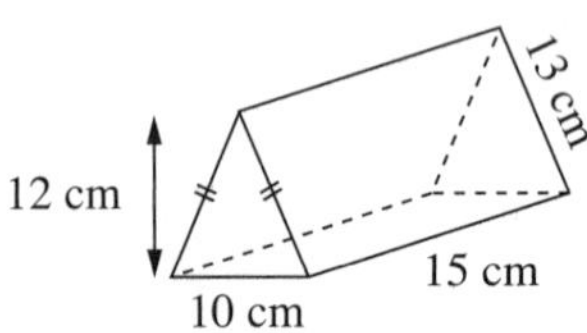

b

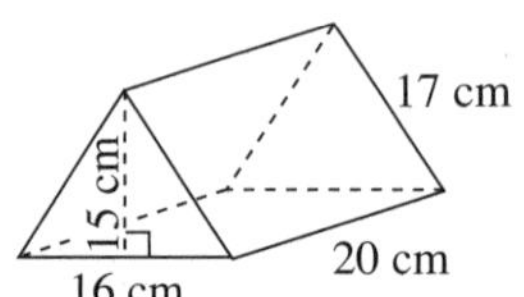

c

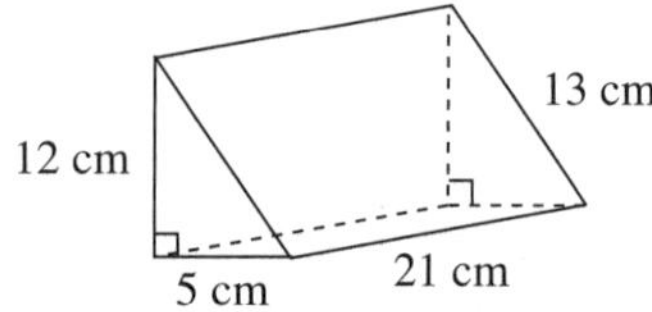

Question 3 Find the surface area of the following trapezoidal prisms.

a

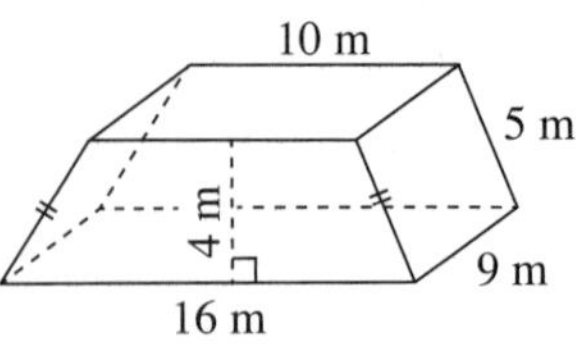

b

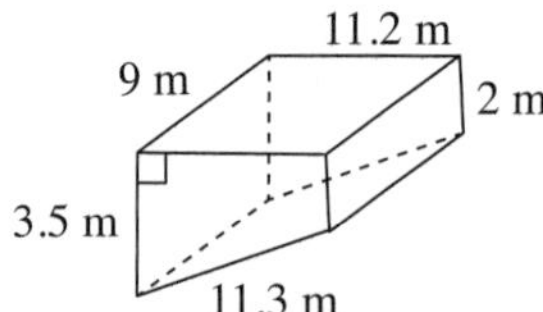

c

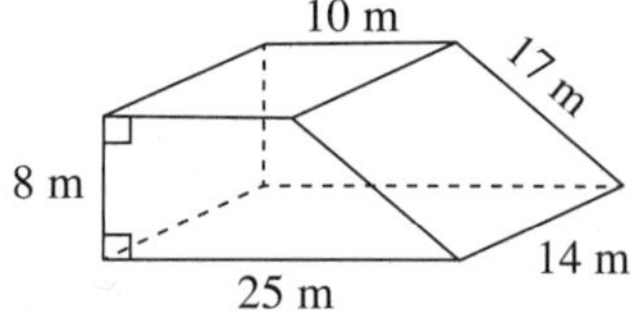

Question 4 Find the surface area of the following closed cylinders correct to one decimal place.

a

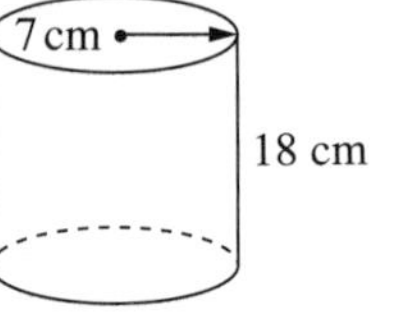

b

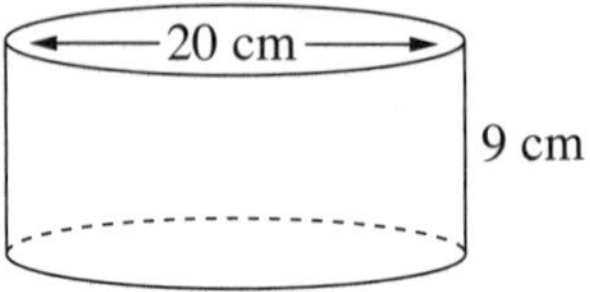

c

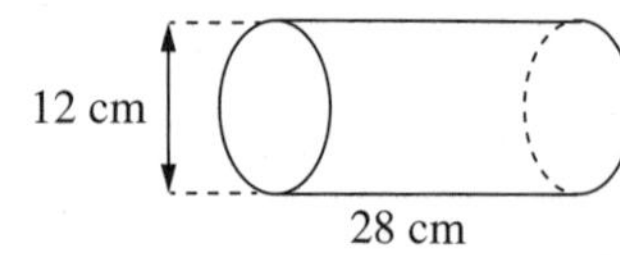

Surface area and volume

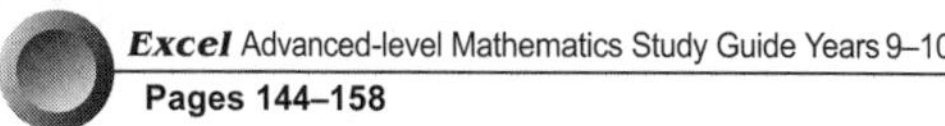

UNIT 2: Perpendicular height and slant height

QUESTION 1 This cone has radius 39 mm and perpendicular height 80 mm. Find the slant height of the cone.

80 mm
39 mm

QUESTION 2 This cone has diameter 32 cm and slant height 28.1 cm. Find the perpendicular height of the cone.

32 cm
28.1 cm

QUESTION 3 The perpendicular height, *OP*, of this rectangular pyramid is 12 cm. *M* is the midpoint of *AB* and *N* is the midpoint of *BC*. *AB* = 32 cm and *BC* = 10 cm. Find the length of:

a *PM* **b** *PN*

P, 12 cm, D, C, O, N, 10 cm, A, M, B, 32 cm

QUESTION 4 In this rectangular pyramid *AB* = 30 m and *BC* = 12 m. *M* is the midpoint of *AB* and *N* is the midpoint of *BC*. *PM* = 10 m. Find:

a the perpendicular height *OP* **b** the slant height *PN*

P, 10 m, D, C, O, N, 12 m, A, M, B, 30 m

QUESTION 5 A square pyramid has perpendicular height, *OP*, of 7 cm and slant height, *AP*, of 21 cm. Find the length of the base of the pyramid.

P, 7 cm, 21 cm, D, C, O, A, B

QUESTION 6 A rectangular pyramid has perpendicular height *OP* of 3.6 m. *M* is the midpoint of *AB* and *N* is the midpoint of *BC*. *PM* = 6 m and *PN* = 3.9 m. Find the length and width of the base.

P, 3.6 m, 6 m, 3.9 m, D, C, O, N, A, M, B

Surface area and volume

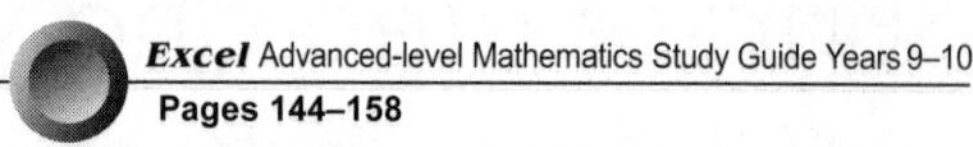

UNIT 3: Surface area of pyramids

QUESTION 1 Calculate the surface area of the following square pyramids correct to four significant figures.

a

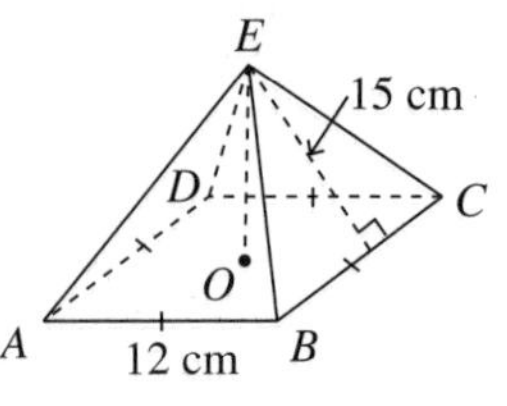

b

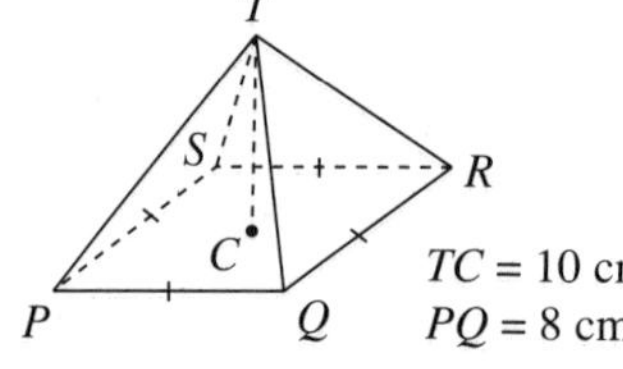

c

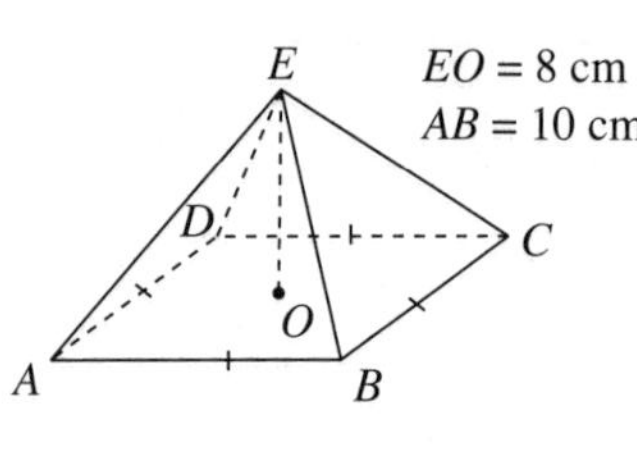

QUESTION 2 Calculate the surface area of the following rectangular pyramids correct to four significant figures.

a

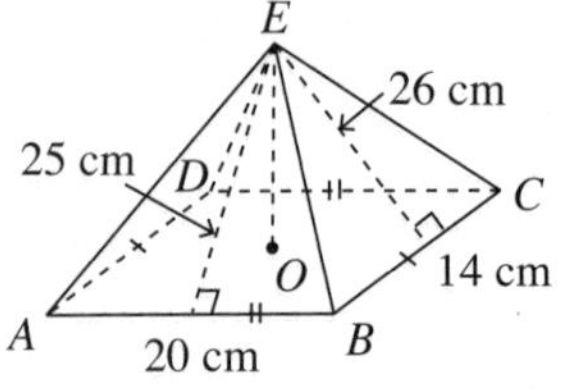

b

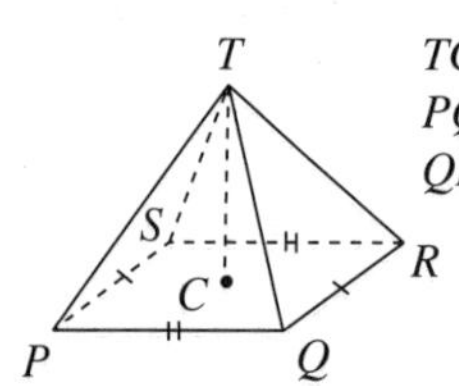

c

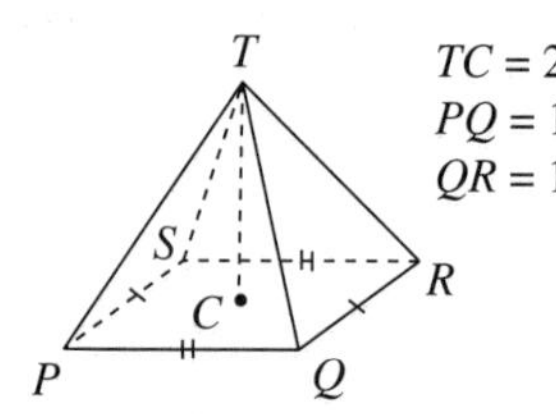

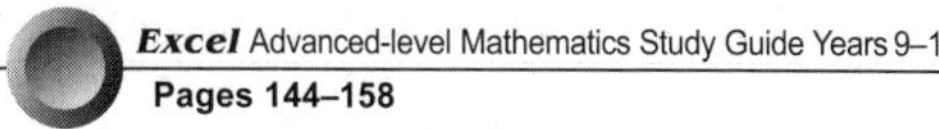

UNIT 4: Surface area of cones

QUESTION 1 Find the **curved** surface area of the following cones correct to one decimal place.

a 12 cm, 8 cm

b 18 cm, 12 cm

c 15 cm, 7 cm

d 34 cm, 28 cm

QUESTION 2 Find the total surface area of a cone with the following measurements. Give answers in terms of π.

a diameter 24 cm, slant height 10 cm

b radius 16 cm, perpendicular height 30 cm

QUESTION 3 The **curved** surface area of a cone is 795 cm^2. The radius of the cone is 11 cm. Find the slant height.

QUESTION 4 The total surface area of a cone is 36π cm^2. The slant height is 5 cm. Find the diameter of the cone.

Surface area and volume

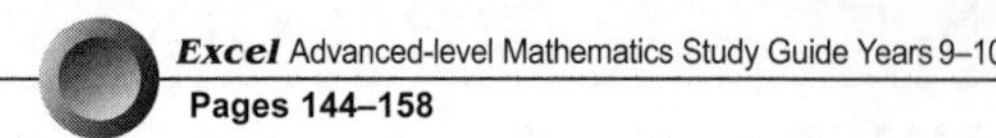

UNIT 5: Surface area of a sphere

QUESTION 1 Find the surface area of the following spheres, giving your answers in terms of π, with:

a radius = 7 cm

b diameter = 18 cm

c radius = 28 cm

d diameter = 42 cm

e radius = 8.3 cm

f diameter = 23.9 cm

QUESTION 2 Calculate the surface area of the following spheres. Leave your answer in terms of π.

a

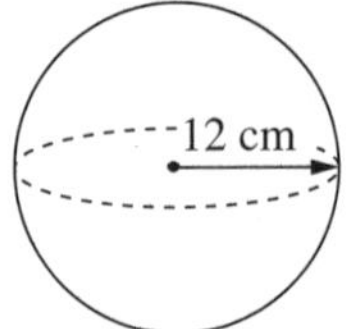

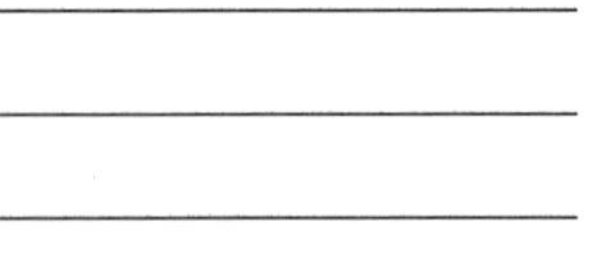

b

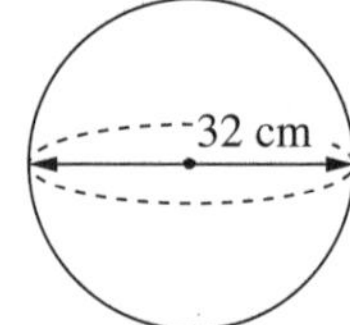

QUESTION 3 Calculate the surface area of the following hemispheres correct to two decimal places.

a

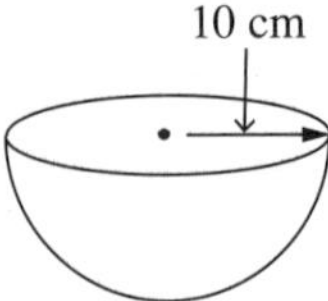

b

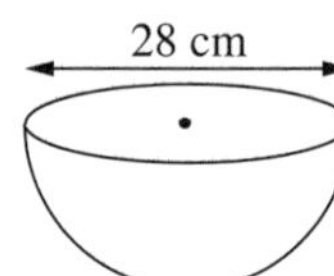

QUESTION 4 Find the external surface area of the following solids correct to three significant figures.

a

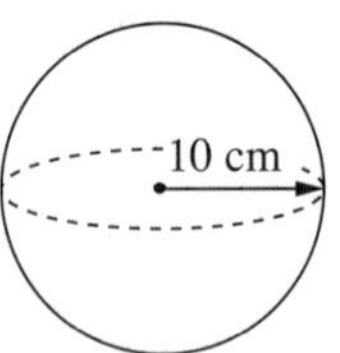

b

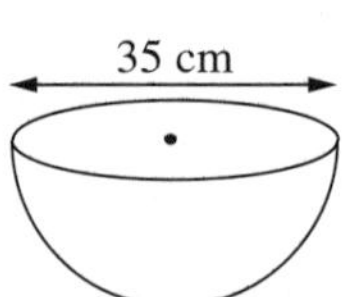

c

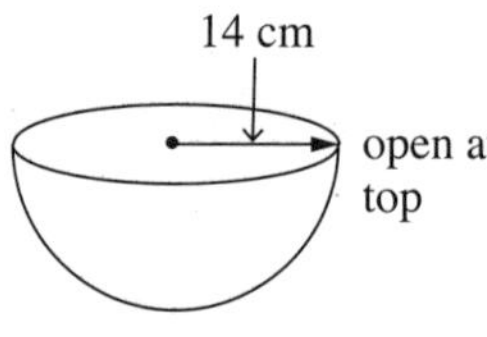

QUESTION 5 A sphere has a surface area of 360 cm^2. Find its radius correct to two decimal places.

Surface area and volume

UNIT 6: Surface area of composite solids

QUESTION **1** Find the total surface area of these solids. Give each answer to the nearest square centimetre.

a

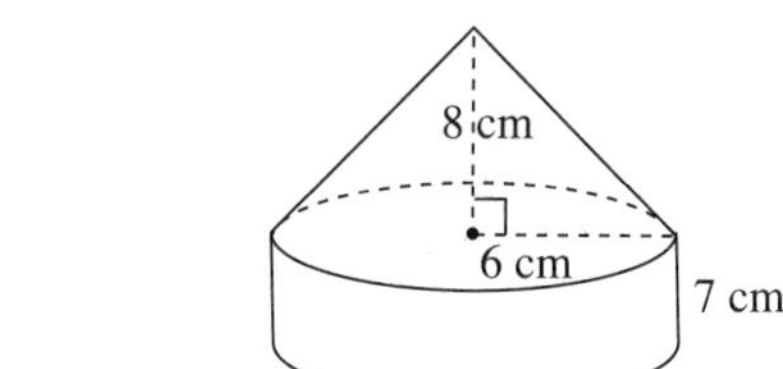

b

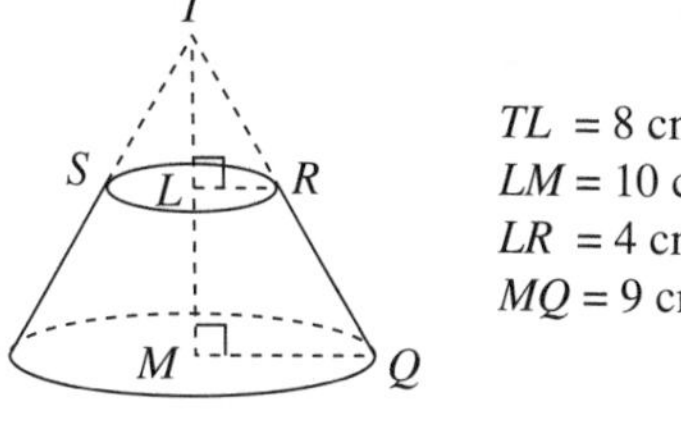

c

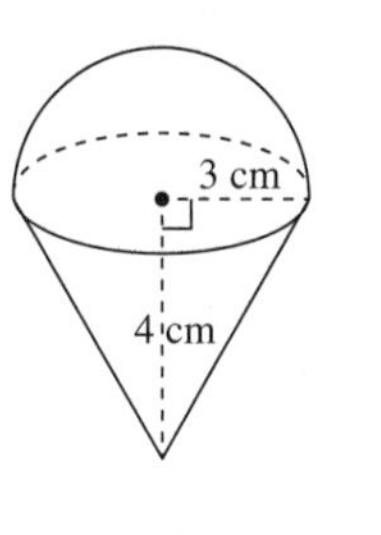

d

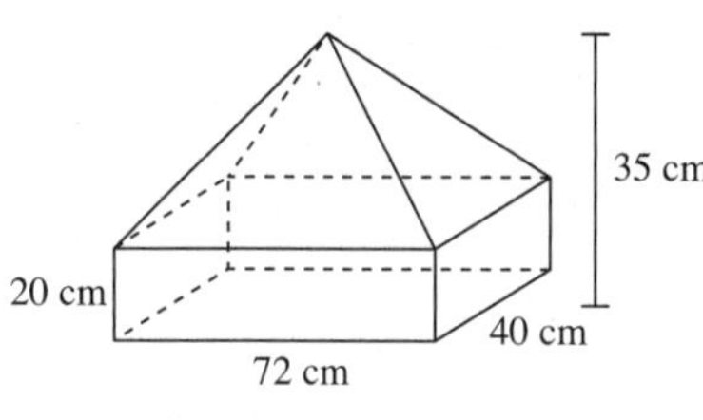

e

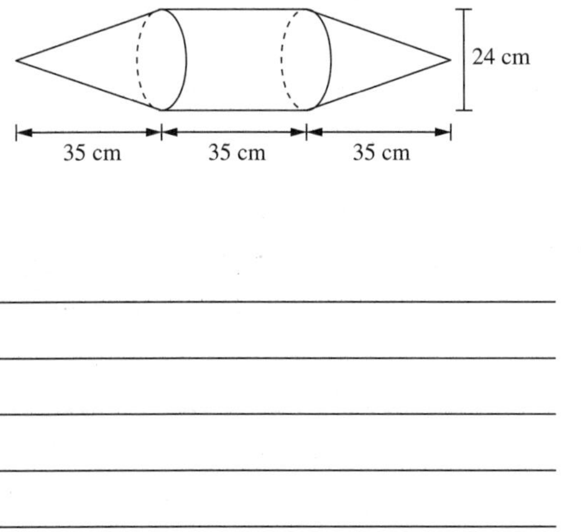

f

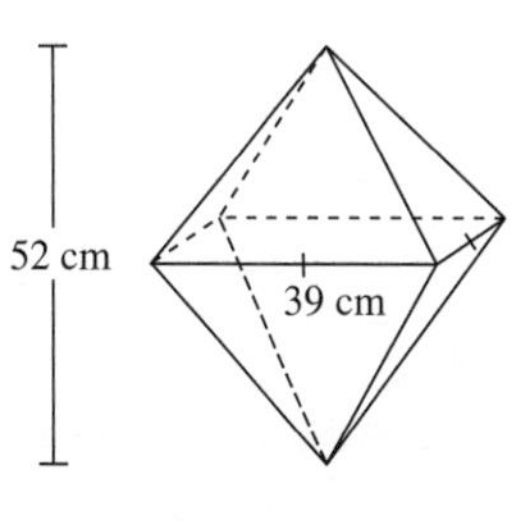

Excel Advanced-level Mathematics Study Guide Years 9–10
Pages 144–158

UNIT 7: Volume of different solids

QUESTION 1 Find the volume of the following rectangular prisms (give answers correct to one decimal place).

a

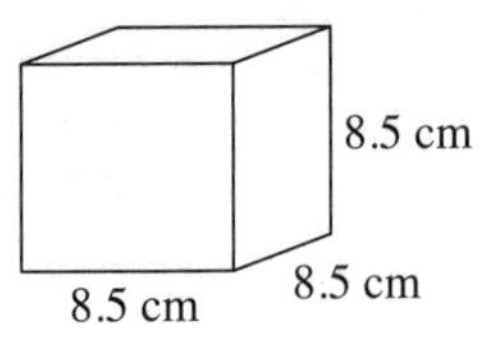

b

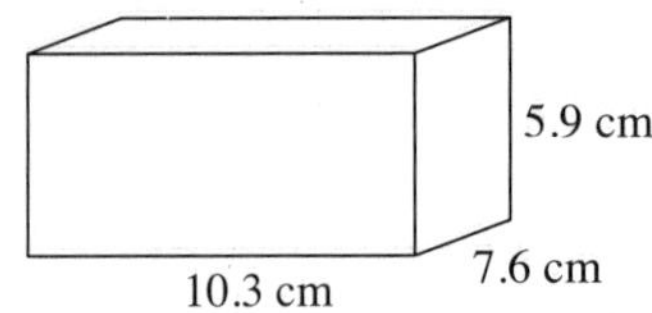

c

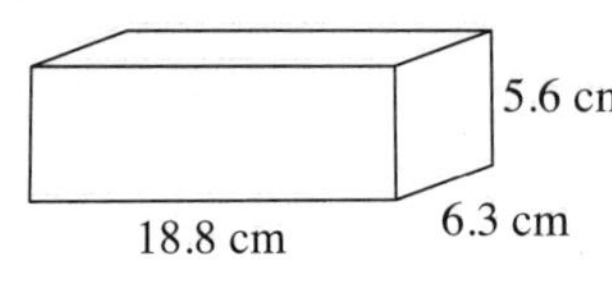

QUESTION 2 Find the volume of the following triangular prisms (give answers correct to four significant figures).

a

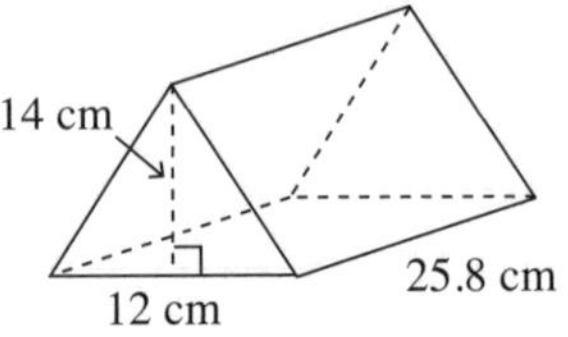

b

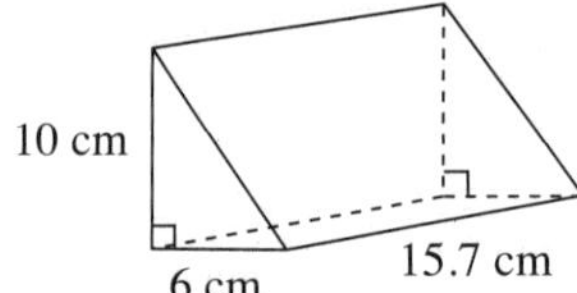

c

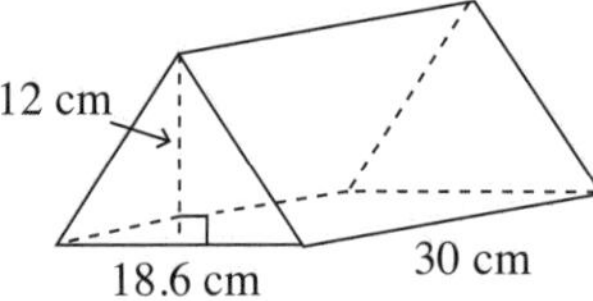

QUESTION 3 Find the volume of the following trapezoidal prisms (give answers correct to two decimal places if necessary).

a

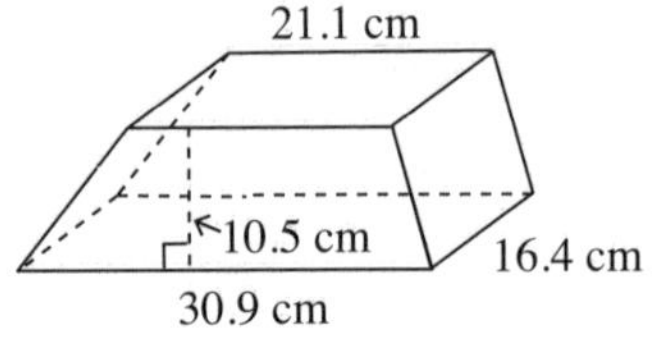

b

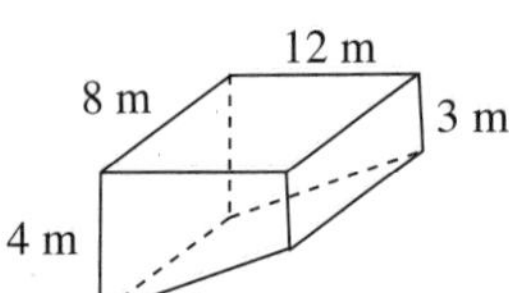

c

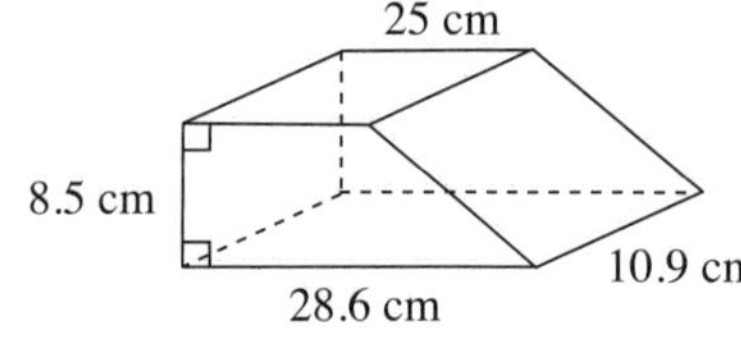

QUESTION 4 Find the volume of the following solids (give answers to one decimal place).

a

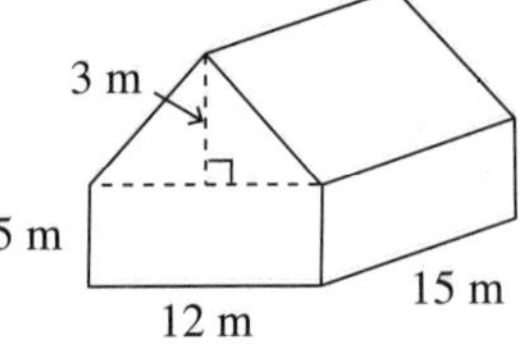

b

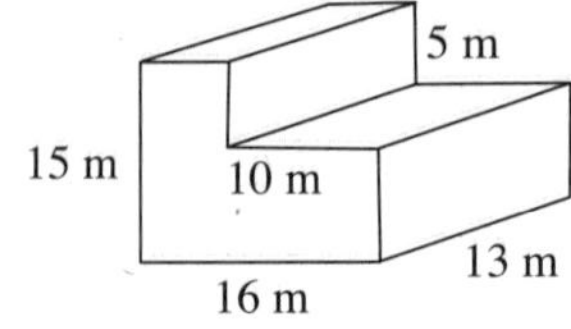

c

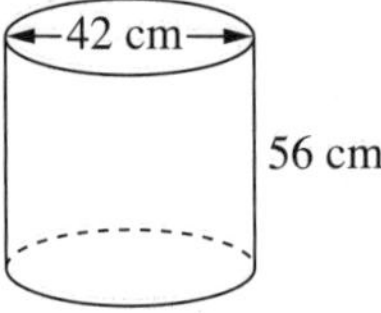

Surface area and volume

Excel Advanced-level Mathematics Study Guide Years 9–10
Pages 144–158

UNIT 8: Volume of a right pyramid

QUESTION 1 Calculate the volume of the following square pyramids correct to one decimal place.

a

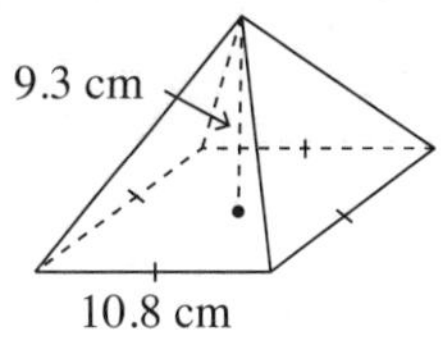

b

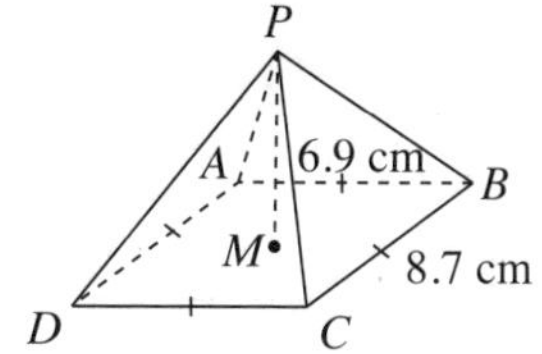

c

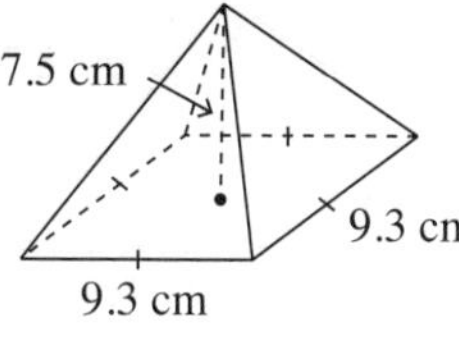

QUESTION 2 Calculate the volume of the following rectangular pyramids correct to two decimal places.

a

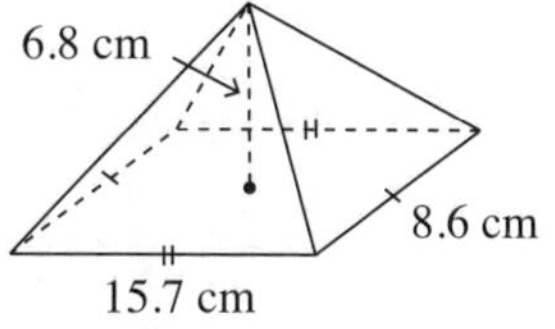

b

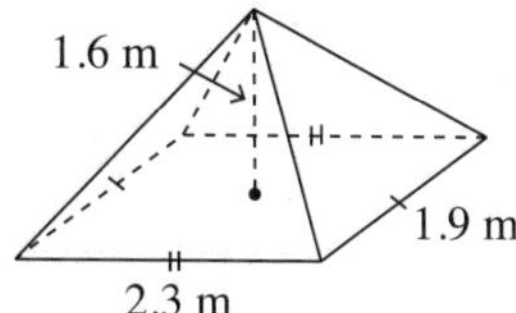

c

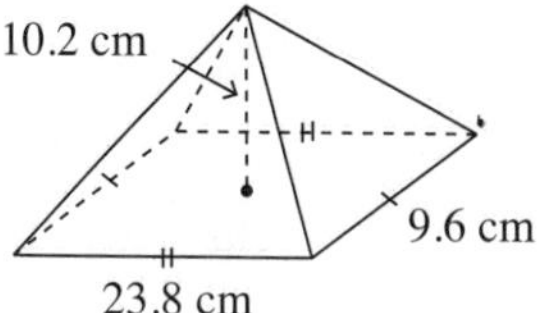

QUESTION 3 Find the volume of the following pyramids, given base area A cm^2 and perpendicular height h cm.

a Octagonal pyramid; $A = 225$, $h = 16.4$

b Hexagonal pyramid; $A = 98$, $h = 12$

QUESTION 4 Find the volume of these pyramids correct to one decimal place.

a

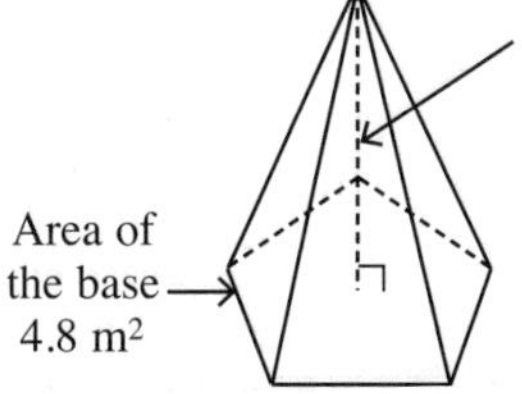

b

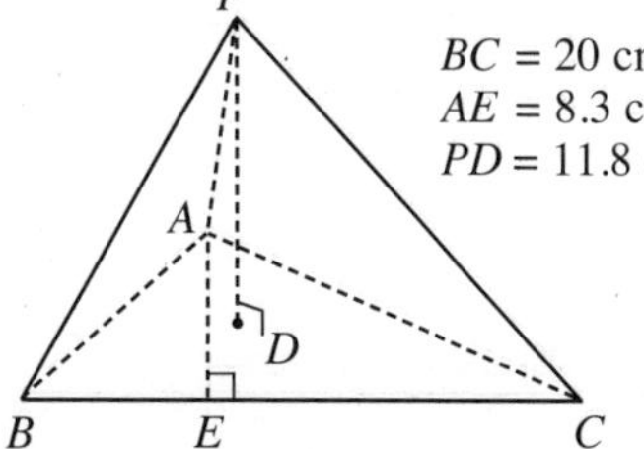

c

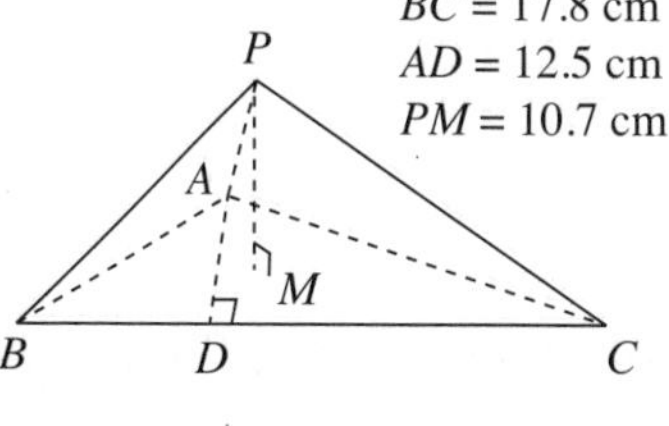

Surface area and volume

Excel Advanced-level Mathematics Study Guide Years 9–10
Pages 144–158

UNIT 9: Volume of a cone

QUESTION **1** Find the volume, correct to one decimal place, of these cones.

a

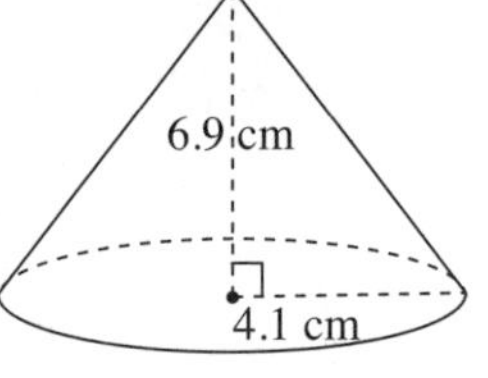

b

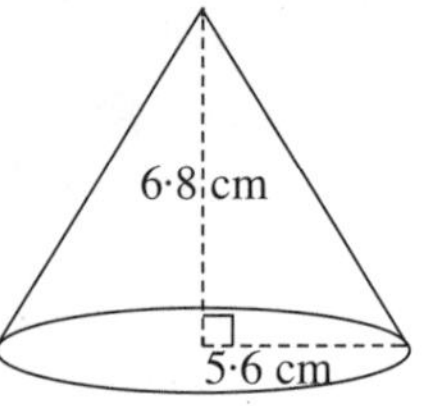

c

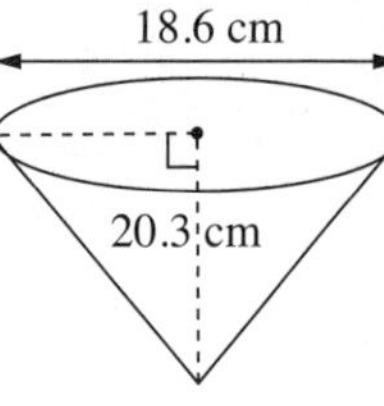

d

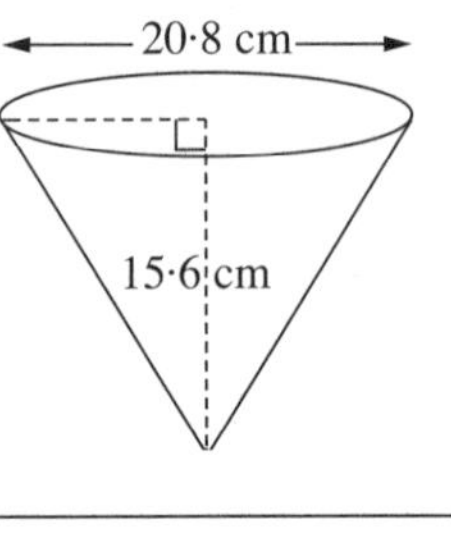

e

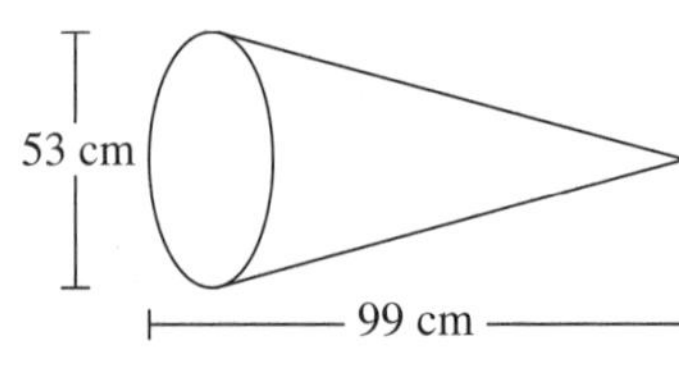

f

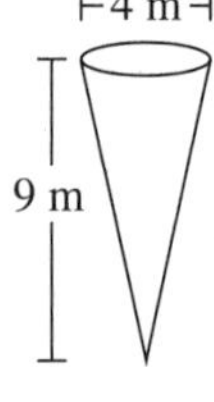

QUESTION **2** Find the volume of a cone with the following dimensions correct to one decimal place.

a radius 8 cm, perpendicular height 15 cm

b diameter 5.2 cm, perpendicular height 7.9 cm

QUESTION **3** Find the volume of these cones correct to two decimal places.

a

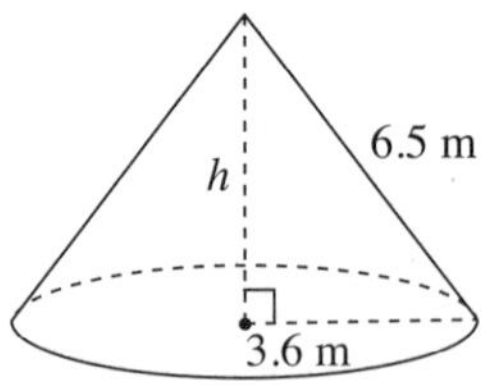

b

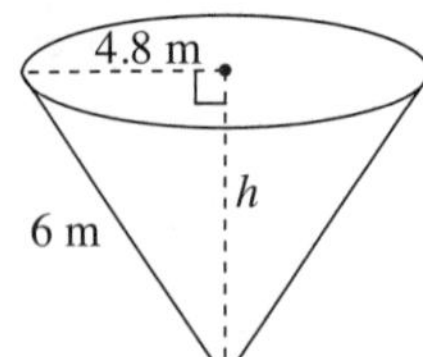

c

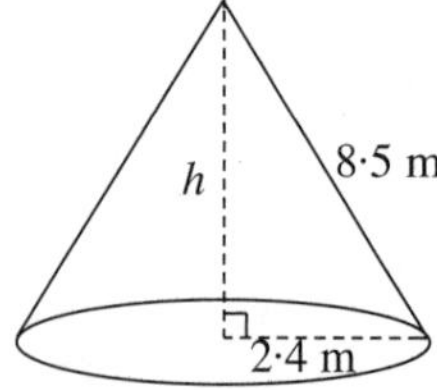

Surface area and volume

UNIT 10: Volume of a sphere

QUESTION 1 Find the volume of the following spheres (correct to one decimal place) with:

a radius = 9 cm

b diameter = 20 cm

c radius = 30 cm

d diameter = 35 cm

e radius = 15.3 cm

f diameter = 56 cm

QUESTION 2 Calculate the volume of the following spheres correct to one decimal place.

a

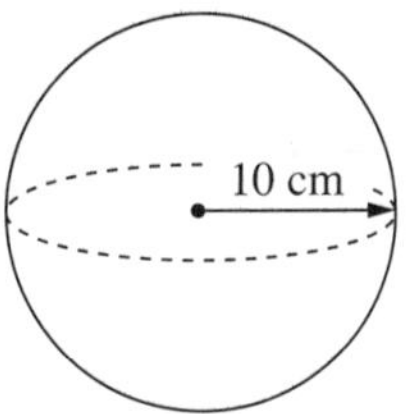

b

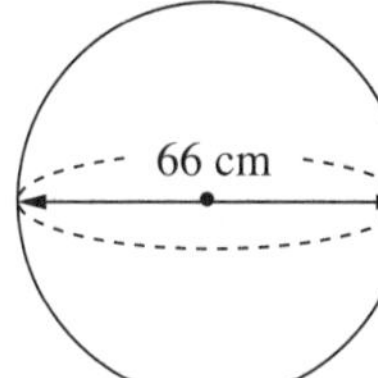

c

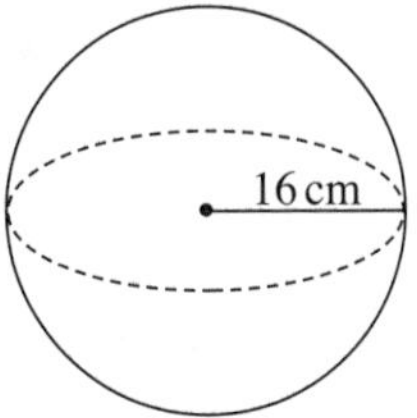

d

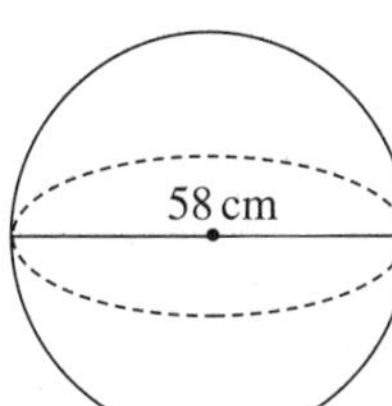

QUESTION 3 Calculate the volume of the following hemispheres correct to one decimal place.

a

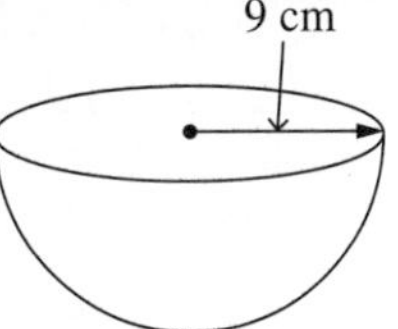

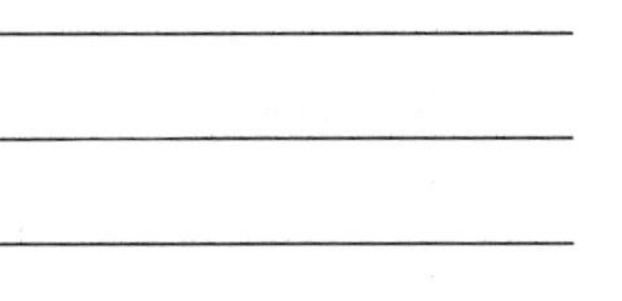

b

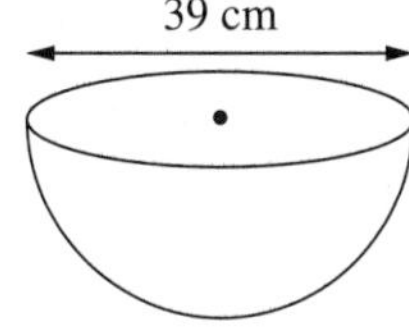

Surface area and volume

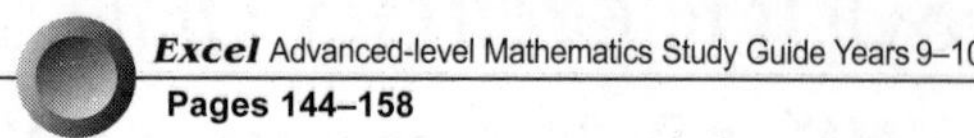

UNIT 11: Volume of composite solids

QUESTION **1** Find the volume of these solids correct to one decimal place.

a

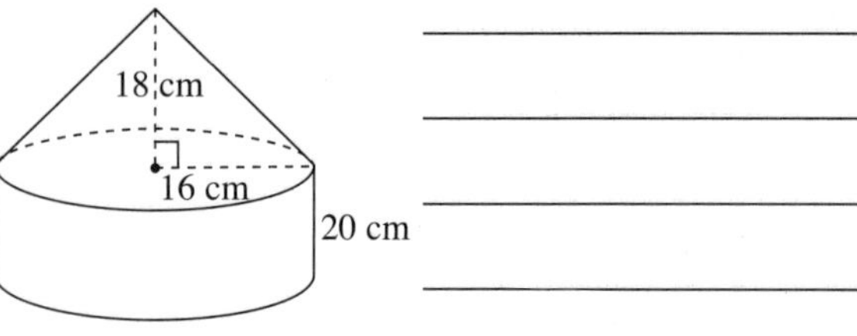

b

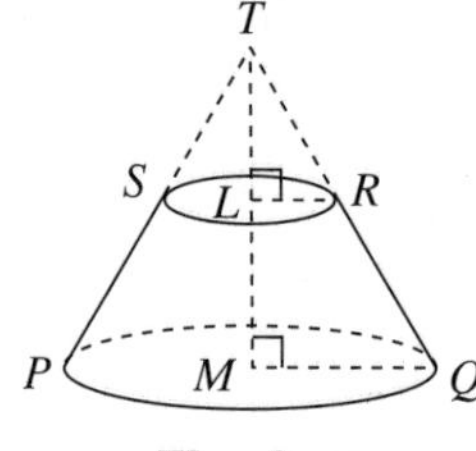

$TL = 8$ cm
$LM = 10$ cm
$LR = 4$ cm
$MQ = 9$ cm

c

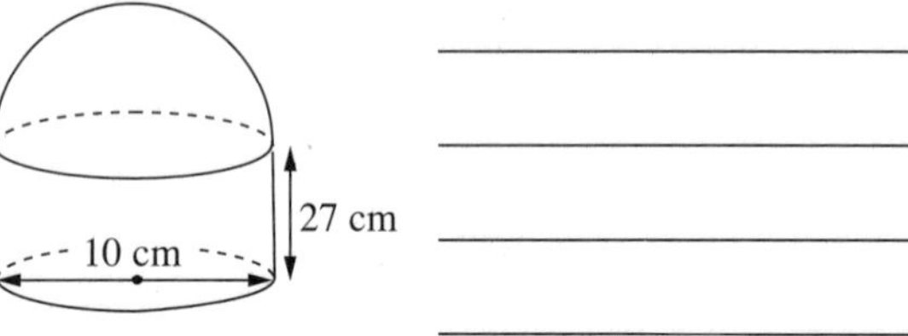

d

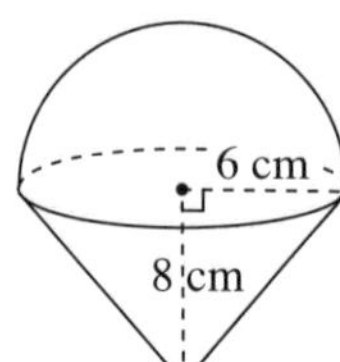

e

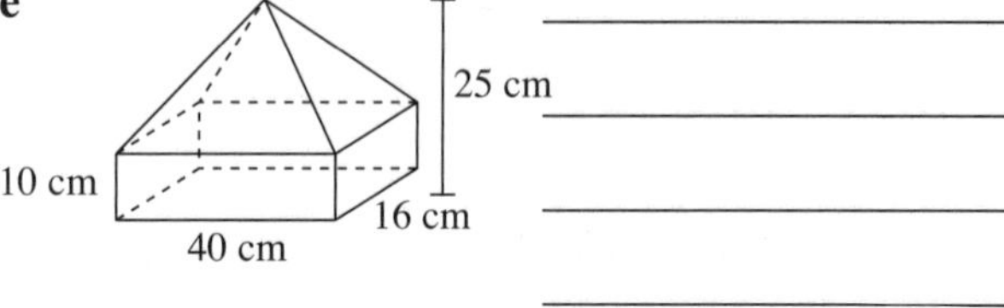

f

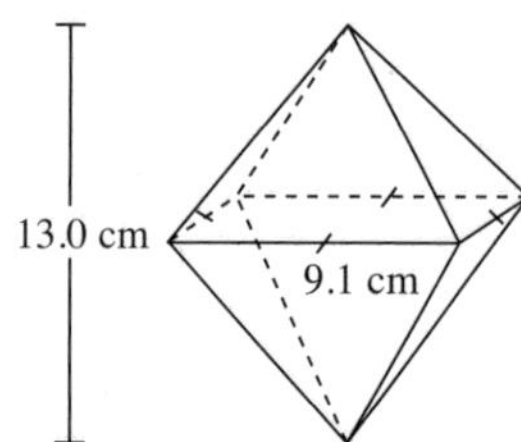

g

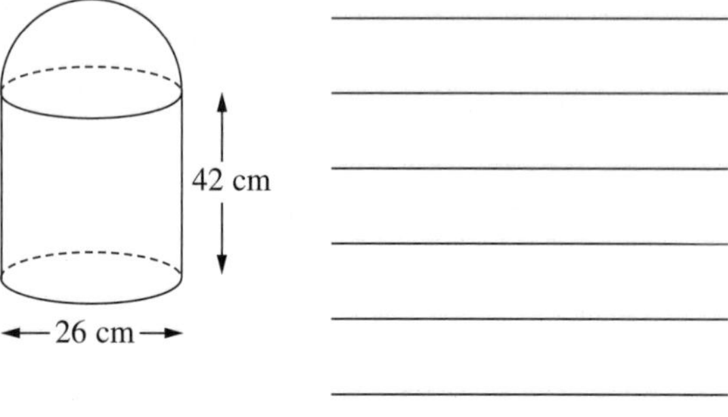

h

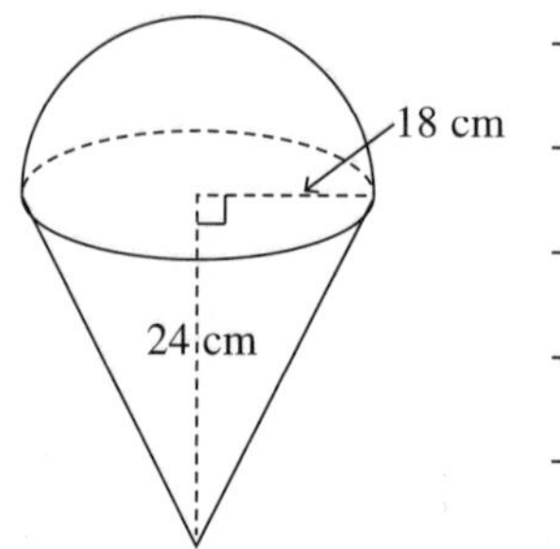

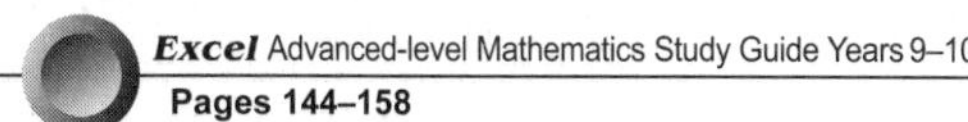

UNIT 12: Surface area and volume of similar figures

QUESTION 1 For each of the following similar figures, find the ratio of the smaller volume to the larger volume. All measurements are in centimetres.

a

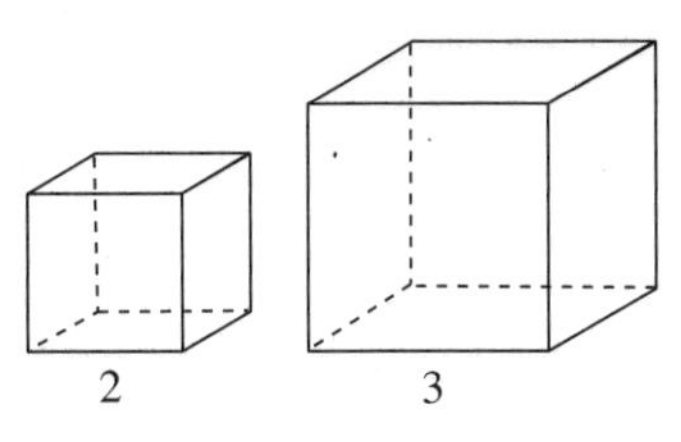

b

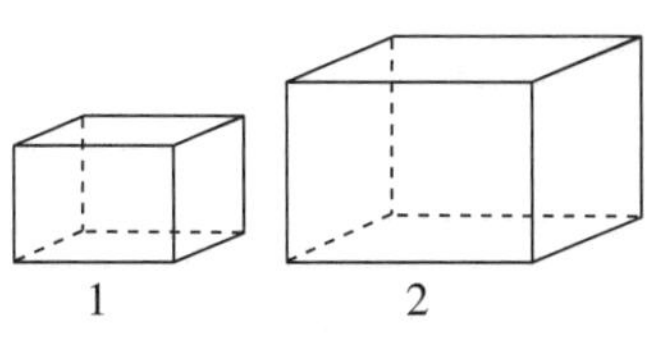

c

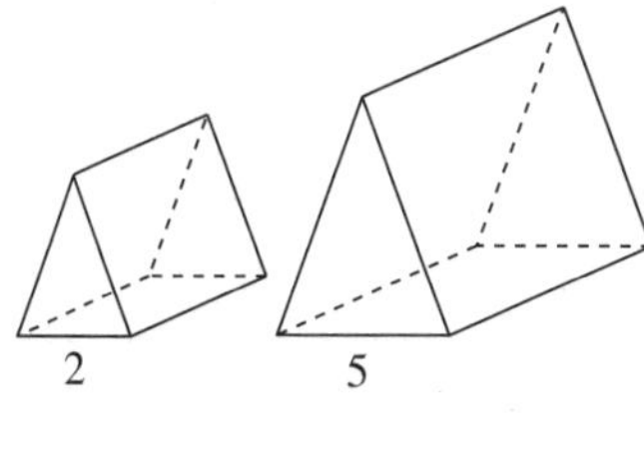

QUESTION 2 For each of the following pairs of similar figures, the ratio of the sides and one volume are given. Find the other volume. All measurements are in centimetres.

a

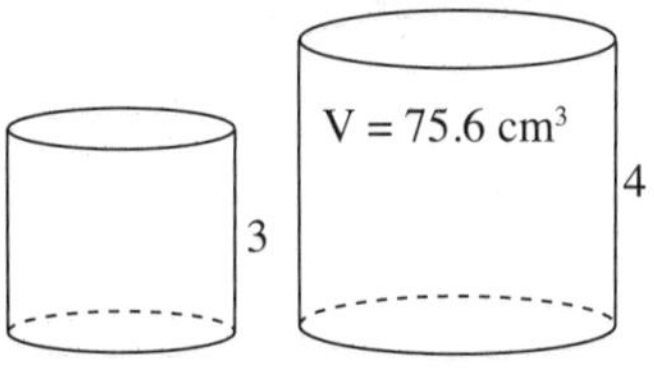

b

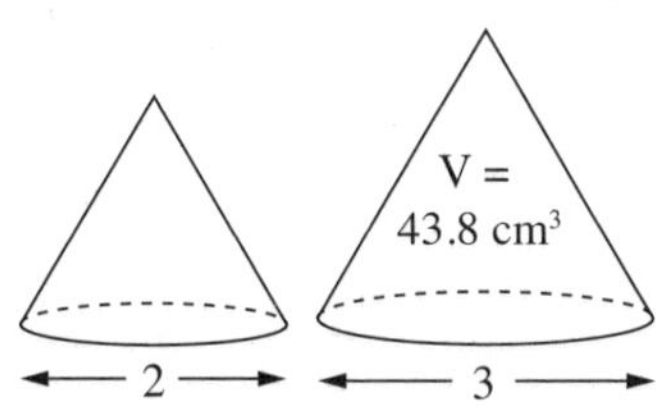

c

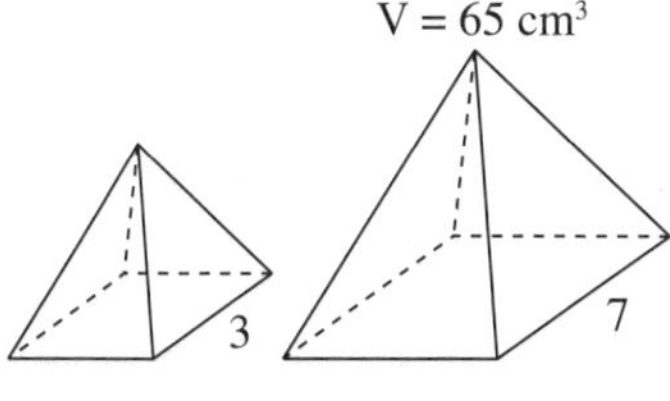

QUESTION 3

a The side lengths of two cubes are in the ratio 5 : 4. Find the ratio of their volumes.

b The surface areas of two spheres are in the ratio 64 : 49.

i What is the ratio of their radii?

ii Find the ratio of their volumes.

c The ratio of the side lengths of two rectangular prisms is 1 : 2. If the smaller prism has a volume of 68 cm^3, what is the volume of the larger prism?

Surface area and volume

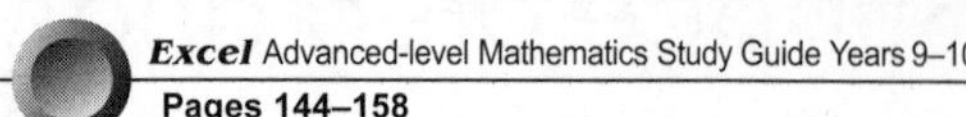
Excel Advanced-level Mathematics Study Guide Years 9–10
Pages 144–158

UNIT 13: Applications of area and volume

QUESTION 1 The radius of the Earth is approximately 6400 km. Calculate:

a its surface area in square kilometres correct to four significant figures.

b its volume correct to four significant figures.

QUESTION 2 A spherical balloon has a radius of 4.56 metres. Calculate:

a its surface area correct to one decimal place.

b its volume correct to two decimal places.

QUESTION 3 A conical tent has a base diameter of 6.5 metres and a slant height of 6 metres. Find the area of canvas used for this tent.

QUESTION 4 The diameter of the base of an oil can in the shape of a cone is 12 cm and its height is 10 cm. Find:

a its volume to the nearest cubic centimetre.

b its capacity to the nearest millilitre.

QUESTION 5 A grain bin is in the shape of a rectangular prism above a rectangular pyramid. Find:

a the area of metal required for the bin.

b the volume of grain the bin can hold.

32 cm
80 cm
70 cm
1 m

Surface area and volume

TOPIC TEST — PART A

Instructions
- This part consists of 10 multiple-choice questions.
- Fill in only ONE CIRCLE for each question.
- Each question is worth 1 mark.

Time allowed: 10 minutes — **Total marks: 10**

Marks

1 A cone has a base diameter of 12 cm and a vertical height of 8 cm. Its volume is

Ⓐ 8π cm^3 Ⓑ 24π cm^3 Ⓒ 72π cm^3 Ⓓ 96π cm^3 — 1

2 The volume of a sphere of radius 5 cm is closest to

Ⓐ 515 cm^3 Ⓑ 524 cm^3 Ⓒ 864 cm^3 Ⓓ 1765 cm^3 — 1

3 Approximately how many spherical balls of diameter 0.5 cm could be made from a melted down cube of side length 5 cm?

Ⓐ 19 Ⓑ 190 Ⓒ 1900 Ⓓ 19 000 — 1

4 The volume of a cone with diameter 7 cm and height 8 cm is closest to

Ⓐ 56 cm^3 Ⓑ 103 cm^3 Ⓒ 392 cm^3 Ⓓ 448 cm^3 — 1

5 A cone has a perpendicular height of 65.1 cm and slant height of 70.1 cm. What is the diameter of the cone?

Ⓐ 26 cm Ⓑ 48 cm Ⓒ 52 cm Ⓓ 96 cm — 1

6 The surface area of a sphere of diameter 28 cm is closest to

Ⓐ 2463 cm^2 Ⓑ 3284 cm^2 Ⓒ 9852 cm^2 Ⓓ 11 494 cm^2 — 1

7 The curved surface area of a cone with radius 9 cm and slant height 15 cm is closest to

Ⓐ 339 cm^2 Ⓑ 424 cm^2 Ⓒ 679 cm^2 Ⓓ 1018 cm^2 — 1

8 What is the surface area of this square-based pyramid?

Ⓐ 1344 cm^2 Ⓑ 1536 cm^2

Ⓒ 2112 cm^2 Ⓓ 2496 cm^2 — 1

20 cm

24 cm

9 A triangular pyramid has base area 72 cm^2 and perpendicular height 16 cm. What is the volume of the pyramid?

Ⓐ 1152 cm^3 Ⓑ 384 cm^3 Ⓒ 576 cm^3 Ⓓ 128 cm^3 — 1

10 The total surface area of a cone with radius 12 cm is 980 cm^2. What is the slant height?

Ⓐ 26 cm Ⓑ 28 cm Ⓒ 13 cm Ⓓ 14 cm — 1

Total marks achieved for PART A /10

Surface area and volume

TOPIC TEST **PART B**

Instructions
- This part consists of 6 questions.
- Each question part is worth 1 mark.
- Show all working.

Time allowed: 20 minutes **Total marks: 15**

Marks

1 Find the following for this cone (to one decimal place).

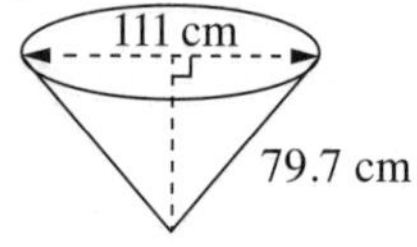

a perpendicular height **b** curved surface area **c** volume

______ ______ ______

3

Find the surface area and the volume of the following to the nearest whole number. All measurements are in centimetres.

2

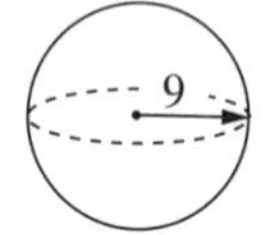

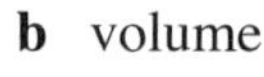

a surface area **b** volume

______ ______

2

3

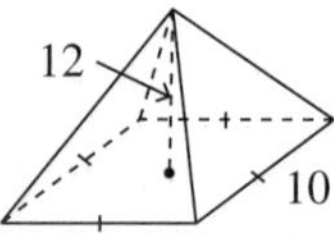

a surface area **b** volume

______ ______

2

4

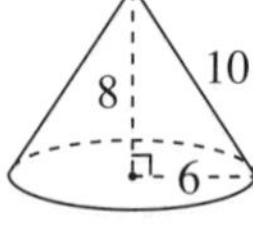

a surface area **b** volume

______ ______

2

5

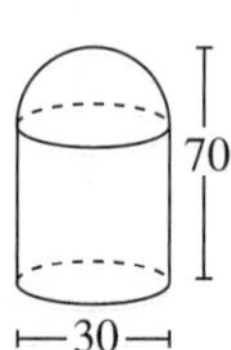

a surface area **b** volume

______ ______

2

6 Consider this rectangular pyramid. Find:

a the length AB **b** the width BC

______ ______

c the surface area **d** the volume

______ ______

P, 24 cm, 25 cm, 26 cm, D, C, O, N, A, M, B

4

Total marks achieved for PART B

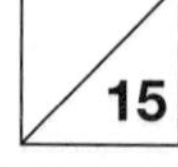

CHAPTER 7

Trigonometry

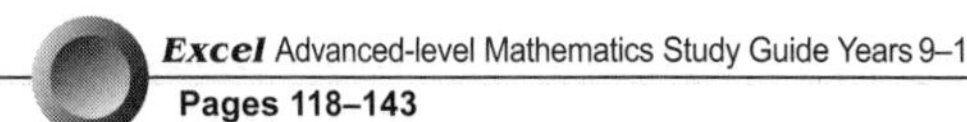

UNIT 1: Review of basic trigonometry

QUESTION 1 Evaluate correct to three decimal places.

a $\sin 75° =$ ________ **b** $\tan 34° =$ ________ **c** $\cos 120° =$ ________

d $\tan 65°07' =$ ________ **e** $\cos 105°36' =$ ________ **f** $\sin 160°23' =$ ________

QUESTION 2 If $0° \le \theta \le 90°$, find θ to the nearest minute.

a $\sin\theta = 0.5$ ________ **b** $\cos\theta = 0.729$ ________ **c** $\tan\theta = 2.715$ ________

d $\cos\theta = 0.89$ ________ **e** $\tan\theta = 1.36$ ________ **f** $\sin\theta = 0.2588$ ________

QUESTION 3 Complete the trigonometric ratios for the following triangles.

a

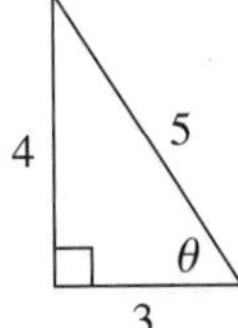

$\sin\theta =$ ________
$\cos\theta =$ ________
$\tan\theta =$ ________

b (triangle with sides 12, 13, 5 and angle θ)

$\sin\theta =$ ________
$\cos\theta =$ ________
$\tan\theta =$ ________

c (triangle with sides 17, 15, 8 and angle θ)

$\sin\theta =$ ________
$\cos\theta =$ ________
$\tan\theta =$ ________

QUESTION 4 Find the value of x. Give each answer correct to one decimal place.

a

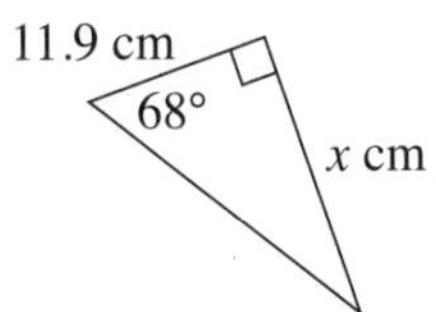

b

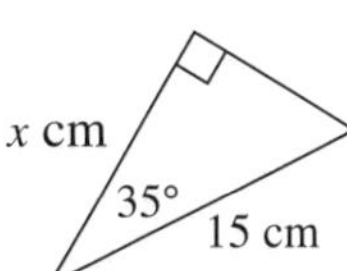

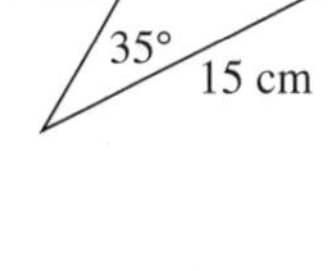

c

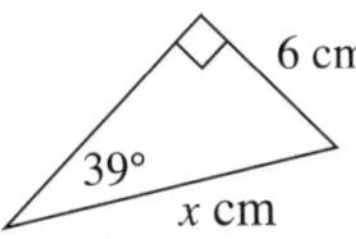

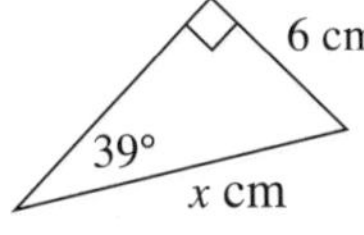

d

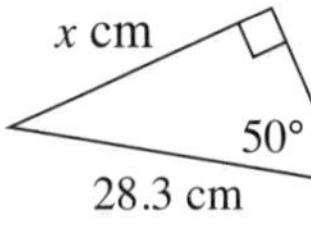

e

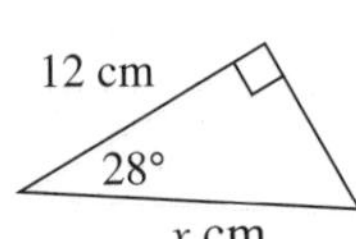

f

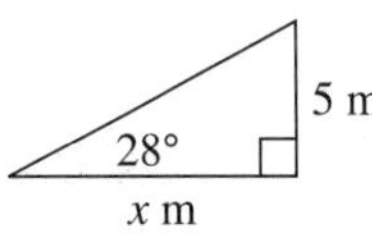

QUESTION 5 Find the size of angle θ to the nearest minute.

a

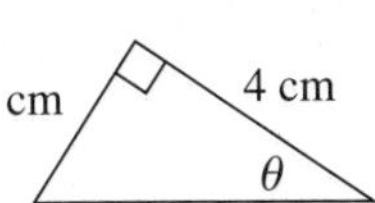

b

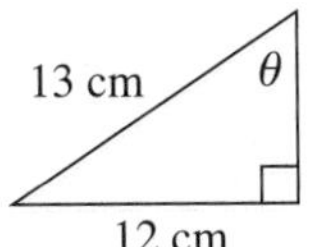

c

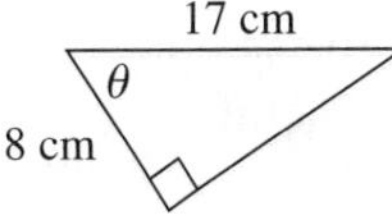

Excel Advanced-level Mathematics Study Guide Years 9–10
Pages 118–143

UNIT 2: Problems involving two right-angled triangles

QUESTION 1 From the diagram given opposite, find:

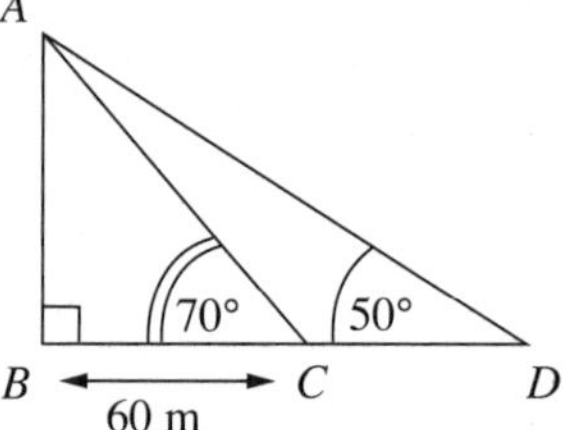

a the length of side AB (to two decimal places).

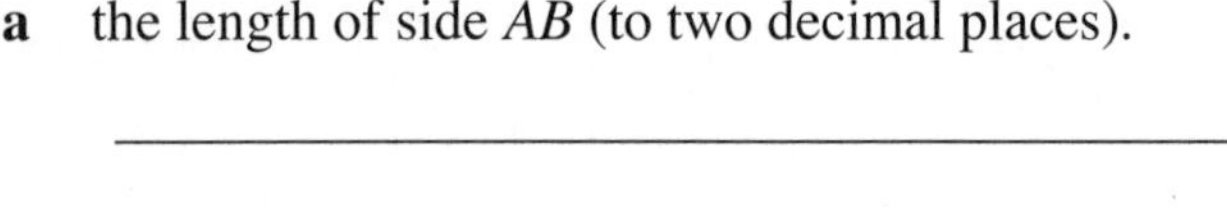

b the size of angle DAC.

c the length of side CD (to two decimal places).

QUESTION 2 In the diagram given opposite, calculate:

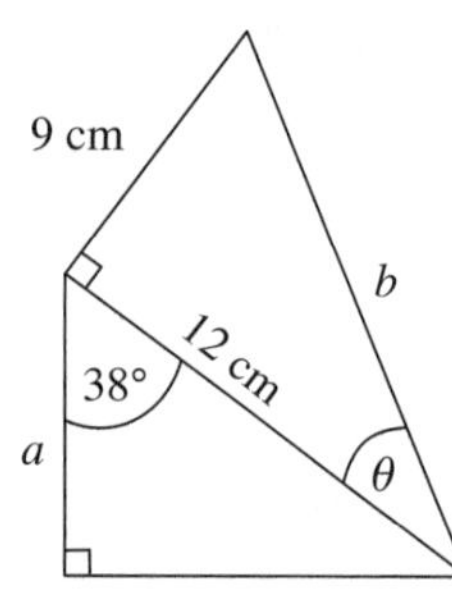

a the length a (correct to one decimal place).

b the size of angle θ.

c the length b.

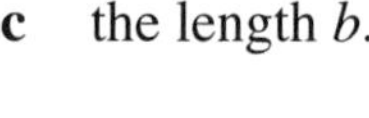

QUESTION 3 Find the value of the p and q in the given diagram.

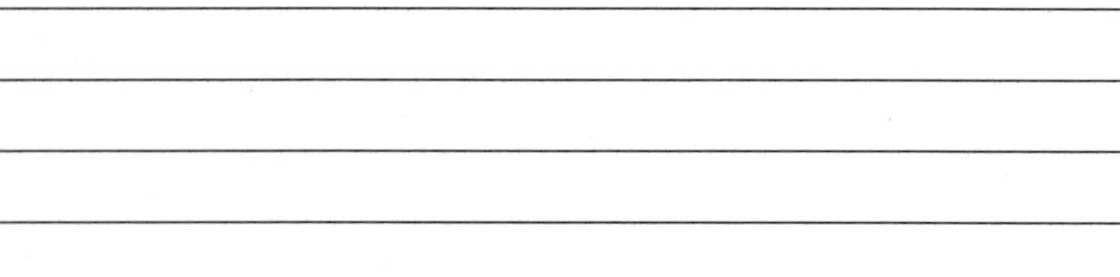

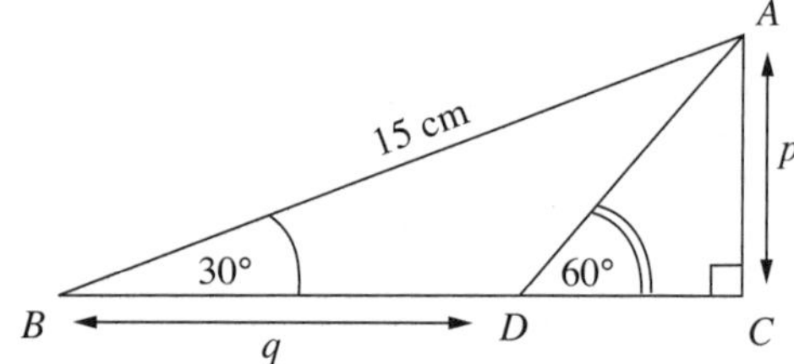

QUESTION 4 The angle of elevation of the top of a cliff from a boat 600 m out to sea is 37°. If the boat then travels a further d metres out to sea, the angle of elevation of the cliff is now 25°. Find:

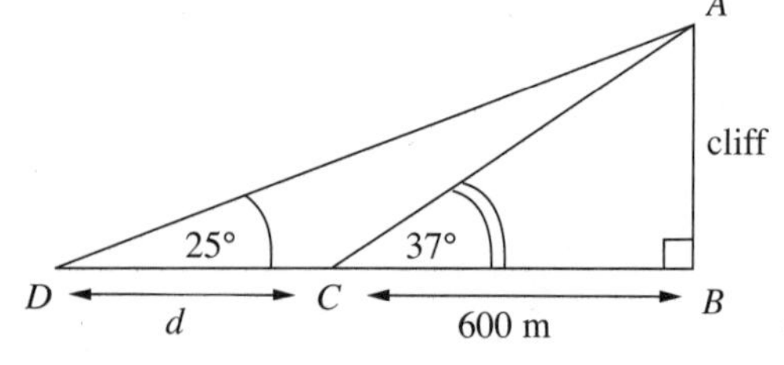

a the height of the cliff above sea level to the nearest metre.

b the value of d to the nearest metre.

Trigonometry

UNIT 3: Review of bearings

QUESTION 1 Write down the bearing of Q from P.

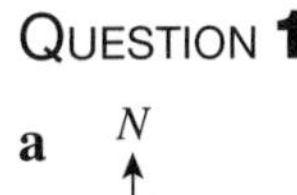

a

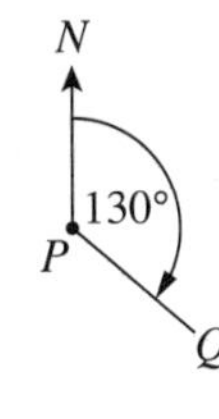

b

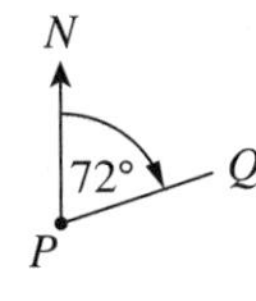

c

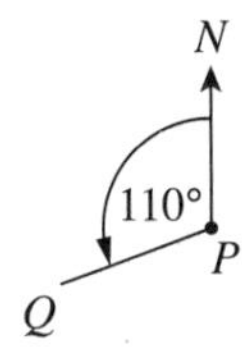

d

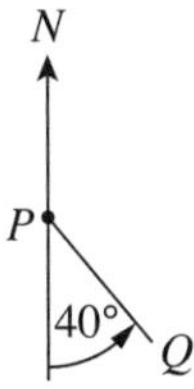

e

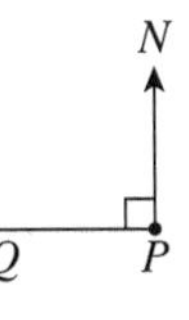

f

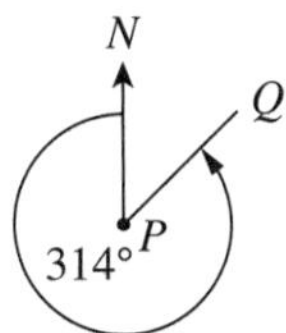

g

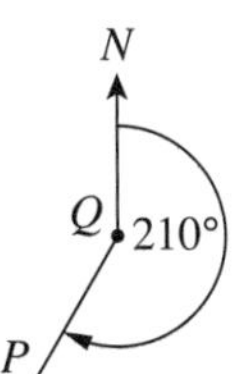

h

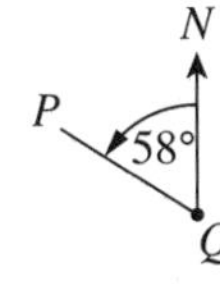

QUESTION 2 Two helicopters leave P. One flies 80 km on a bearing of 127°. The other flies 39 km on a bearing of 217°. How far apart are the helicopters after the flights?

QUESTION 3 A plane flies from A on a bearing of 064° to B a distance of 113 km. From B it flies on a bearing of 154° to C, which is due east of A.

a Draw a diagram to show this information.

b Find the size of $\angle ABC$.

c Find the size of $\angle BAC$.

d Find the distance from A to C.

e On what bearing should the plane fly from C back to A?

QUESTION 4 P is 80 m due east of O and 50 m due north of Q. What is the bearing of Q from O?

Excel Advanced-level Mathematics Study Guide Years 9–10
Pages 118–143

UNIT 4: Trigonometry in three dimensions

QUESTION 1 A rectangular prism drawn below has its length 15 cm, width 8 cm and height 6 cm. Find:

a the length of the diagonal of the base.

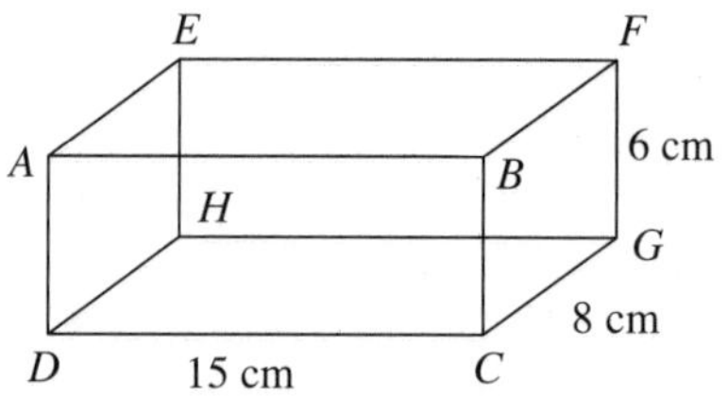

b the length of the diagonal of the prism (leave answer in surd form).

E F
A B
H
6 cm
G
8 cm
D 15 cm C

c the size of the angle a diagonal of the prism makes with the base.

QUESTION 2 A cone has a base diameter of 10 cm and a perpendicular height of 12 cm, find the vertical angle of the cone.

QUESTION 3 From a point P which is 100 m due east of a tower, the angle of elevation of its top is 30°. Find the angle of elevation of the top of the tower from a point Q which is 70 metres due south of it.

Trigonometry

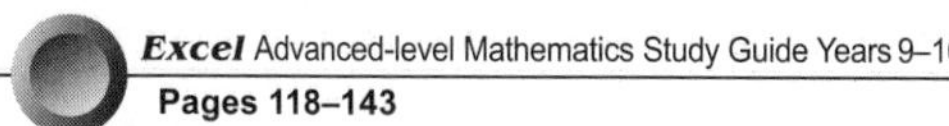

UNIT 5: Solving three-dimensional problems

QUESTION 1 $AB = 7$ m. $\angle AQB = 30°$. $\angle APB = 40°$. AB is perpendicular to both PB and QB. PB is perpendicular to QB. Find the length, to one decimal place, of:

a BQ **b** PB **c** PQ

QUESTION 2 $BC = 30$ m. $\angle ADB = 38°$. $\angle ACB = 25°$. AB is perpendicular to both BC and BD. BC is perpendicular to BD. Find the length, to one decimal place, of:

a AB **b** BD **c** CD

QUESTION 3 A is due south of a tower and B is due east of the tower on level ground. B is 75 m from A on a bearing of 056°. From B the angle of elevation of the top of the tower is 34°. Find:

a the distance from B to the foot of the tower to one decimal place.

b the height of the tower to the nearest metre.

QUESTION 4 From a point, P, due south of a tower, the angle of elevation of the top of the tower is 27°. From a point Q due east of the tower the angle of elevation of the top of the tower is 39°. If the tower is 85 m tall and the ground is level, find the distance from P to Q to the nearest metre.

QUESTION 5 From B, the base of a cliff, the bearing of point P is 123° and the bearing of Q is 180°. Q is 100 m due west of P. From Q the angle of elevation of the top of the cliff, A, is 48°. Find the height of the cliff to the nearest metre.

Trigonometry

UNIT 6: The unit circle

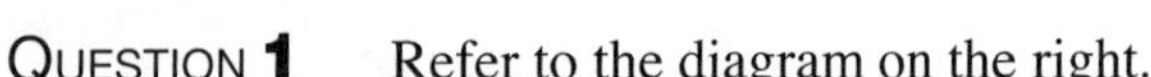

QUESTION 1 Refer to the diagram on the right.

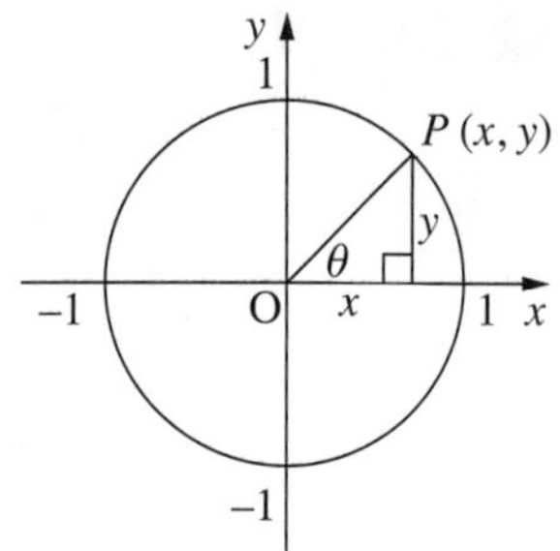

a What is the radius of the circle? ____________

b What is the length of OP? ____________

c Write in terms of x and y.

i $\sin\theta =$ ________ **ii** $\cos\theta =$ ________ **iii** $\tan\theta =$ ________

QUESTION 2 Refer to the diagram on the right.

a Write in terms of x and y.

i $\sin\theta =$ ________ **ii** $\cos\theta =$ ________ **iii** $\tan\theta =$ ________

b Complete the following statement by adding either of the words positive or negative.

When x is negative but y is positive, $\sin\theta$ is ________ , $\cos\theta$ is ________ and $\tan\theta$ is ________ .

QUESTION 3 Refer to the diagram on the right.

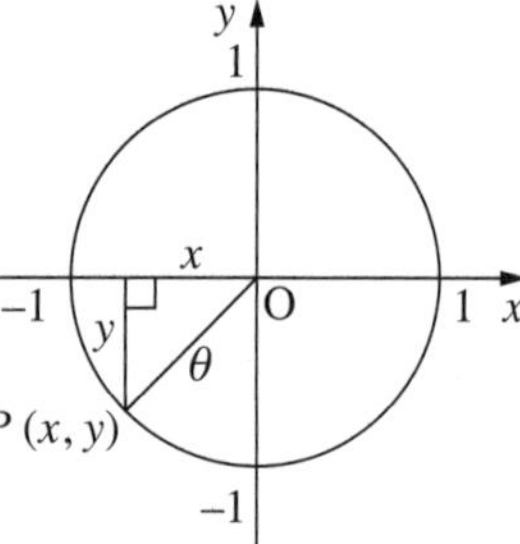

a Write in terms of x and y.

i $\sin\theta =$ ________ **ii** $\cos\theta =$ ________ **iii** $\tan\theta =$ ________

b Complete the following statement by adding either of the words positive or negative.

When x and y are both negative, $\sin\theta$ is ________ , $\cos\theta$ is ________ and $\tan\theta$ is ________ .

QUESTION 4 Refer to the diagram on the right.

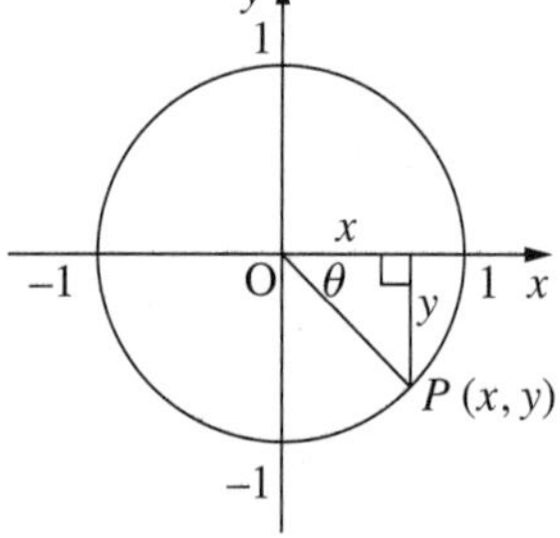

a Write in terms of x and y.

i $\sin\theta =$ ________ **ii** $\cos\theta =$ ________ **iii** $\tan\theta =$ ________

b Complete the following statement by adding either of the words positive or negative.

When x is positive but y is negative, $\sin\theta$ is ________ , $\cos\theta$ is ________ and $\tan\theta$ is ________ .

QUESTION 5

a In terms of $\sin\theta$ and $\cos\theta$, $\tan\theta =$ ____________________

b If $\sin\theta = 0.6$ and $\cos\theta = 0.8$, what is the value of $\tan\theta$? ____________________

QUESTION 6 Use a calculator to find, to 4 decimal places, the value of:

a sin 125° ____________ **b** cos 200° ____________ **c** tan 320° ____________

d cos 140° ____________ **e** tan 95° ____________ **f** sin 248° ____________

g tan 195° ____________ **h** sin 302° ____________ **i** cos 356° ____________

UNIT 7: Graphs of trigonometric ratios

QUESTION 1 Sketch the graph, for $0° \le x° \le 360°$, of:

a $y = \sin x°$

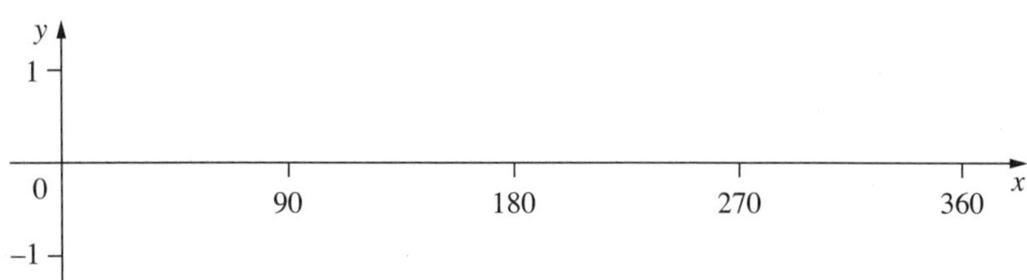

b $y = \cos x°$

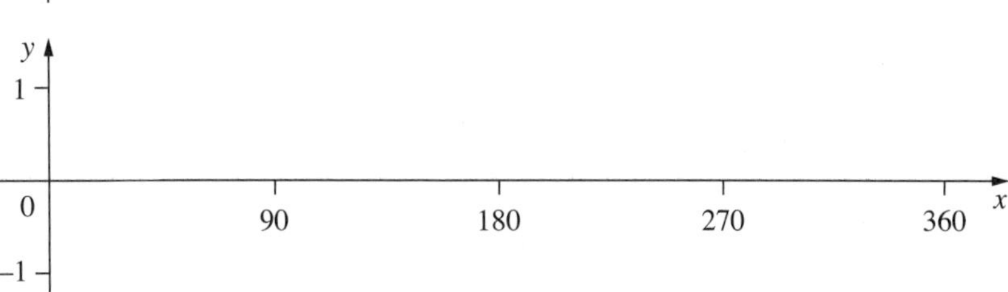

c $y = \tan x°$

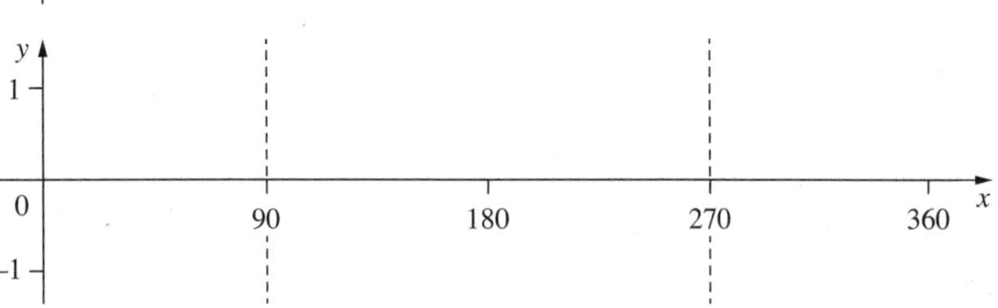

QUESTION 2 Each of the graphs drawn in question 1 is periodic. After how many degrees does each graph repeat?

a $y = \sin x°$ ______________ **b** $y = \cos x°$ ______________ **c** $y = \tan x°$ ______________

QUESTION 3 For the curve $y = \sin x°$ what is the:

a maximum y value ______________ **b** minimum y value ______________

QUESTION 4 For the curve $y = \cos x°$ what is the:

a maximum y value ______________ **b** minimum y value ______________

QUESTION 5

a Does the graph of $y = \tan x°$ have a maximum or minimum value? ______________

b Briefly explain why or why not. ______________

QUESTION 6 Briefly discuss any similarities between the curves.

____

____

____

____

____

UNIT 8: Obtuse angles

QUESTION 1 Determine whether the following ratios will be positive or negative.

a $\sin 160°$ ______ **b** $\cos 99°$ ______ **c** $\tan 124°$ ______

d $\cos 175°$ ______ **e** $\sin 108°$ ______ **f** $\tan 156°$ ______

QUESTION 2 Write the value, correct to four decimal places, of:

a $\sin 115°$ ______ **b** $\cos 167°$ ______ **c** $\tan 105°$ ______

d $\tan 168°$ ______ **e** $\sin 93°$ ______ **f** $\cos 121°$ ______

QUESTION 3 Express as the ratio of an acute angle.

a $\sin 110°$ ______ **b** $\tan 115°$ ______ **c** $\sin 158°$ ______ **d** $\cos 130°$ ______

e $\cos 112°$ ______ **f** $\tan 128°$ ______ **g** $\cos 117°$ ______ **h** $\sin 98°$ ______

QUESTION 4 Find the obtuse angle, θ, to the nearest degree, if:

a $\sin \theta = 0.5723$ **b** $\cos \theta = -0.2145$ **c** $\tan \theta = -1.786$ **d** $\cos \theta = -0.6983$

______ ______ ______ ______

QUESTION 5 Find, to the nearest degree, two possible values for θ, (one acute and the other obtuse).

a $\sin \theta = 0.5$ **b** $\sin \theta = 0.7823$ **c** $\sin \theta = 0.8945$ **d** $\sin \theta = 0.2394$

______ ______ ______ ______

______ ______ ______ ______

QUESTION 6 Use a calculator to solve each equation for $0° \le \theta \le 180°$.
Give each answer correct to the nearest minute.

a $\sin \theta = 0.125$ **b** $\tan \theta = 0.386$ **c** $\cos \theta = 0.279$ **d** $\cos \theta = -0.3$

______ ______ ______ ______

______ ______ ______ ______

e $\sin \theta = 0.4$ **f** $\tan \theta = -0.5$ **g** $\sin \theta = 0.6$ **h** $\cos \theta = -0.89$

______ ______ ______ ______

______ ______ ______ ______

QUESTION 7 Use the fact that the gradient $m = \tan \theta$ to find the angle of inclination of each line.

a

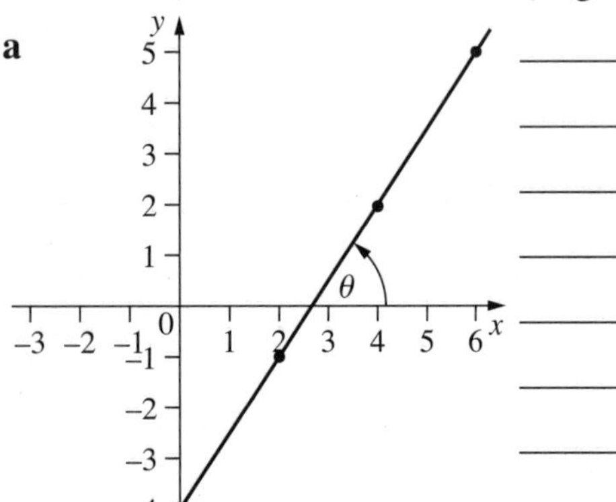

______

b

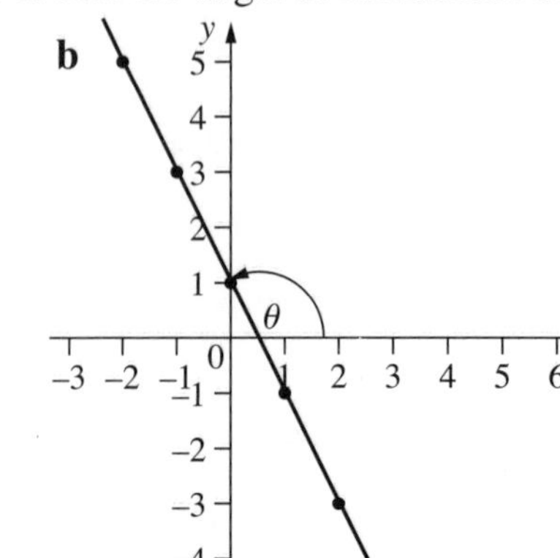

______

Trigonometry

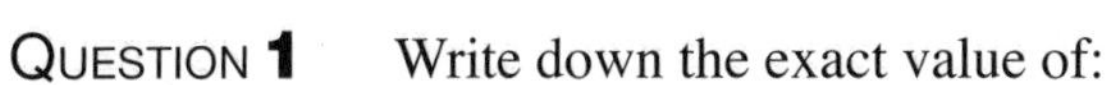

UNIT 9: Exact ratios

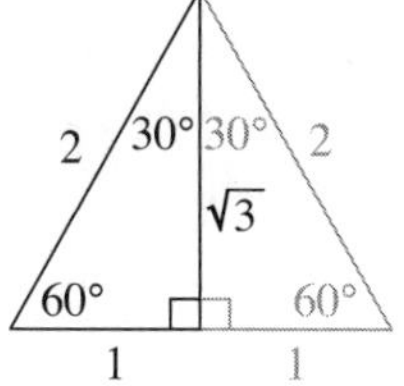

QUESTION 1 Write down the exact value of:

a sin 60° ________ **b** cos 30° ________ **c** tan 30° ________

d tan 60° ________ **e** sin 30° ________ **f** cos 60° ________

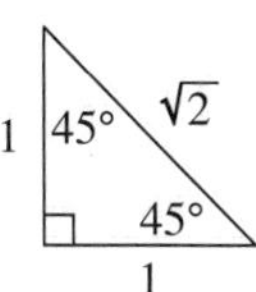

QUESTION 2 Write down the exact value of:

a tan 45° ________ **b** cos 45° ________ **c** sin 45° ________

QUESTION 3 Write down the value of:

a sin 0° ________ **b** cos 0° ________ **c** tan 0° ________ **d** sin 90° ________ **e** cos 90° ________

QUESTION 4 Find the exact value of:

a sin 150° **b** cos 135° **c** tan 120°

d tan 150° **e** cos 120° **f** sin 135°

g sin 120° **h** tan 135° **i** cos 150°

QUESTION 5 Show that:

a $\sin 30° \times \cos 60° = \frac{1}{4}$ **b** $\sin 60° + \cos 30° = \sqrt{3}$ **c** $\frac{\sin 45°}{\cos 45°} = \tan 45°$

QUESTION 6 Find the exact value of *x*.

a

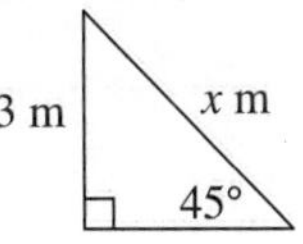

b

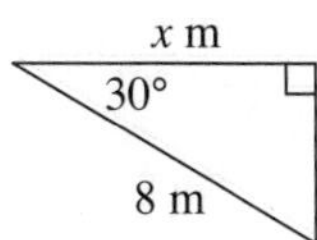

c

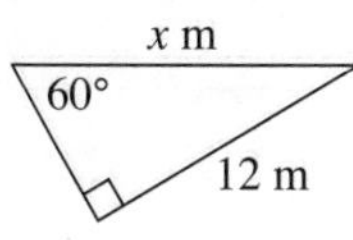

Excel Advanced-level Mathematics Study Guide Years 9–10
Pages 118–143

UNIT 10: Simple trigonometric equations

QUESTION 1 Solve the following trigonometric equations for the domain $0° \le \theta \le 180°$. Give answers correct to the nearest minute.

a $\sin \theta = 0.4$ **b** $\tan \theta = 0.3$ **c** $\cos \theta = 0.5$

d $\sin \theta = 0.7$ **e** $\tan \theta = -0.4$ **f** $\sin \theta = 0.62$

g $\cos \theta = 0.83$ **h** $\tan \theta = 0.5$ **i** $\cos \theta = -0.25$

QUESTION 2 Find the size of the angle θ ($0° \le \theta \le 90°$) for which:

a $\tan \theta = \sqrt{3}$ **b** $\sin \theta = 1$ **c** $\cos \theta = \frac{1}{\sqrt{2}}$

d $\sin \theta = \frac{1}{2}$ **e** $\tan \theta = 1$ **f** $\cos \theta = \frac{\sqrt{3}}{2}$

g $2 \cos \theta = 1$ **h** $\sqrt{3} \tan \theta = 1$ **i** $\sqrt{2} \sin \theta = 1$

QUESTION 3 If $0° \le \theta \le 180°$, find the value of θ for which:

a $\cos \theta = -\frac{1}{2}$ **b** $\tan \theta = -\sqrt{3}$ **c** $\cos \theta + 1 = 0$

QUESTION 4 If $0° \le \theta \le 180°$, find two possible values of θ for which:

a $\sin \theta = 0$ **b** $2 \sin \theta = \sqrt{3}$ **c** $\sin \theta = \frac{\sqrt{2}}{2}$

QUESTION 5 If $0° \le \theta \le 180°$, find the value(s) of θ for which:

a $3 \tan \theta + \sqrt{3} = 0$ **b** $\sqrt{2} \cos \theta + 1 = 0$ **c** $2 \sin \theta - 1 = 0$

QUESTION 6 Fill in the blanks.

a $\sin 70° =$ ________ $20°$ **b** $\cos 50° = \sin$ ________

c $\sin 28° = \cos$ ________ **d** $\cos 15° = \sin$ ________

Excel Advanced-level Mathematics Study Guide Years 9–10
Pages 118–143

UNIT 11: Using the sine rule to find a side

QUESTION 1 Use the sine rule to find the value of x to two decimal places.

a

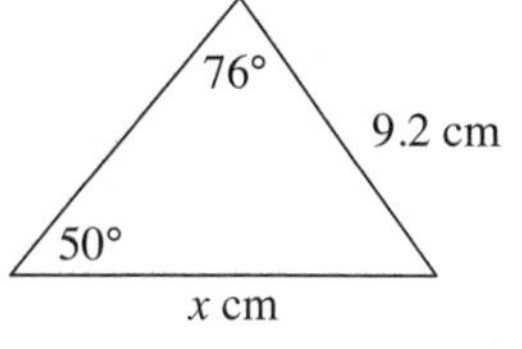

b

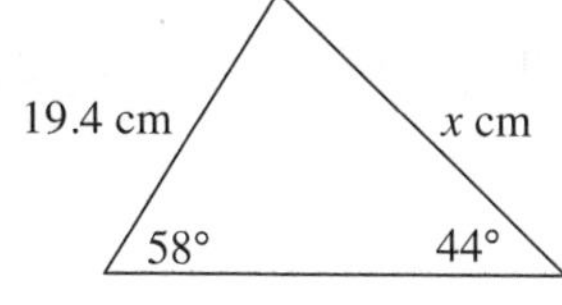

c

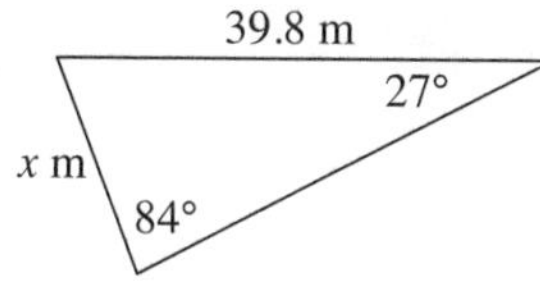

d

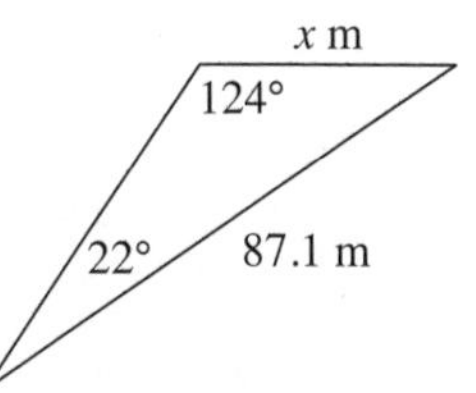

e

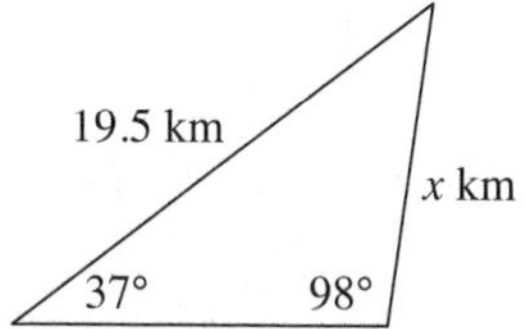

f

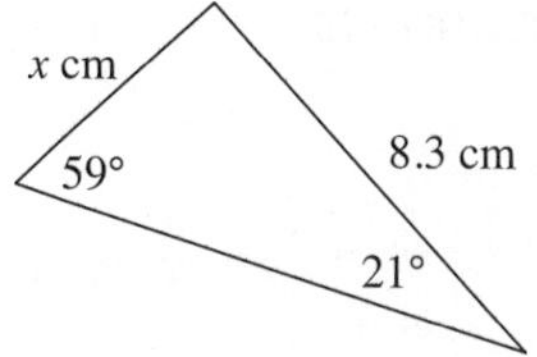

QUESTION 2 Find.

a

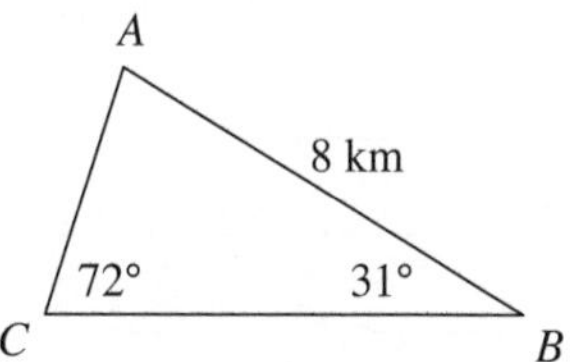

i the size of $\angle BAC$

ii the length of side BC

b

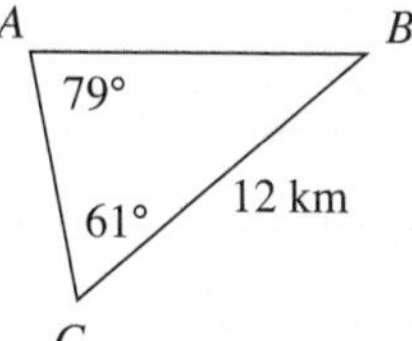

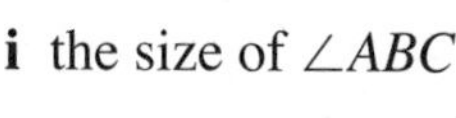

i the size of $\angle ABC$

ii the length of AC

c

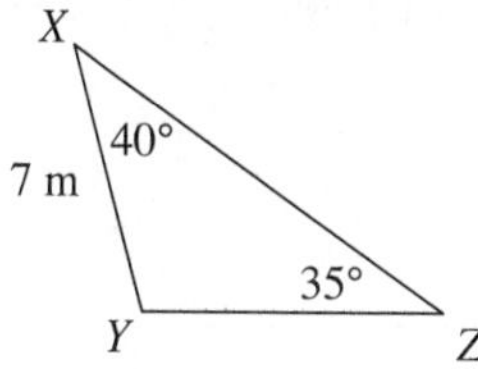

i the size of $\angle XYZ$

ii XZ

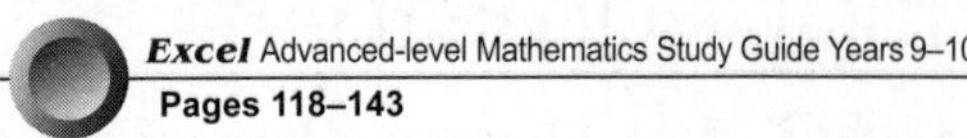

UNIT 12: Using the sine rule to find an angle

QUESTION 1 Given that θ is acute, find its value to the nearest degree.

a

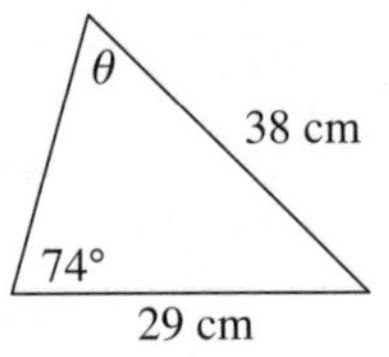

b

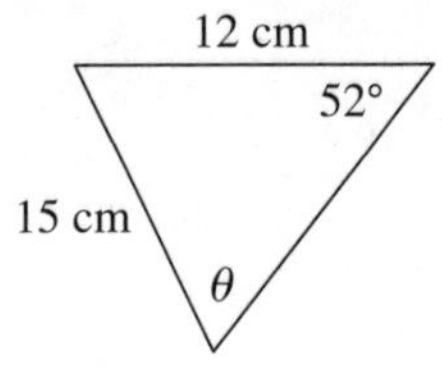

c

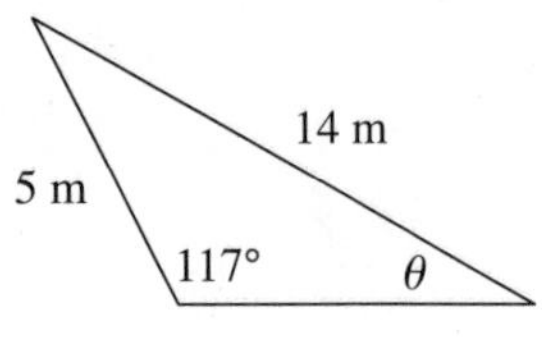

QUESTION 2 Given that θ is obtuse, find its value to the nearest degree.

a

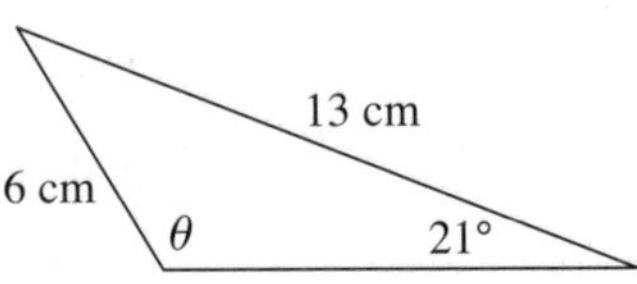

b

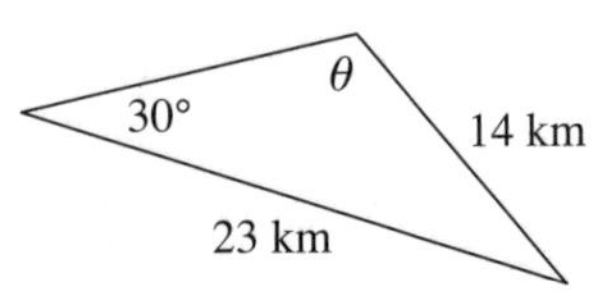

c

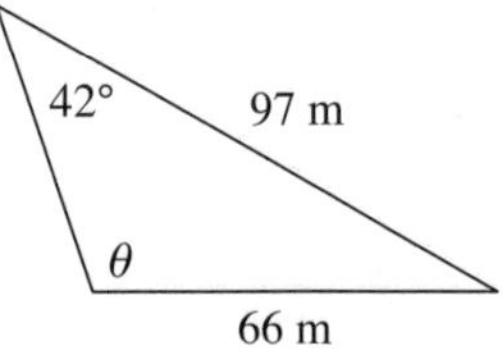

QUESTION 3 Use the sine rule to find the two possible sizes for $\angle A$ if:

a $a = 10, b = 7, \angle B = 36°$

b $a = 67, b = 53, \angle B = 47°$

c $a = 39, b = 27, \angle B = 24°$

QUESTION 4 Use the sine rule to find the two possible values for $\angle A$. Then show that only one value is physically possible.

a $a = 8, b = 12, \angle B = 40°$

b $a = 49, b = 58, \angle B = 61°$

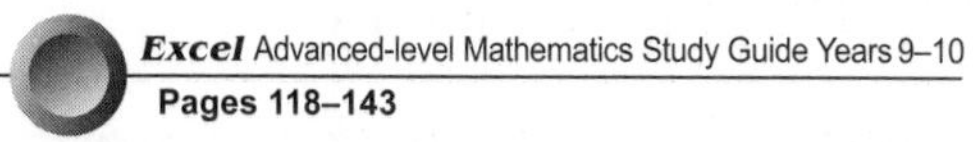

UNIT 13: Further use of the sine rule

QUESTION 1 Find all possible sizes of angle C (The diagrams are not to scale.).

a

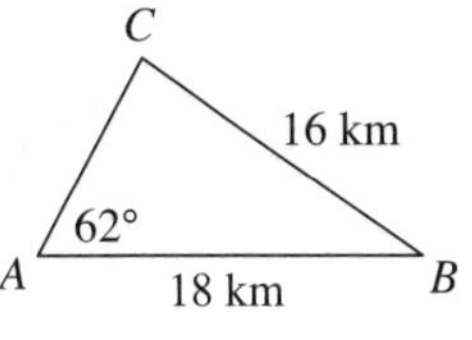

b

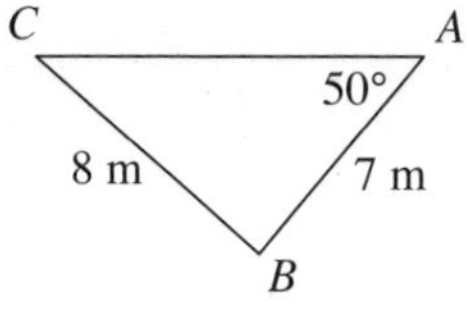

c

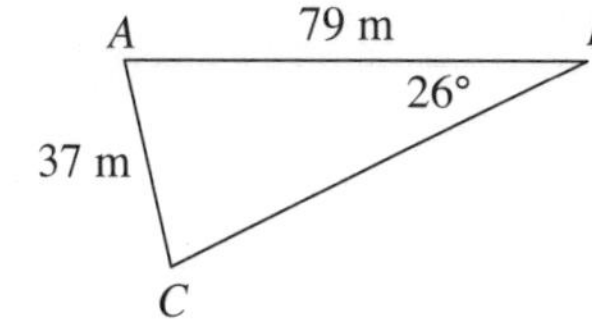

QUESTION 2

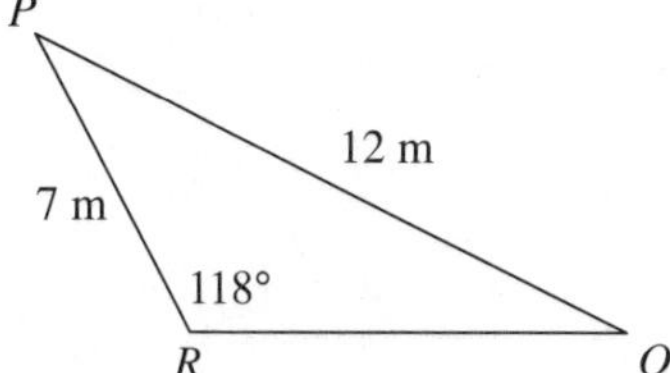

a Use the sine rule to find the size of $\angle PQR$.

b Find the size of $\angle QPR$.

QUESTION 3 Find the size of $\angle XYZ$.

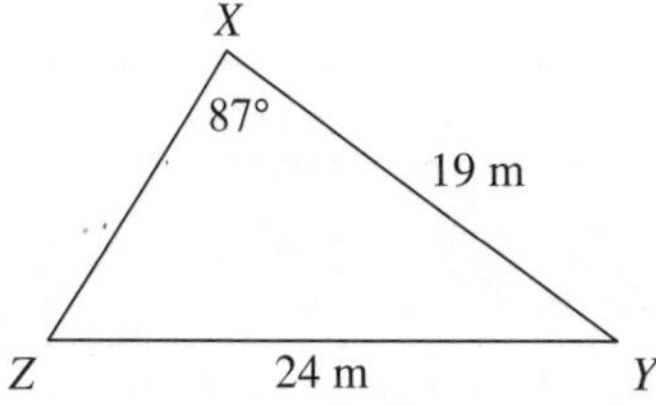

QUESTION 4

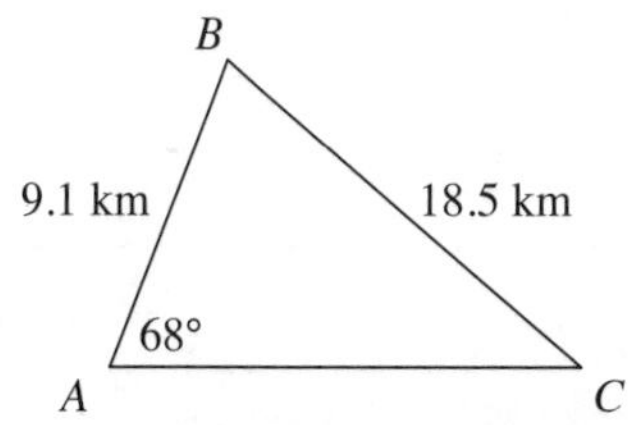

a Find the size of $\angle ACB$ to the nearest degree.

b Find the size of $\angle ABC$.

c Use the sine rule to find the length of AC to the nearest kilometre.

Excel Advanced-level Mathematics Study Guide Years 9–10
Pages 118–143

UNIT 14: Using the cosine rule to find a side

QUESTION 1 Find the value of x correct to one decimal place.

a

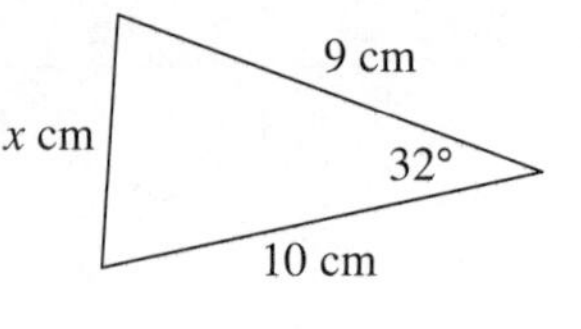

b

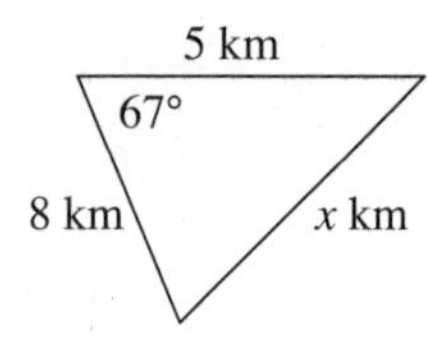

c

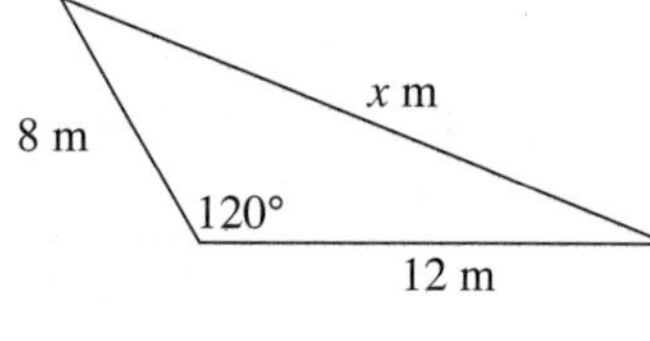

d

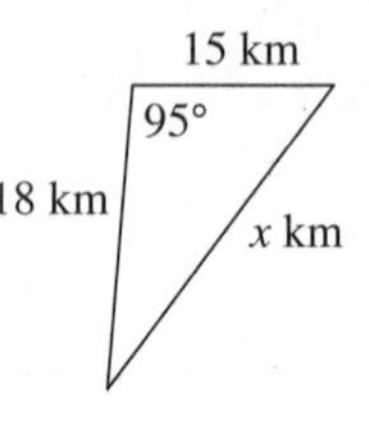

e

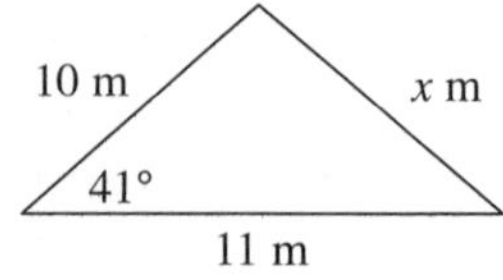

f

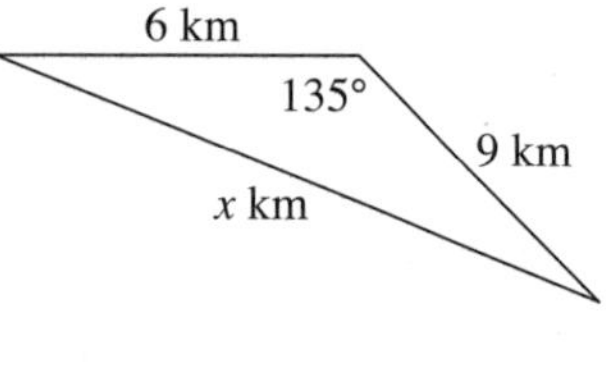

QUESTION 2

a Find the length of side BC to the nearest metre.

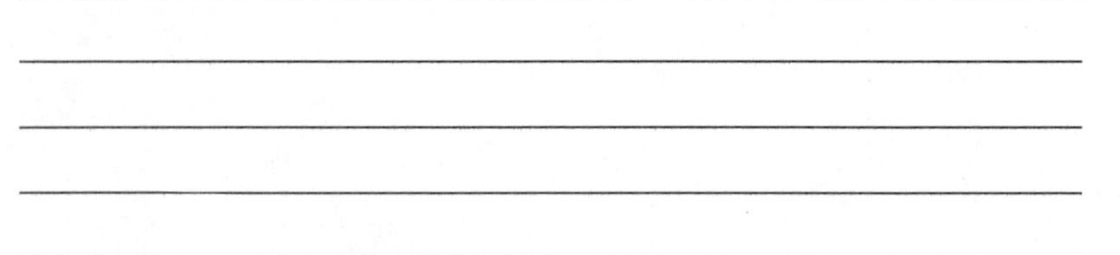

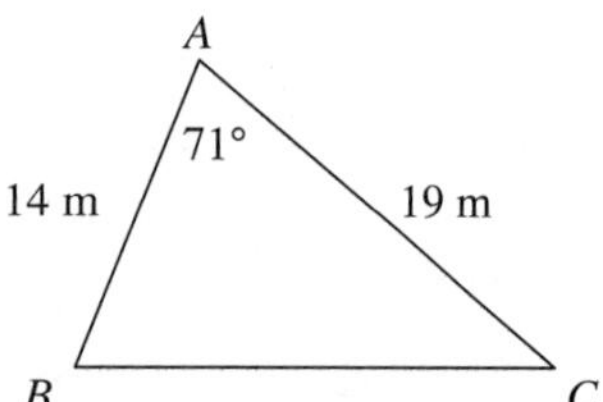

b What is the perimeter of the triangle?

QUESTION 3 Find, correct to one decimal place, the length of:

a AC

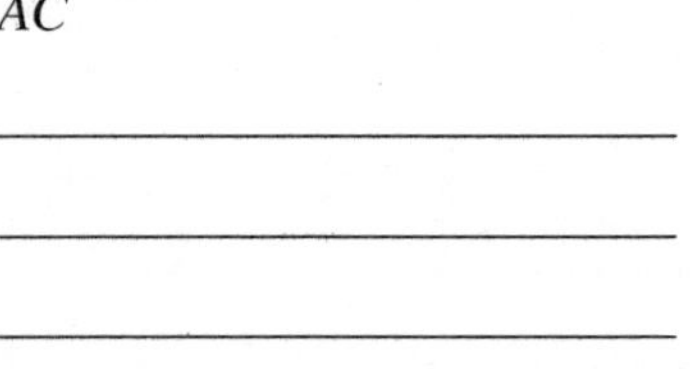

b AD

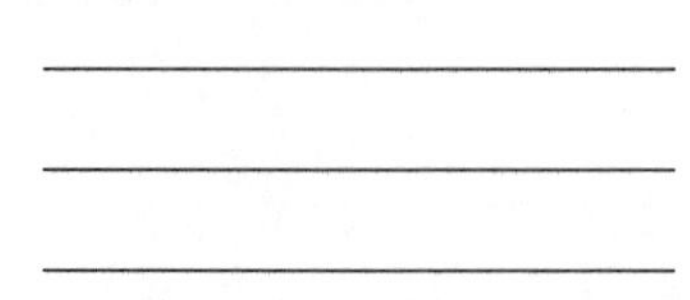

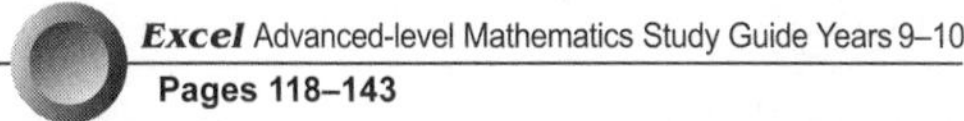

UNIT 15: Using the cosine rule to find an angle

Question 1 Use the cosine rule to find the value of θ to the nearest degree.

a

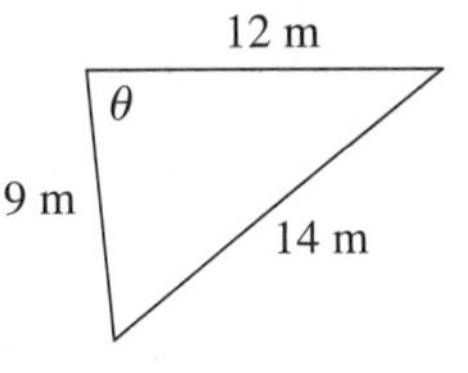

b

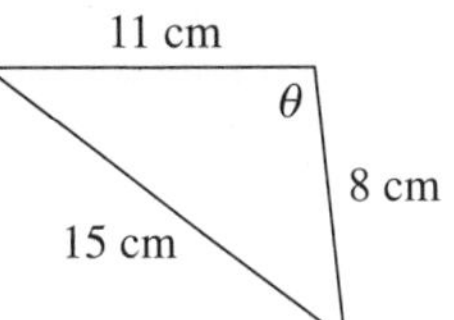

c

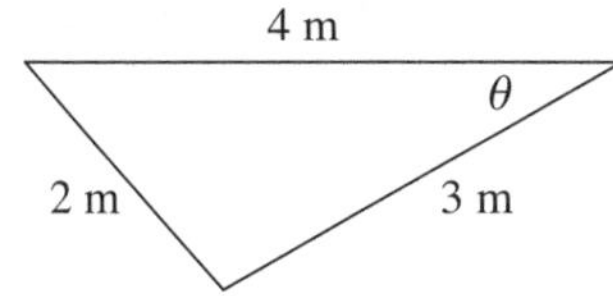

d

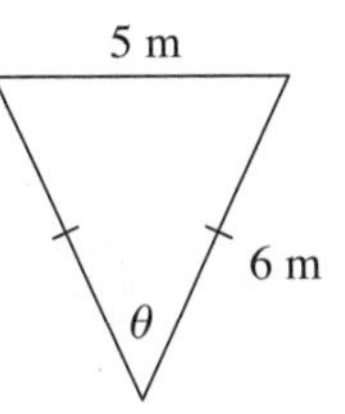

e

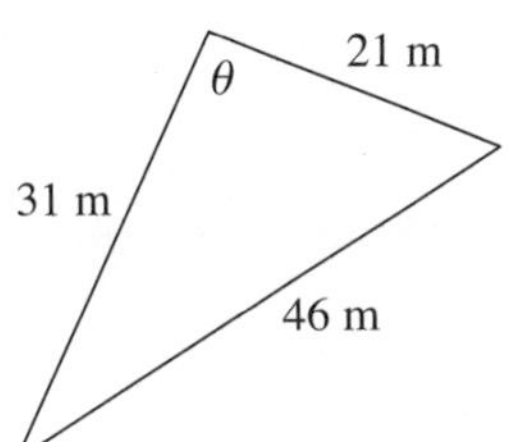

f

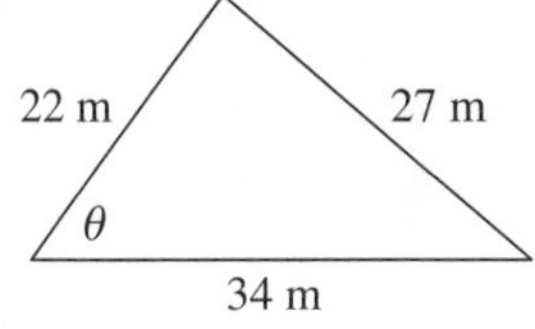

Question 2 Use the cosine rule in ABC to find the size, to the nearest minute, of:

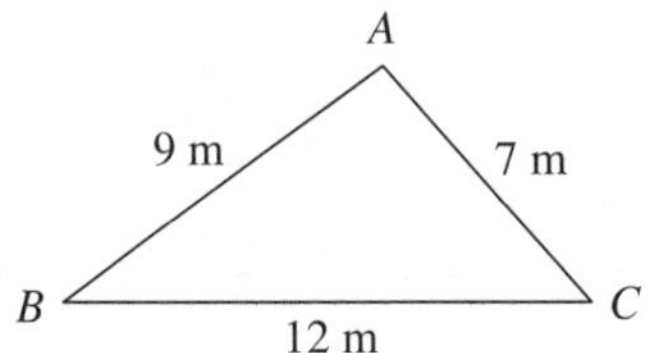

a $\angle ABC$

b $\angle CAB$

c $\angle BCA$

d Find the sum of the three angles. Is it exactly 180°? 

Excel Advanced-level Mathematics Study Guide Years 9–10
Pages 118–143

UNIT 16: Mixed exercises on the sine and cosine rules

QUESTION 1 Find the value of x correct to one decimal place.

a

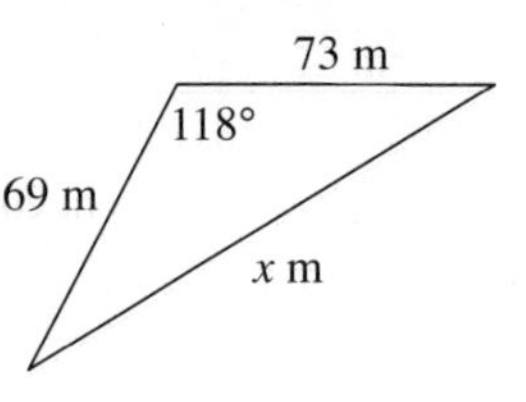

b

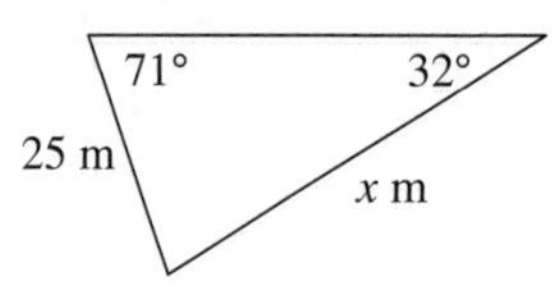

c

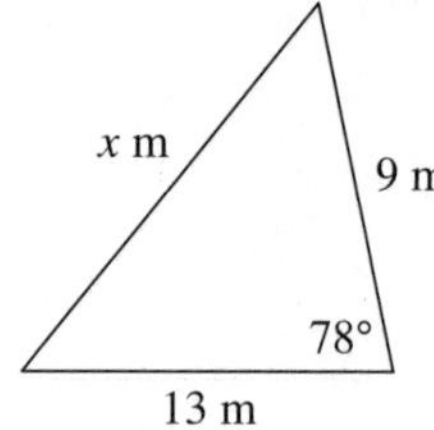

d

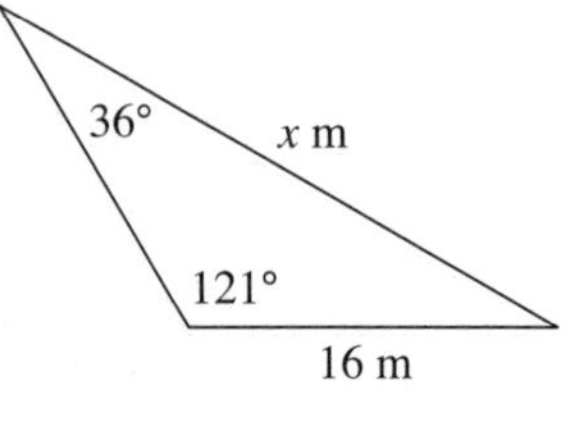

e

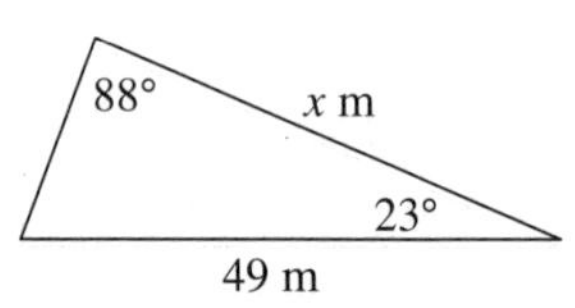

f

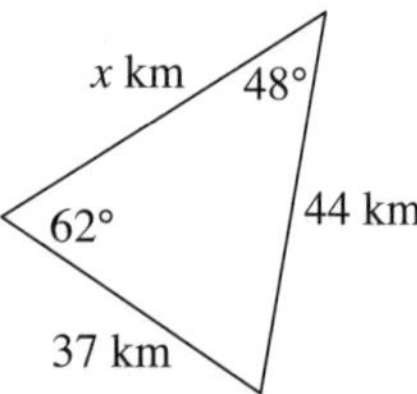

QUESTION 2 Find the value(s) of θ to the nearest degree.

a

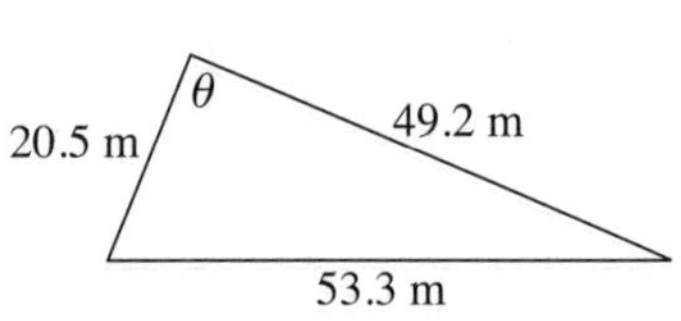

b

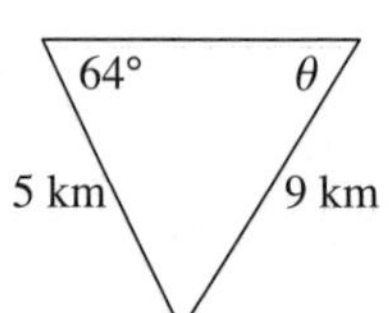

c

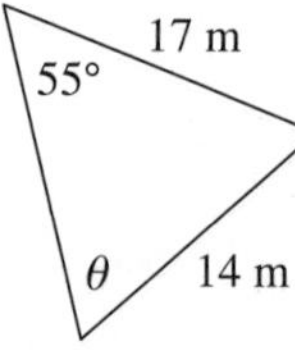

d

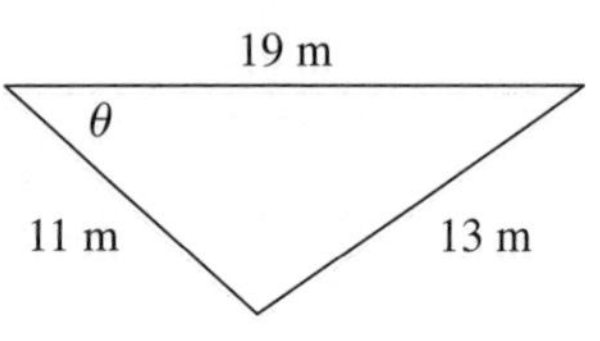

e

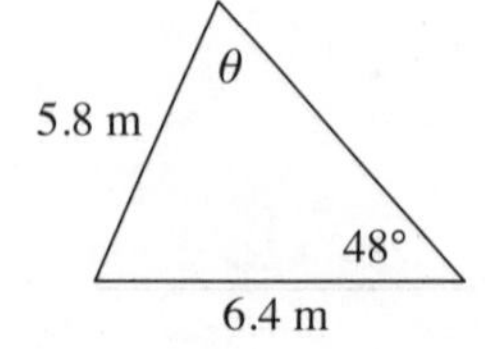

f

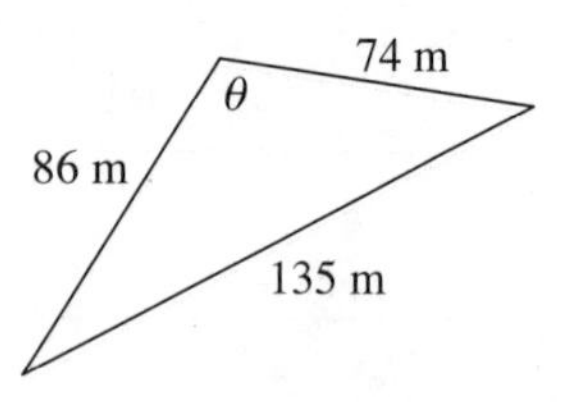

Trigonometry

UNIT 17: Area of a triangle

Question 1 Find the area of each triangle to the nearest square metre.

a

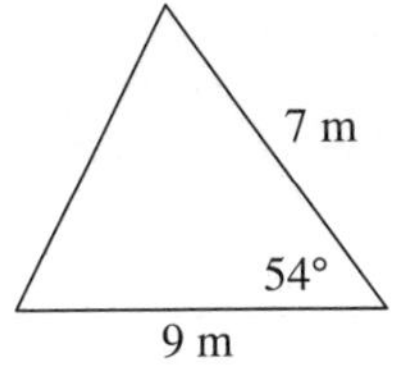

b

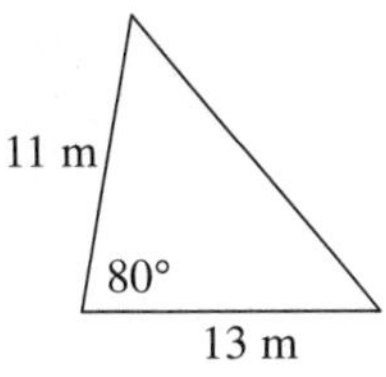

c

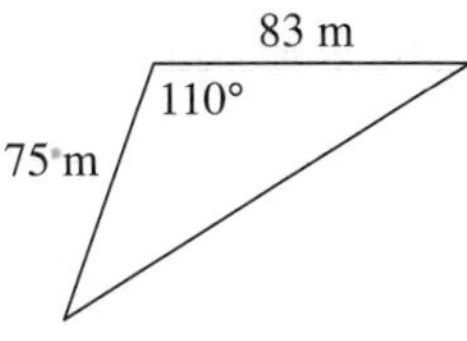

d

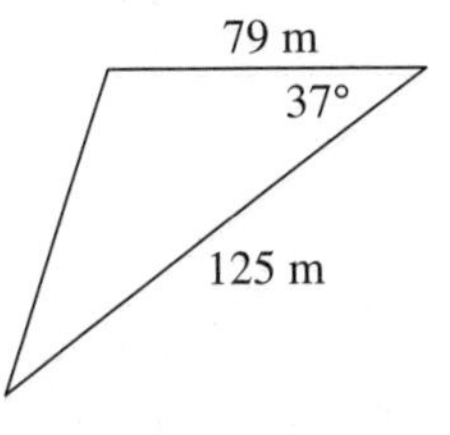

e

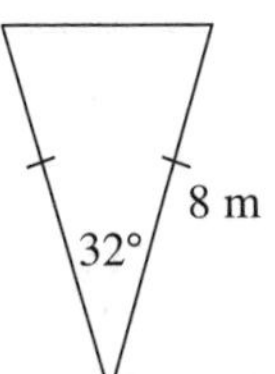

f

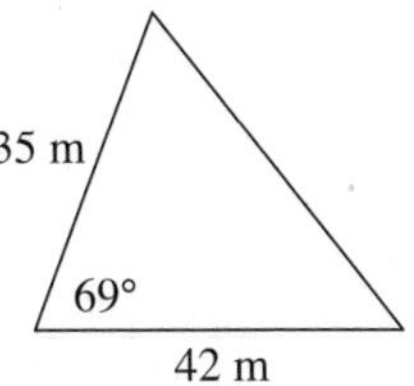

Question 2 Find the area of these triangles to the nearest square centimetre.

a

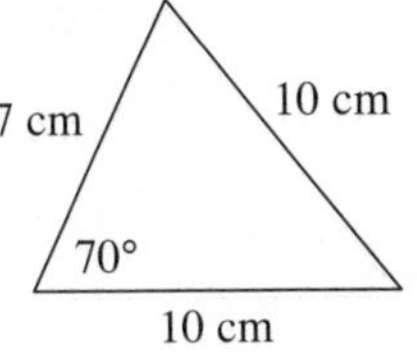

b

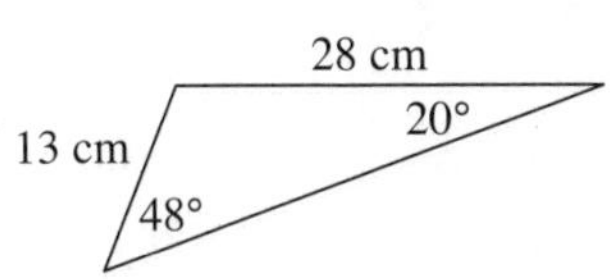

c

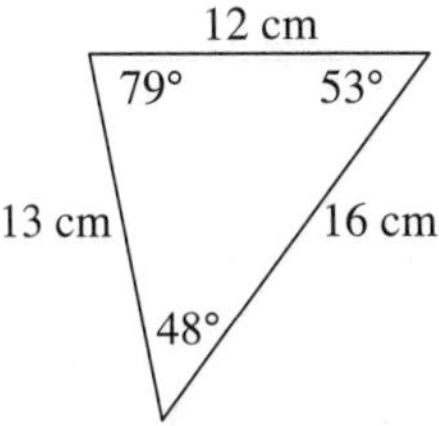

Question 3 Each triangle below has an area of 114 m^2.

a Find the value of x.

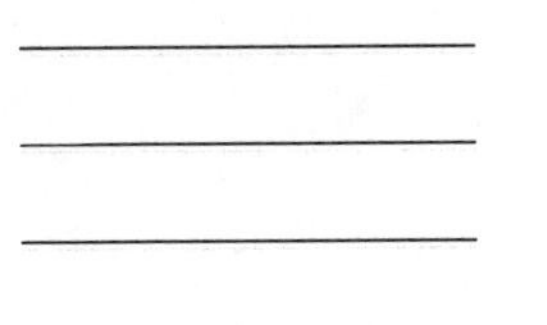

b Find θ.

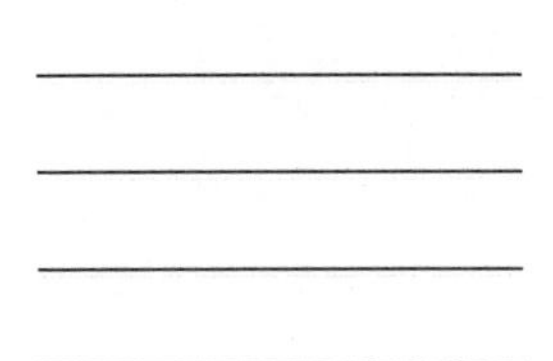

UNIT 18: Solving problems (1)

QUESTION 1 A triangular paddock has boundaries of length 59 m, 75 m and 112 m. Find:

a the size of the smallest angle.

b the area of the land to the nearest square metre.

QUESTION 2 A ship sails from P on a bearing of 075° for 125 km to Q. It then turns and sails on a bearing of 155° to R. The bearing of R from P is 115°.

a How far did the ship sail from Q to R?

b How far is it from R to P?

c On what bearing should the ship sail to return directly to P from R?

QUESTION 3 Find.

B
θ
6 m
5 m
C
2 m
A
11 m
D

a the value of θ to the nearest degree.

b the distance from A to C.

QUESTION 4 Two helicopters leave a town. One flies 293 km on a bearing of 317°. The other flies 562 km on a bearing of 212°. How far apart are the two helicopters at the end of the flights?

Trigonometry

UNIT 19: Solving problems (2)

QUESTION 1 P is 57 km from R and 78 km from Q. The bearing of P from R is 065° and the bearing of Q from R is 160°.

a What is the size of $\angle PRQ$? ______________________

b Find the size of $\angle QPR$.

c What is the bearing of P from Q?

QUESTION 2 A survey was taken of a piece of land ($ABCD$).

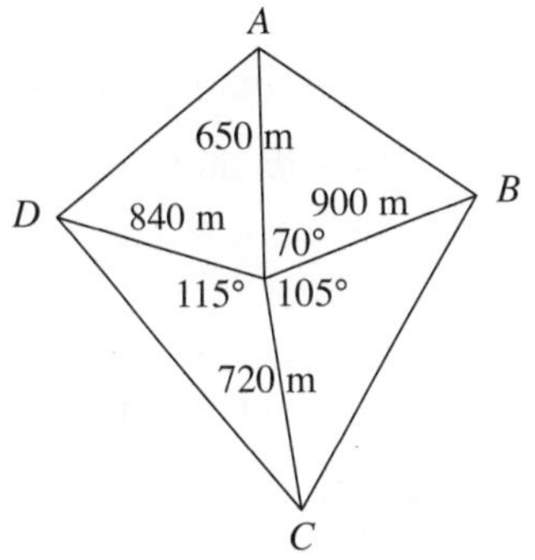

a What is the area of the land in hectares, to two decimal places?

A, B, C, D, 650 m, 840 m, 900 m, 70°, 115°, 105°, 720 m

b What is the length of boundary DC, to the nearest metre?

QUESTION 3 From point A the angle of elevation of the top (D) of a building is 39°. From a point B, 100 m closer to the base (C) of the building and along line AC, the angle of elevation of D is 54°.

a What is the size of $\angle ADB$? ______________________

b Use the sine rule in $\triangle ADB$ to find the length of DB.
Give the answer correct to two decimal places.

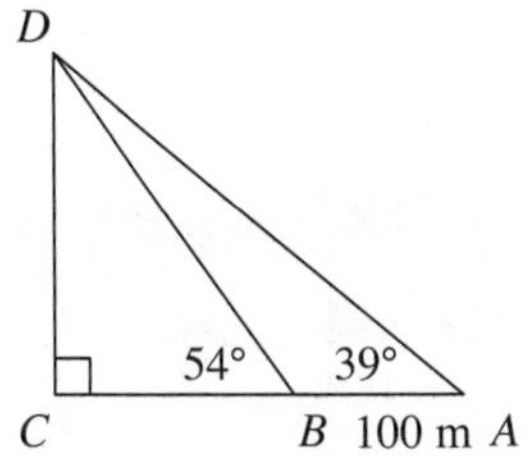

D, 54°, 39°, C, B 100 m A

c Use the answer to part b and $\triangle DBC$ to find the height of the building. Give the answer to the nearest metre.

Trigonometry

TOPIC TEST **PART A**

Instructions
- This part consists of 10 multiple-choice questions.
- Fill in only ONE CIRCLE for each question.
- Each question is worth 1 mark.

Time allowed: 10 minutes **Total marks: 10**

Marks

1 $\cos 150° =$

Ⓐ $-\dfrac{1}{\sqrt{2}}$ Ⓑ $-\dfrac{\sqrt{3}}{2}$ Ⓒ $-\dfrac{1}{2}$ Ⓓ $-\dfrac{1}{\sqrt{3}}$ 1

2 If $\dfrac{a}{\sin 70°} = \dfrac{9}{\sin 55°}$, then the value of a to one decimal place is

Ⓐ 7.8 Ⓑ 9.9 Ⓒ 10.3 Ⓓ 11.4 1

3 A triangle has sides 5 cm, 6 cm and 7 cm long. What is the size of the smallest angle?

Ⓐ 25° Ⓑ 37° Ⓒ 44° Ⓓ 57° 1

4 Which formula is NOT correct?

Ⓐ $A = \dfrac{1}{2}ab \sin C$ Ⓑ $\dfrac{a}{\sin A} = \dfrac{b}{\sin B}$

Ⓒ $c^2 = a^2 + b^2 - 2ab \cos C$ Ⓓ $\cos A = \dfrac{a^2 + b^2 - c^2}{2ab}$

5 Which is a solution to $2 \sin \theta - 1 = 0$?

Ⓐ $\theta = 30°$ Ⓑ $\theta = 60°$ Ⓒ $\theta = 45°$ Ⓓ $\theta = 90°$ 1

6 P is due south of Q. The bearing of R from Q is 230° and the bearing of R from P is 320°. What is the bearing of Q from R?

Ⓐ 040° Ⓑ 050°

Ⓒ 060° Ⓓ 070° 1

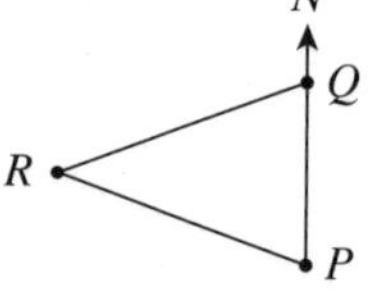

7 Which is closest to the area of an equilateral triangle of side length 5 cm?

Ⓐ 6.25 cm^2 Ⓑ 10.8 cm^2 Ⓒ 12.5 cm^2 Ⓓ 13 cm^2 1

8 If $x^2 = 11^2 + 15^2 - 2 \times 11 \times 15 \times \cos 76°$, then x, to one decimal place, is

Ⓐ 3.9 Ⓑ 15.8 Ⓒ 5.1 Ⓓ 16.3 1

9 If $\cos A = \dfrac{8^2 + 9^2 - 7^2}{2 \times 8 \times 9}$, then the size of $\angle A$ to the nearest degree is

Ⓐ 48° Ⓑ 42° Ⓒ 58° Ⓓ 52° 1

10 Which expression will give the length of AB?

Ⓐ $\dfrac{16 \sin 40°}{\sin 85°}$ Ⓑ $\dfrac{16 \sin 85°}{\sin 40°}$

Ⓒ $\dfrac{16 \sin 55°}{\sin 85°}$ Ⓓ $\dfrac{16 \sin 85°}{\sin 55°}$ 1

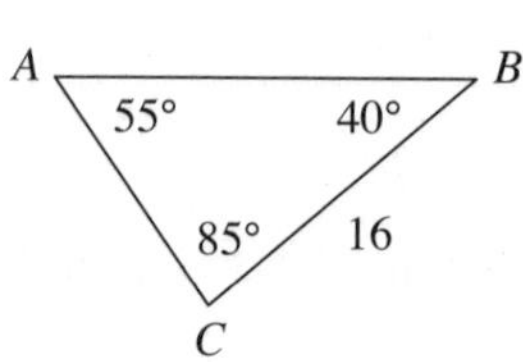

Total marks achieved for PART A

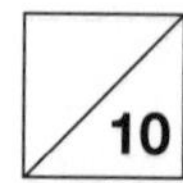

Trigonometry

TOPIC TEST — PART B

Instructions
- This part consists of 7 questions.
- Each question part is worth 1 mark, except question 6.
- Show all working.

Time allowed: 20 minutes — **Total marks: 15**

Marks

1 Find the value of θ if:

a $\cos\theta = -0.6428$ **b** $\tan\theta = \dfrac{1}{\sqrt{3}}$

2

2 The diagram shows the curves $y = \sin x°$ and $y = \cos x°$.

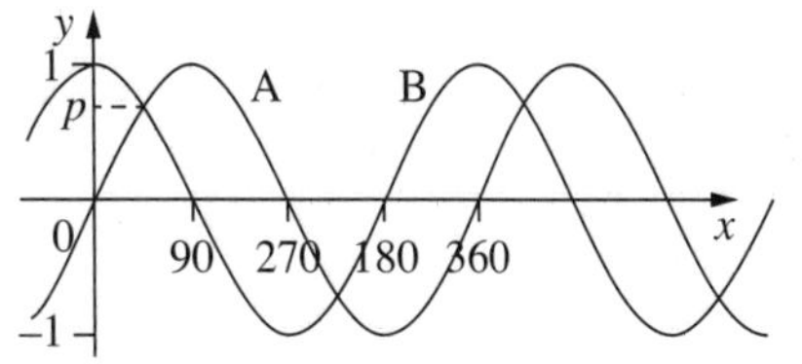

a Which graph (A or B) is the graph of $y = \cos x°$?

b What is the exact y-value of p?

2

3 A tower stands 236 m high on level ground. P is due south of the tower and Q is due east of it. From P the angle of elevation of the top of the tower, A, is 25° and from Q the angle of elevation of A is 32°. Find (to one decimal place):

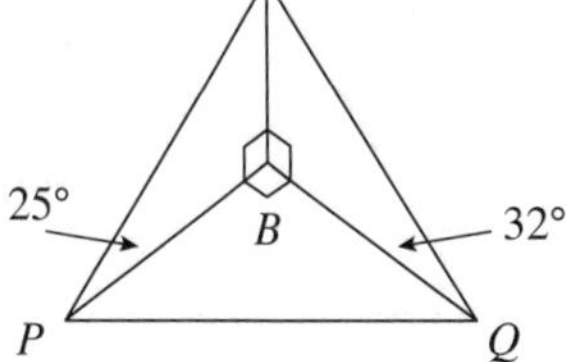

a PB **b** BQ **c** PQ

3

4 Find the value of x to one decimal place.

a

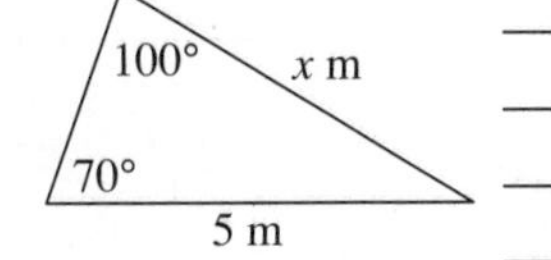

b

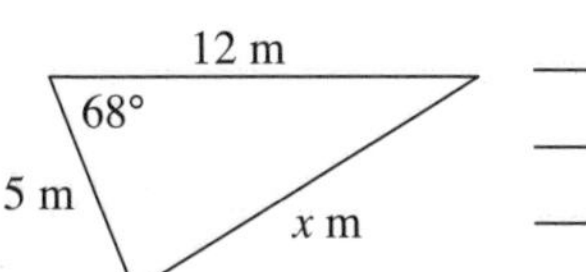

2

5 Find the area, to the nearest square metre, of each triangle in question 4.

a **b**

2

6 Find the two possible values of θ (to the nearest degree).

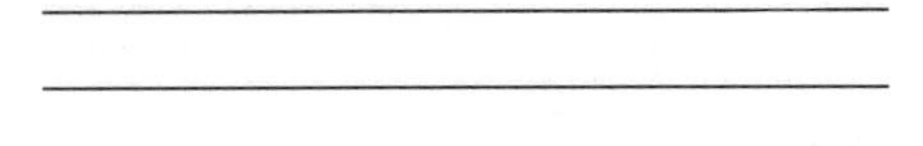

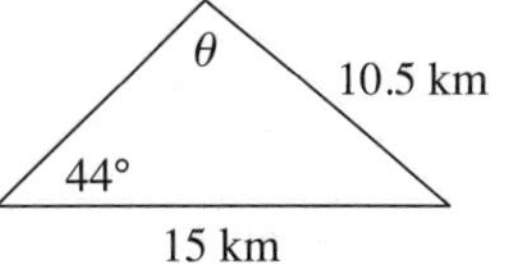

2

7 **a** Find the size of $\angle PQR$ to the nearest degree.

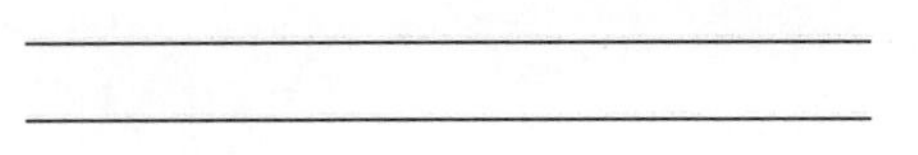

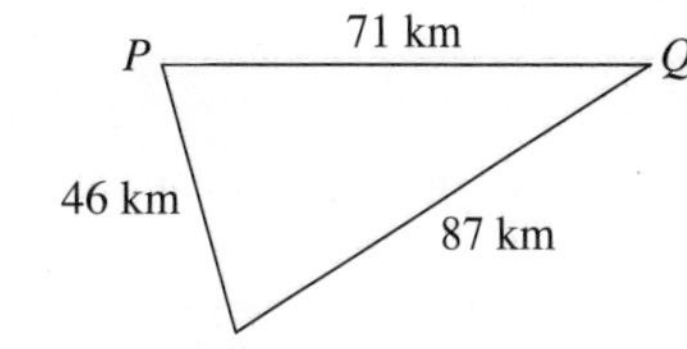

b If Q is due east of P, find the bearing of R from Q.

2

Total marks achieved for PART B

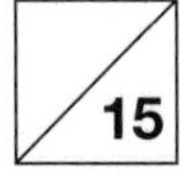

CHAPTER 8

Circle geometry

Excel Advanced-level Mathematics Study Guide Years 9–10
Pages 263–286

UNIT 1: Terminology

QUESTION 1 Name the part of the circle that is drawn, or shaded, in grey.

a

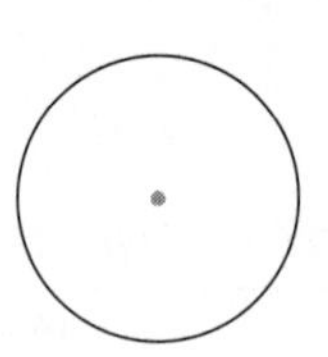

____

b

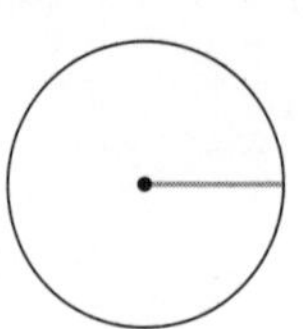

____

c

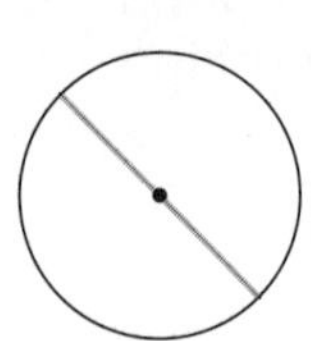

____

d

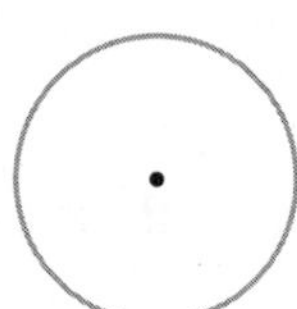

____

e

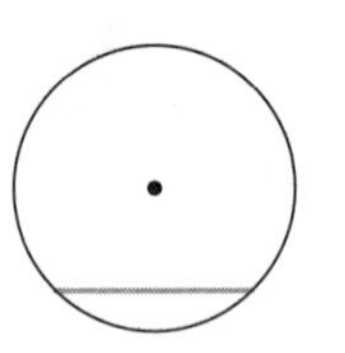

____

f

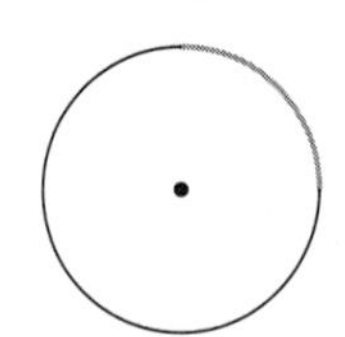

____

g

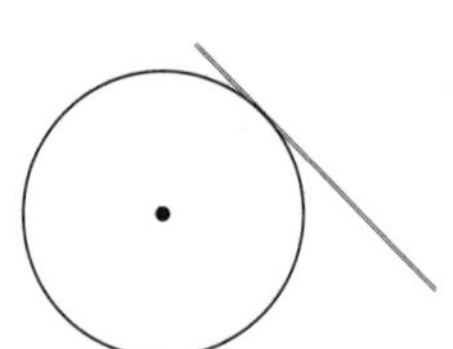

____

h

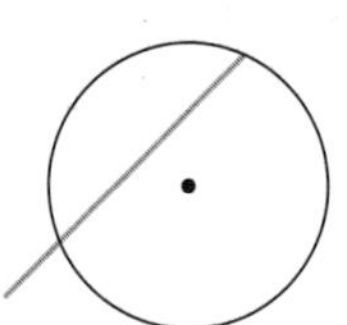

____

i

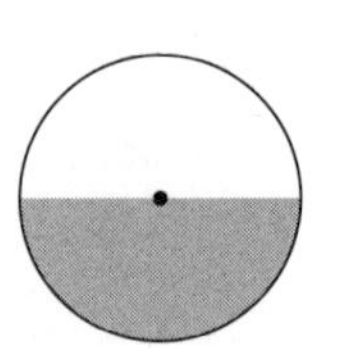

____

j

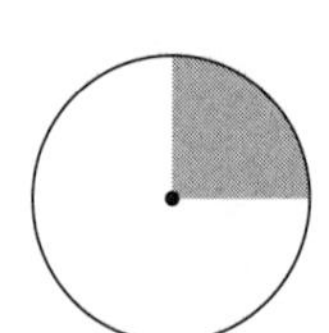

____

k

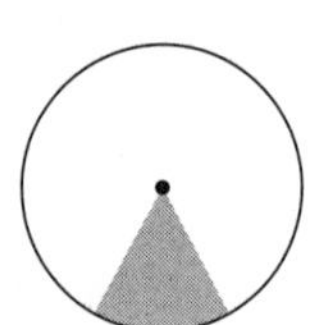

____

l

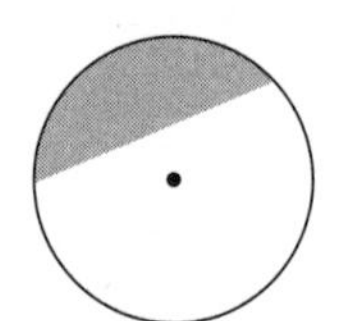

____

QUESTION 2 Name the angle(s) in the same segment standing on the same arc as:

a $\angle ABD$ ______________

b $\angle FAC$ ______________

c $\angle BED$ ______________

d $\angle CDA$ ______________

e $\angle EBD$ ______________

f $\angle DAF$ ______________

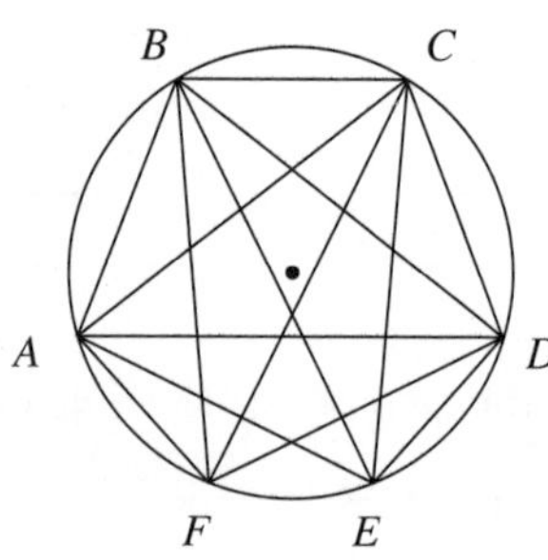

QUESTION 3

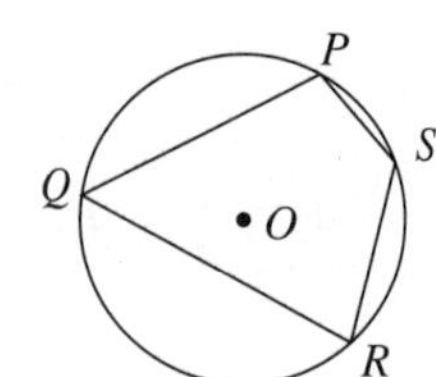

a How many different tangents can be drawn to the circle at P?

____

b What type of quadrilateral is $PQRS$?

____

c Name the angle at the centre subtended by arc PQ.

____

Circle geometry

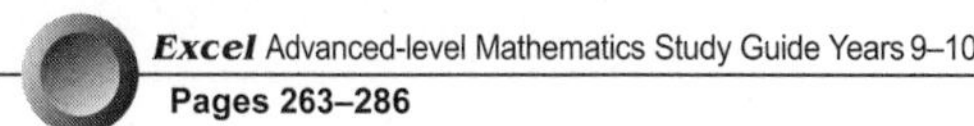

UNIT 2: Chord properties

QUESTION 1 Complete the following statements.

a Chords of equal length in a circle subtend ______________ ______________ at the centre.

b Chords of equal length in a circle are ______________________________ from the centre.

c The line from the centre of a circle to the midpoint of a chord is __________________ to that chord.

d The perpendicular from the centre of a circle to a chord ______________ the chord.

e The perpendicular bisector of a chord of a circle passes through the ______________ of the circle.

f When two circles intersect, the line joining their centres ______________ their common chord at ______________ ______________.

g For any three points that form a triangle, the point of intersection of the perpendicular bisectors of any two sides is the ____________________ of the circle through all three points.

QUESTION 2 Find the value of the pronumeral in the following questions.

a

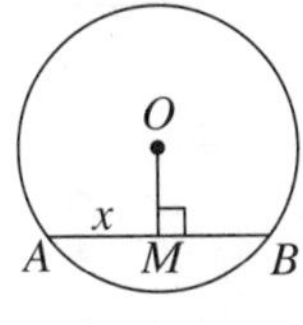

$AB = 18$ cm

____

b

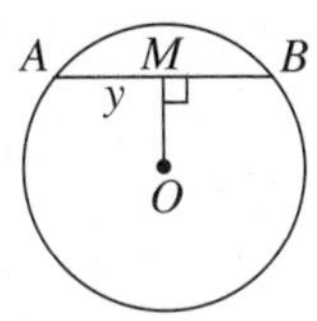

$AB = 36$ cm

____

c

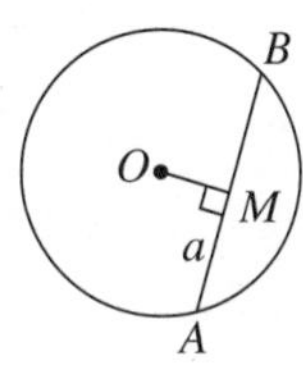

$AB = 12.6$ cm

____

d

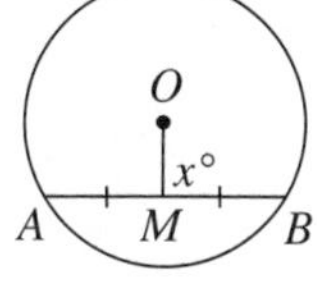

$AM = BM = 5$ cm

____

e

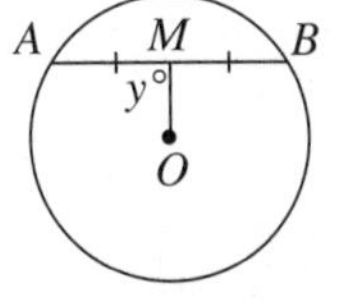

$AM = BM = 8$ cm

____

f

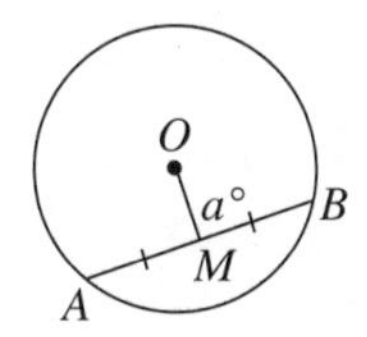

$AM = BM = 12$ cm

____

g

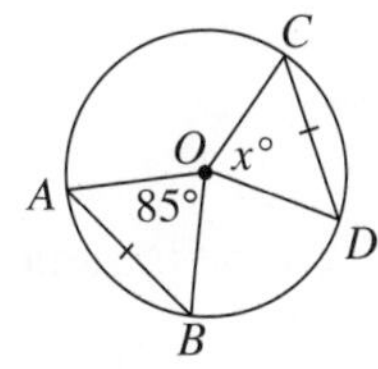

$AB = CD = 9$ cm

____

h

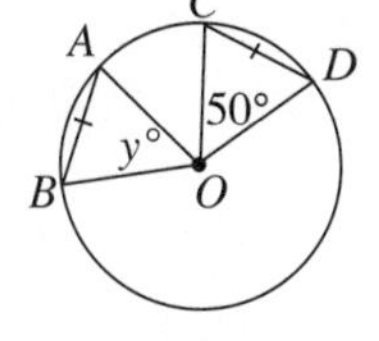

$AB = CD = 6$ cm

____

i

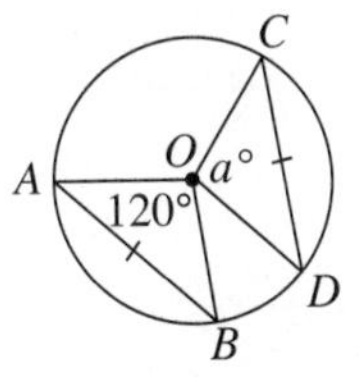

$AB = CD = 12$ cm

____

j

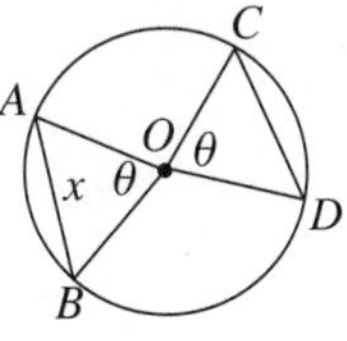

$CD = 5$ cm

____

k

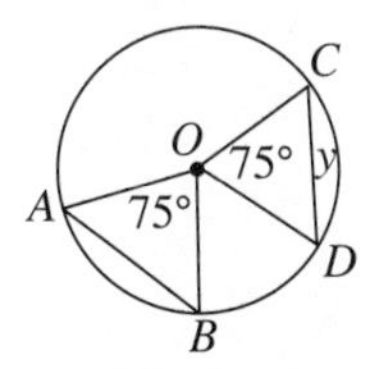

$AB = 7$ cm

____

l

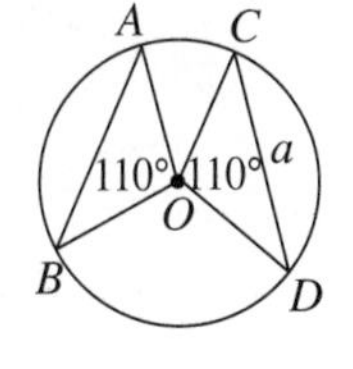

$AB = 12$ cm

____

QUESTION 3 Find the required length.

a

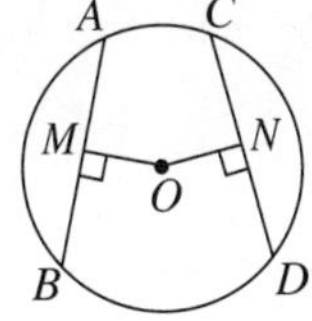

$AB = CD$
$OM = 6$ cm
Find ON.

____

b

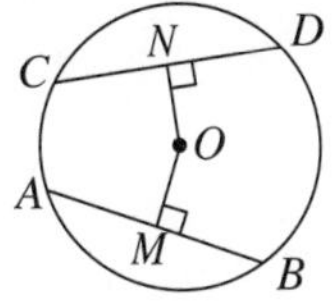

$AB = CD$
$ON = 9$ cm
Find OM.

____

c

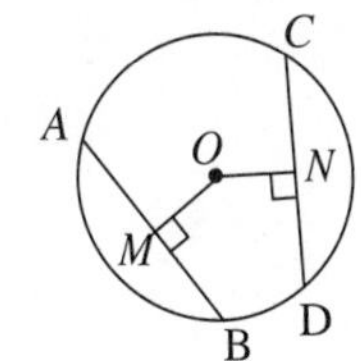

$OM = ON = 7$ cm
$CD = 21$ cm
Find AB.

____

Circle geometry

Excel Advanced-level Mathematics Study Guide Years 9–10
Pages 263–286

UNIT 3: Angle properties

QUESTION 1 Complete the following statements.

a The angle at the centre of a circle is twice the angle at the ___________ standing on the same ___________ .

b The angle in a semicircle is a ___________ ___________ .

c Angles at the _________________, standing on the same ___________ , are ___________ .

QUESTION 2 Find the value of the pronumeral in each of the following.

a

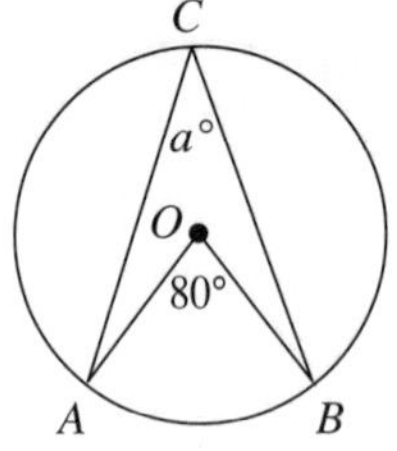

____

b

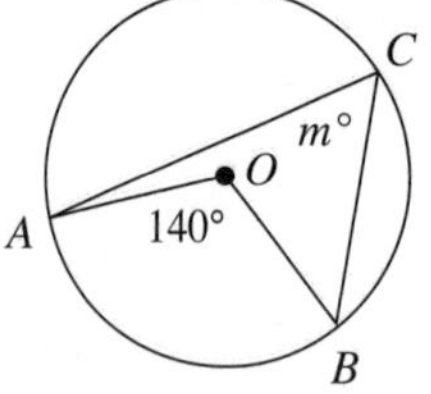

____

c

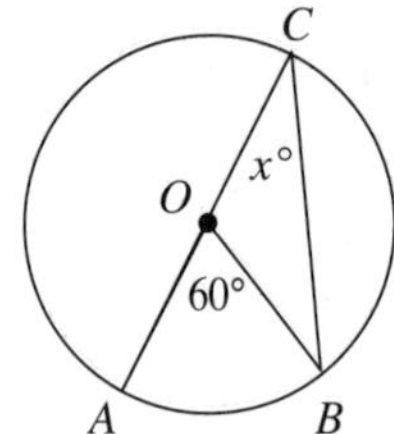

____

d

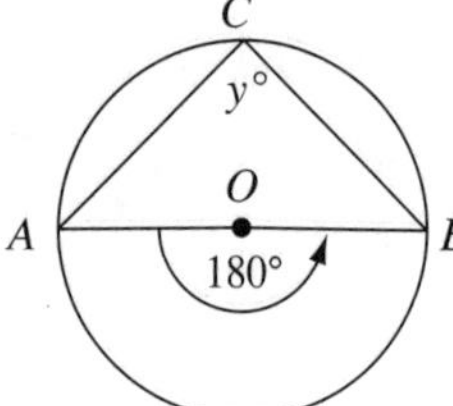

____

e

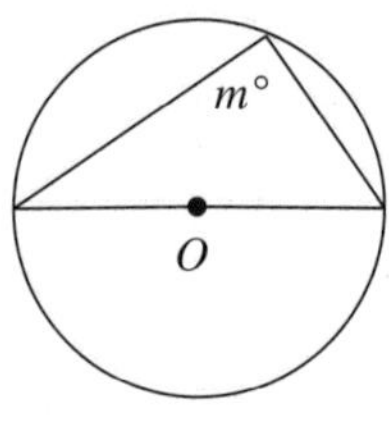

____

f

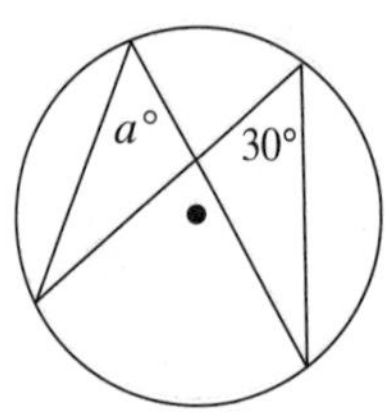

____

g

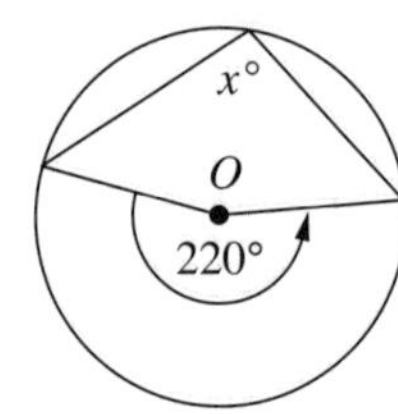

____

h

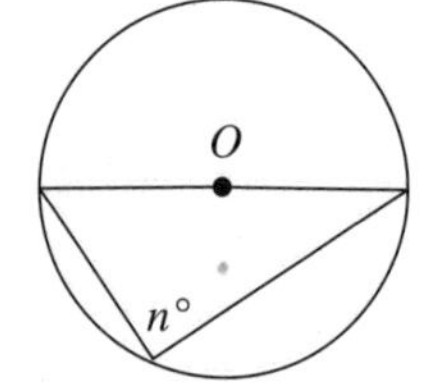

____

QUESTION 3 Find the value of each pronumeral.

a

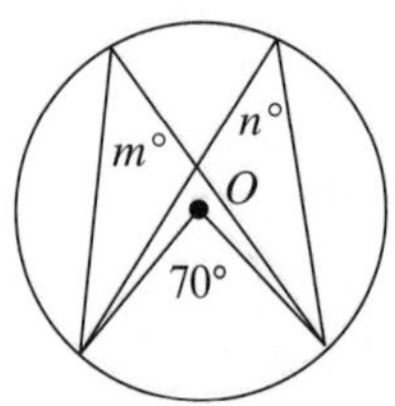

____

____

b

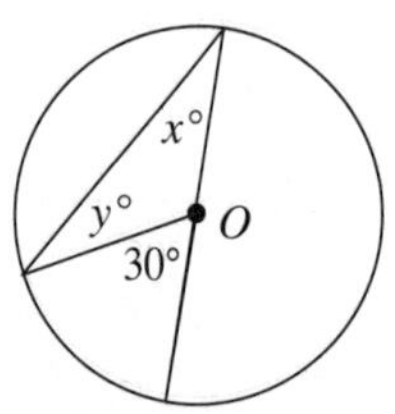

____

____

c

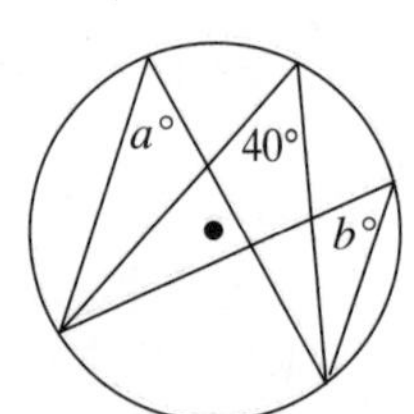

____

____

d

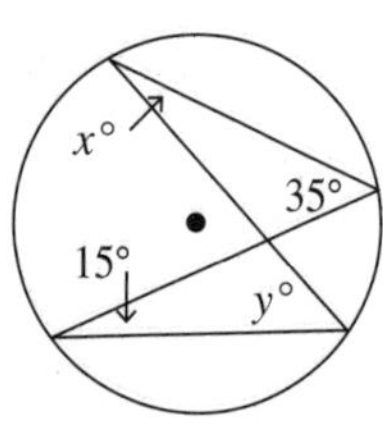

____

____

e

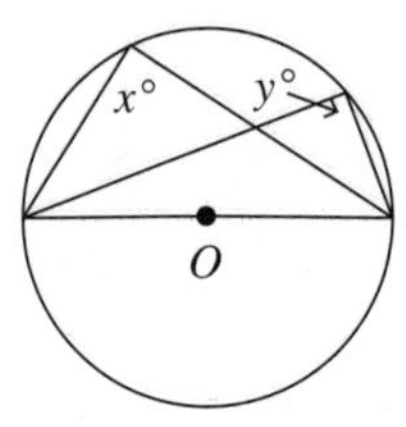

____

____

f

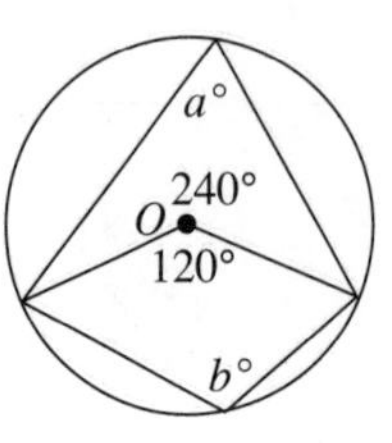

____

____

g

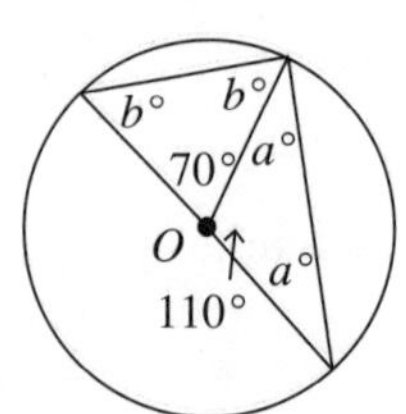

____

____

h

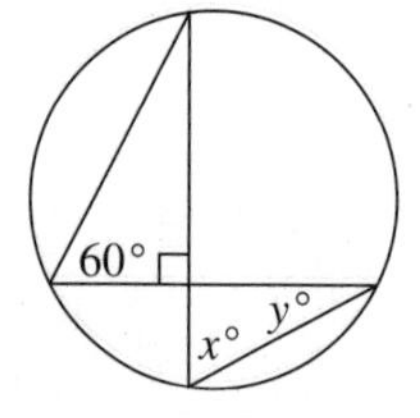

____

____

Circle geometry

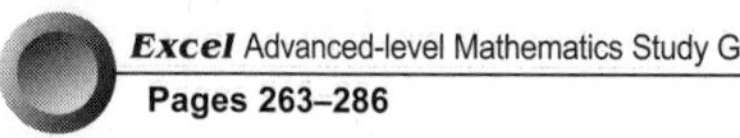

Excel Advanced-level Mathematics Study Guide Years 9–10
Pages 263–286

UNIT 4: Cyclic quadrilaterals

QUESTION 1 Complete the following statements.

a The ______________ angles of cyclic quadrilaterals are __________________.

b An ______________ angle at a vertex of a cyclic quadrilateral is _________ to the interior opposite angle.

QUESTION 2 Find the value of the pronumeral.

a

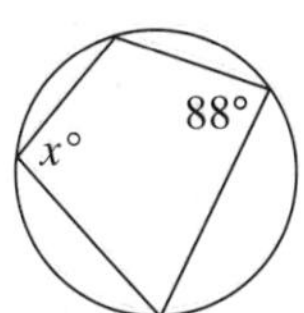

b

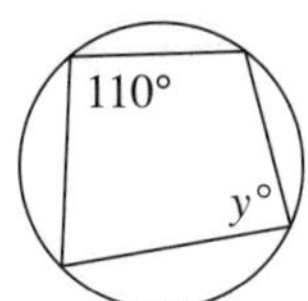

c

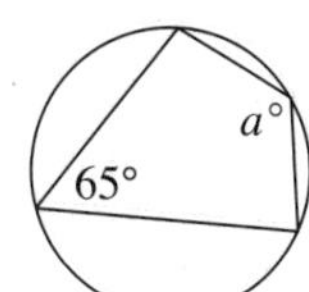

d

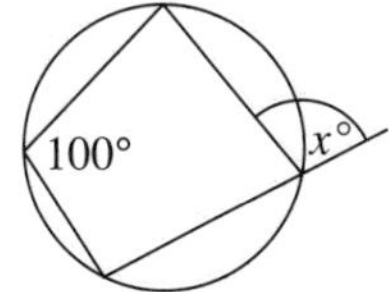

QUESTION 3 Find the value of each pronumeral.

a

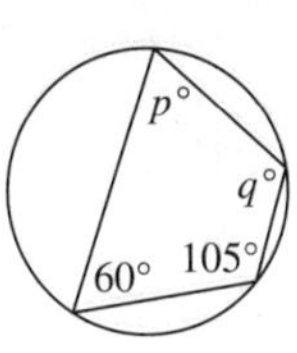

b

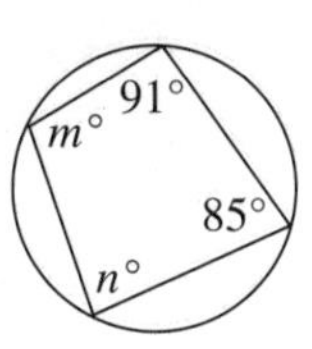

c

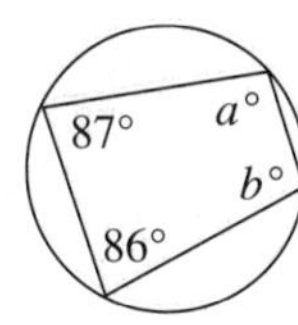

d

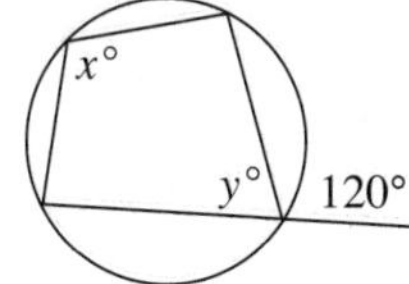

e

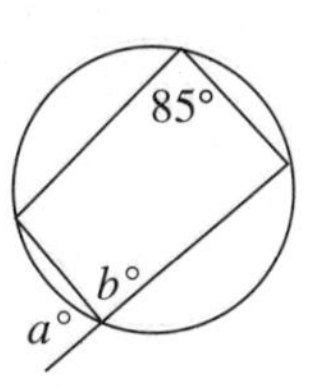

f

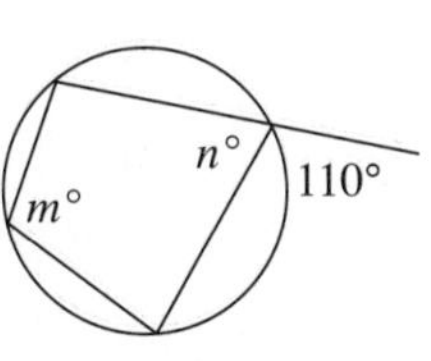

g

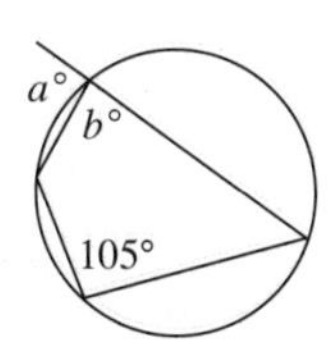

h

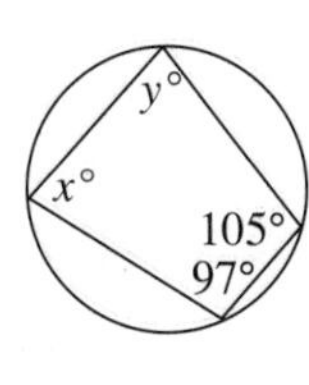

i

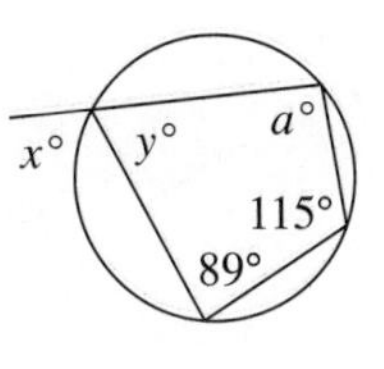

j

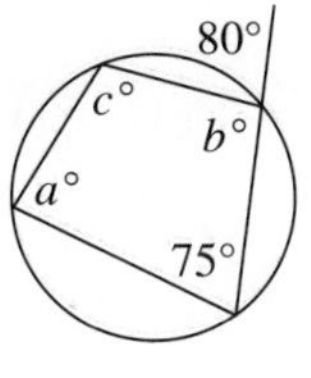
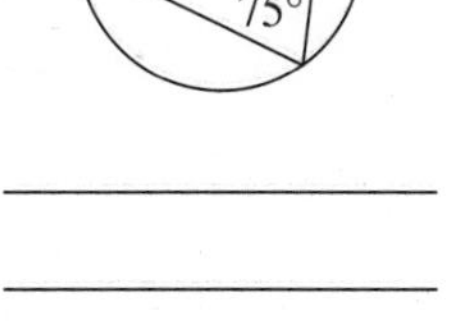

k

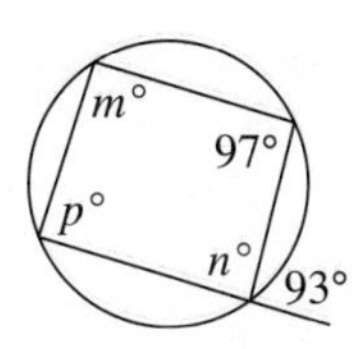

l

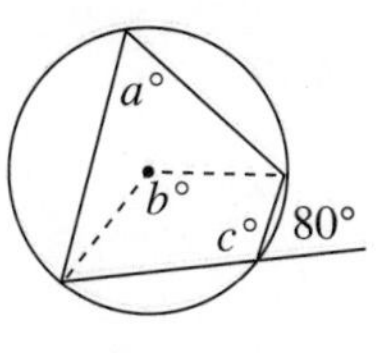

Circle geometry

UNIT 5: Tangent properties

Question 1 Complete the following statements.

a Any tangent to a circle is ____________________ to the ________________ at the point of contact.

b The two tangents drawn to a circle from an external point are ______________ .

c When two circles touch, their centres and the point of contact are ________________ .

Question 2 Find the value of the pronumeral.

a

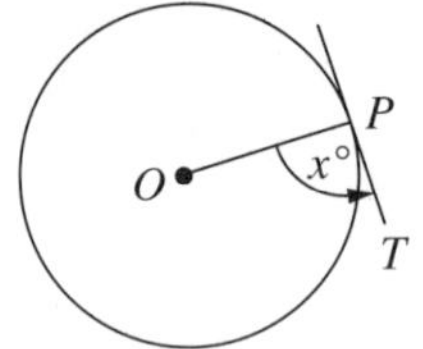

____

b

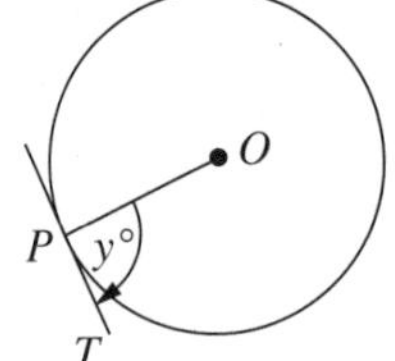

____

c

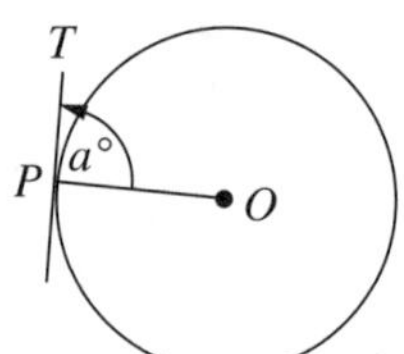

____

d

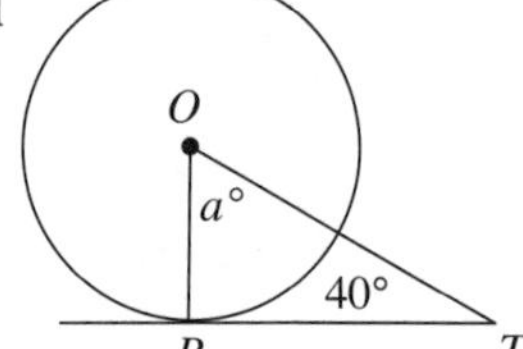

____

e

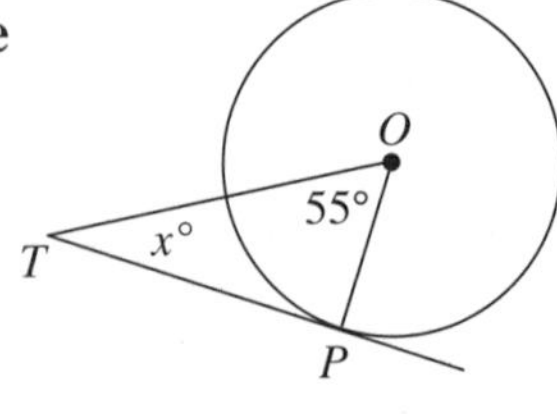

____

f

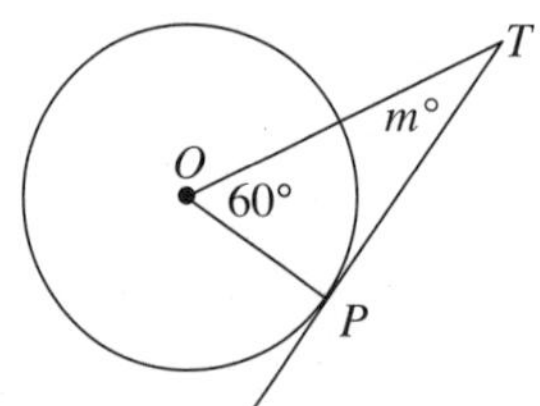

____

g

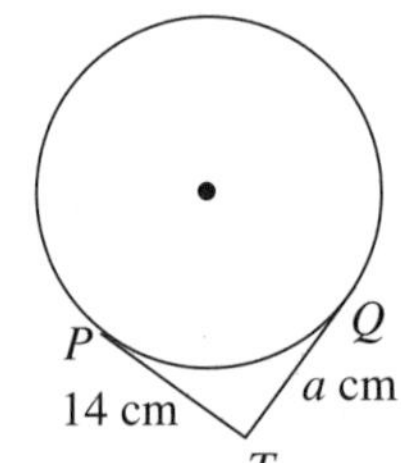

____

h

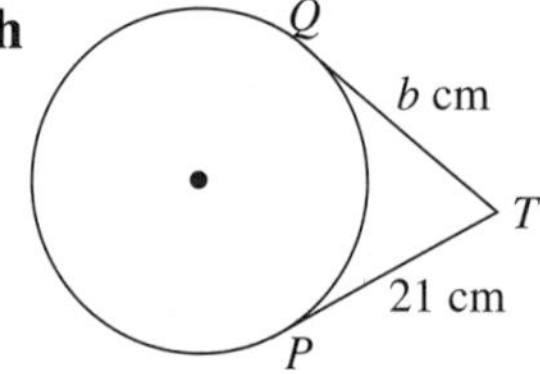

____

Question 3 Find the value of each pronumeral.

a

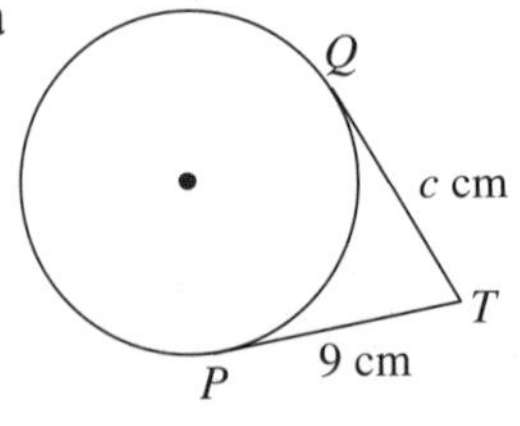

____
____

b

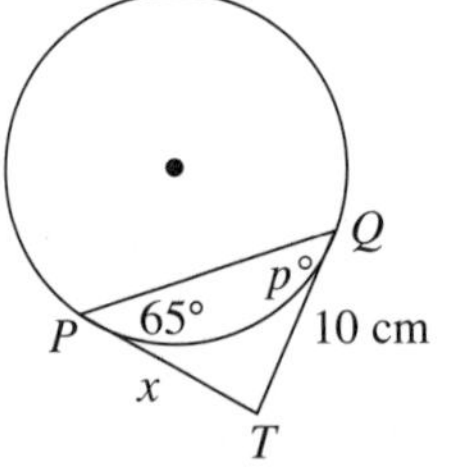

____
____

c

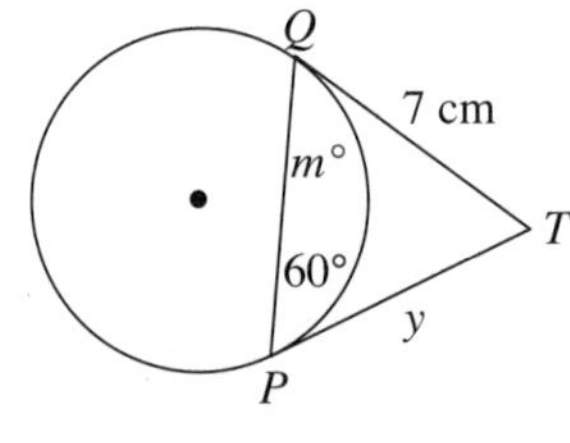

____
____

d

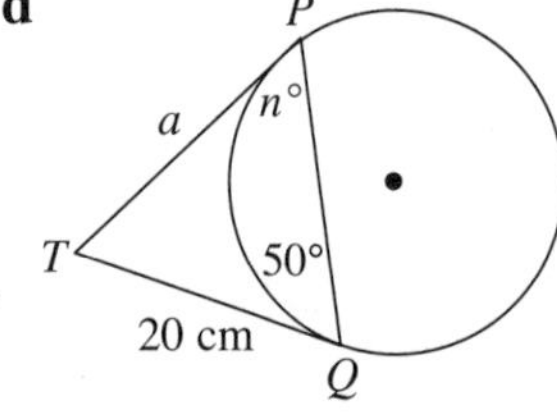

____
____

e

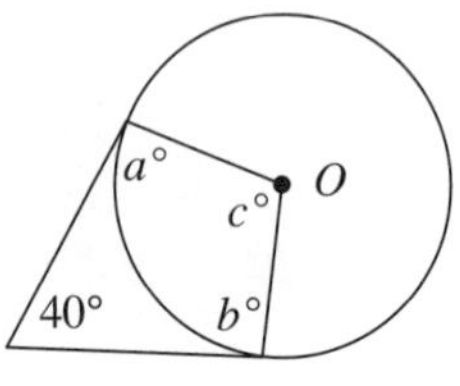

____
____
____

f

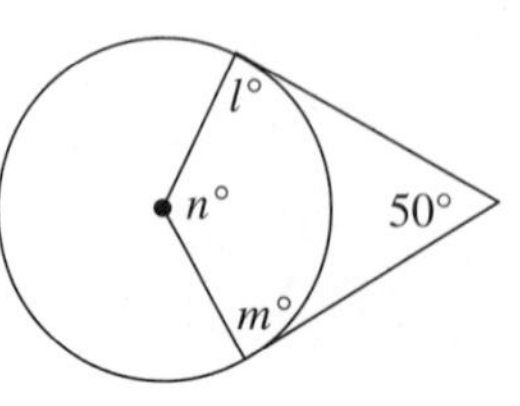

____
____
____

g

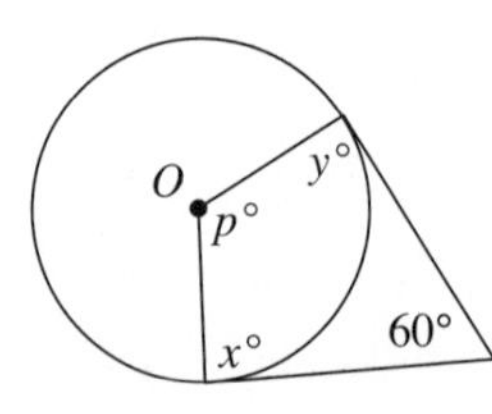

____
____
____

h

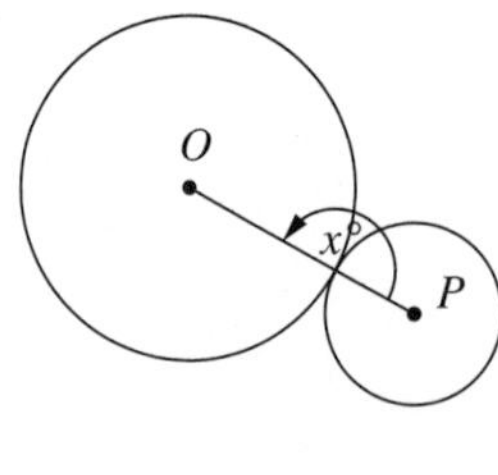

____
____
____

Excel Advanced-level Mathematics Study Guide Years 9–10
Pages 263–286

UNIT 6: Angle in the alternate segment

QUESTION 1 Complete the following statement.

The angle between a tangent and a chord drawn to the point of contact is ______________ to the angle in the ____ .

QUESTION 2 Find the value of the pronumerals.

a

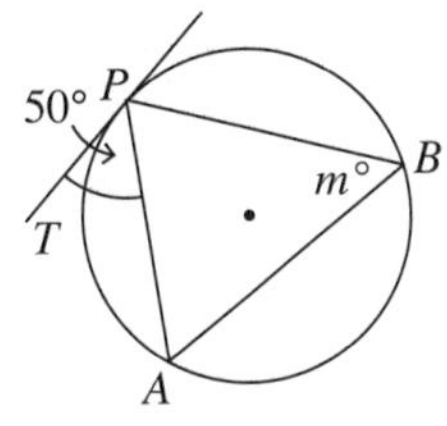

b

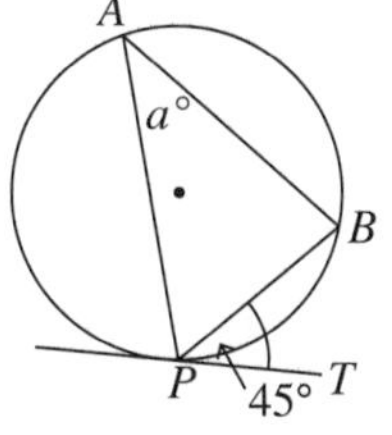

c

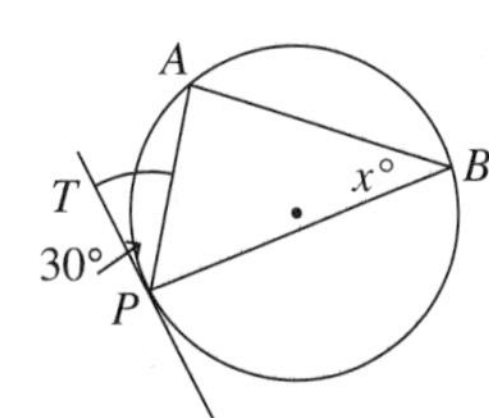

d

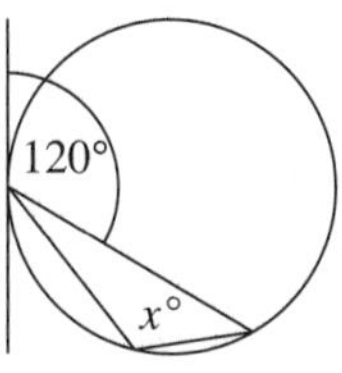

e

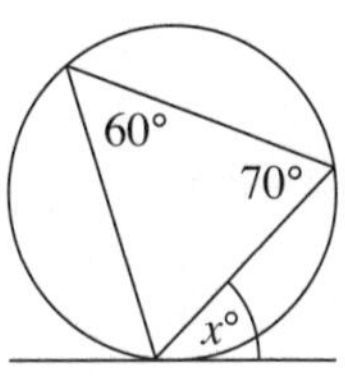

f

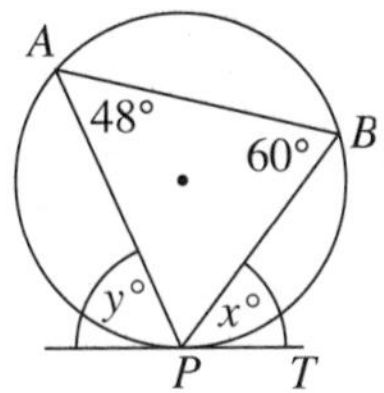

g

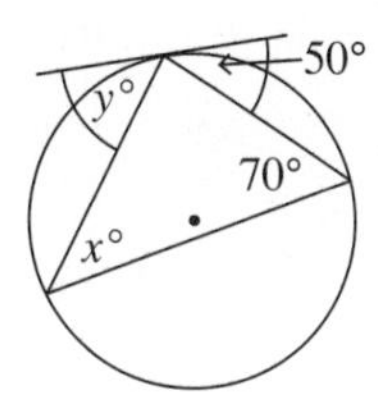

h

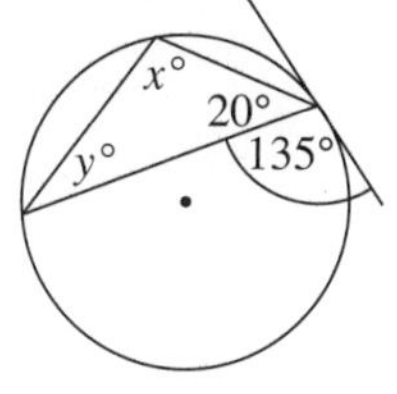

i

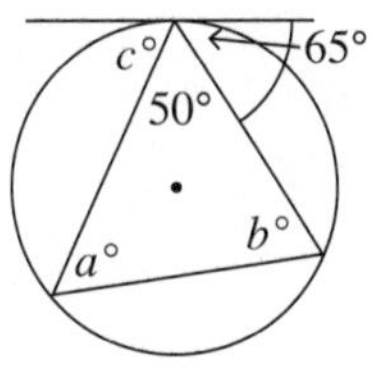

j

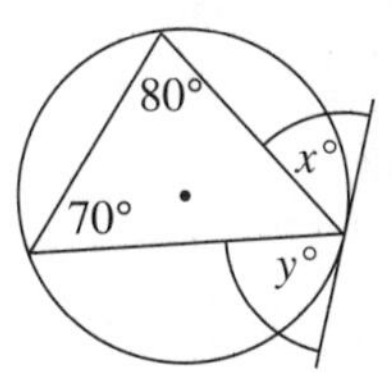

k

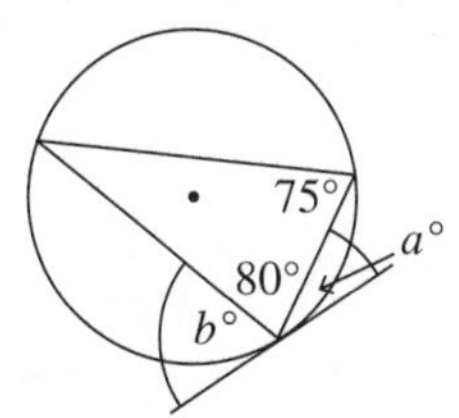

l

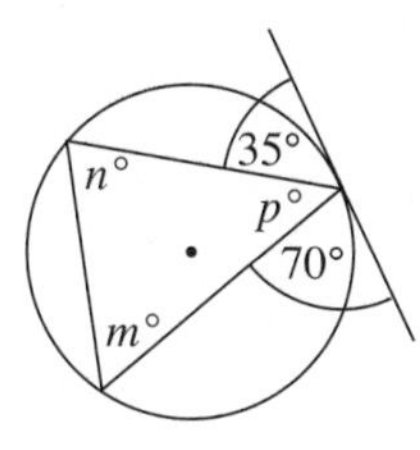

m

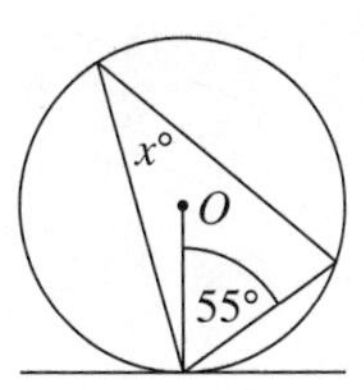

n

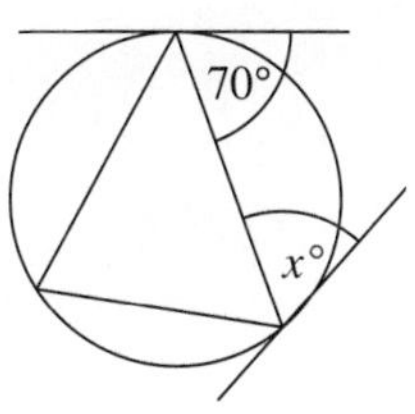

o

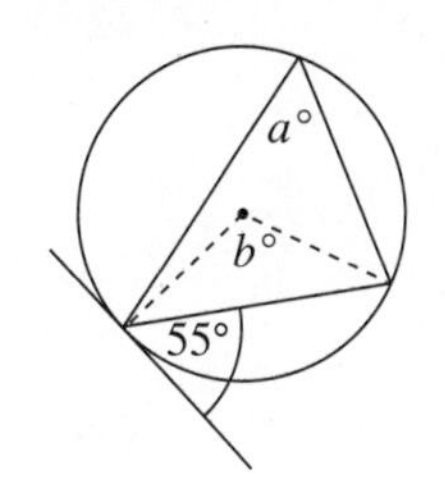

p

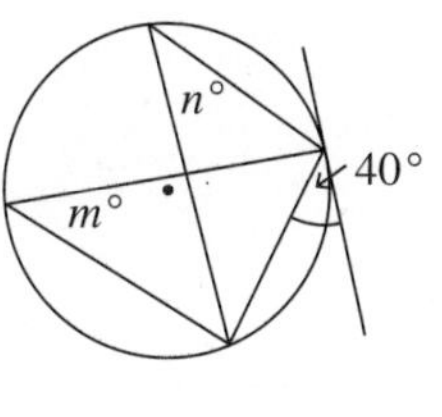

Circle geometry

Excel Advanced-level Mathematics Study Guide Years 9–10
Pages 263–286

UNIT 7: Secant properties

QUESTION 1 Complete the following statements.

a The __________ of the intercepts of two equal chords are __________ .

b The __________ of the intercepts of two intersecting secants to a circle from an external point are ________.

c The ___________ of a tangent to a circle from an external point __________ the product of the intercepts of any __________ from that point.

QUESTION 2 Find the value of the pronumeral. All measurements are in cm.

a

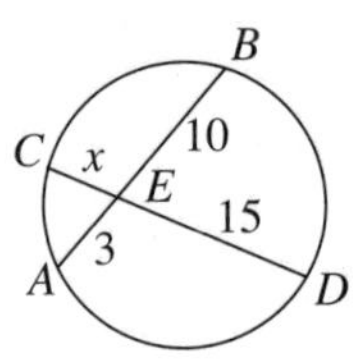

b

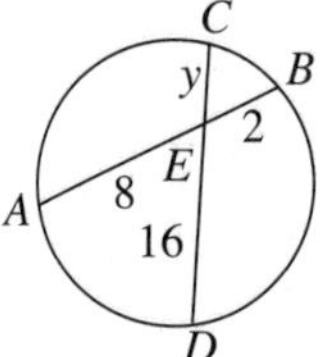

c

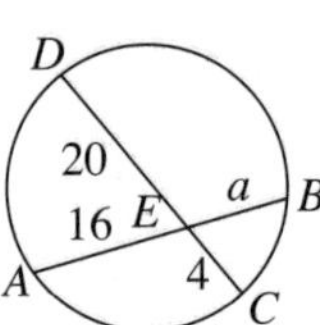

d

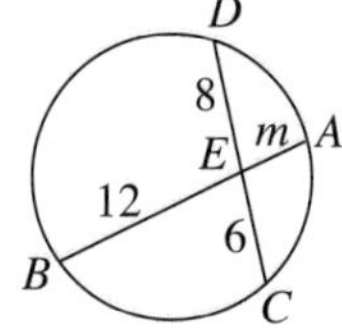

QUESTION 3 Find the value of the pronumeral. All measurements are in cm.

a

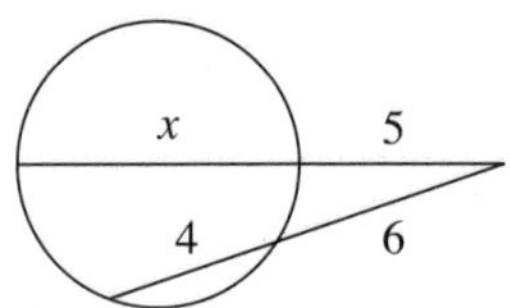

b

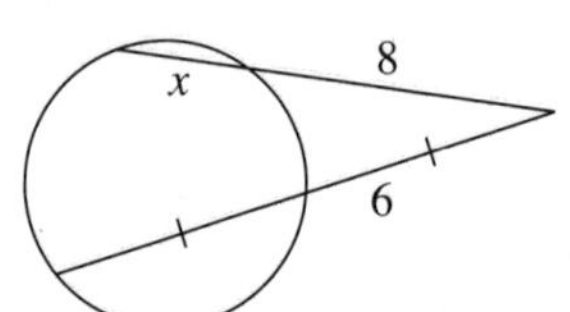

c

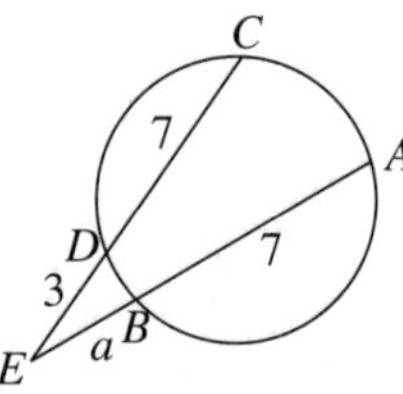

d

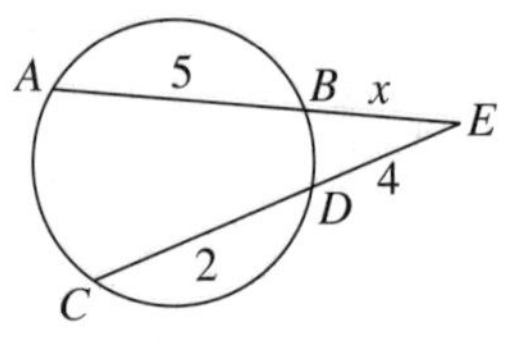

e

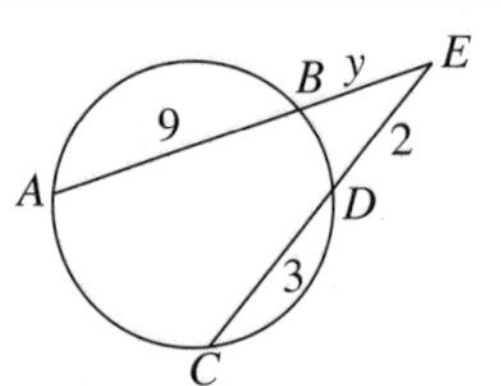

f

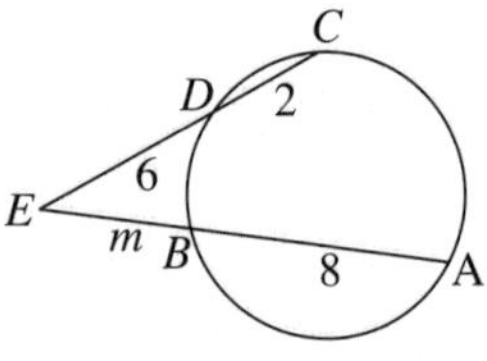

QUESTION 4 Find the value of the pronumeral. All measurements are in cm.

a

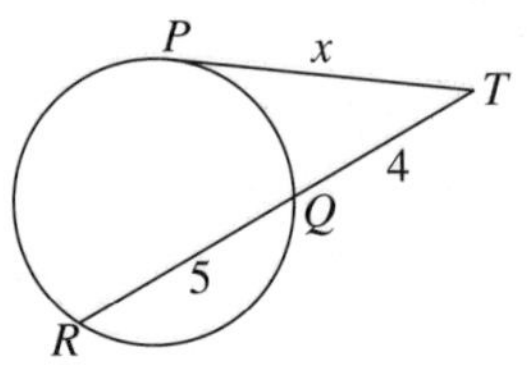

b

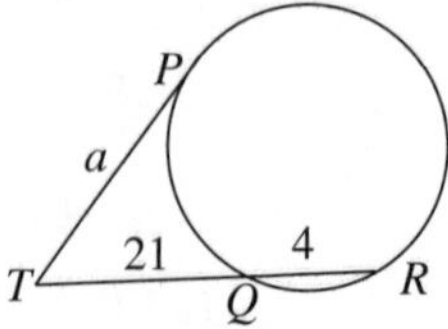

c

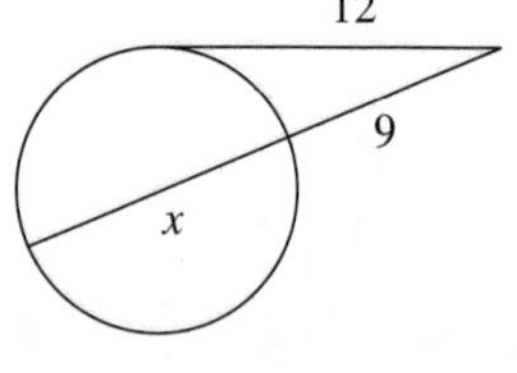

Circle geometry

UNIT 8: Mixed questions

QUESTION 1 Find the value of the pronumeral.

a

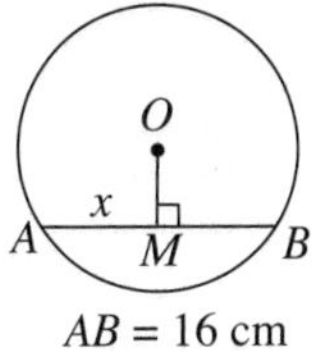

$AB = 16$ cm

b

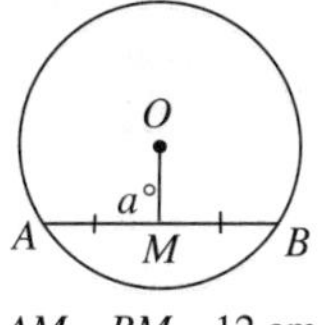

$AM = BM = 12$ cm

c

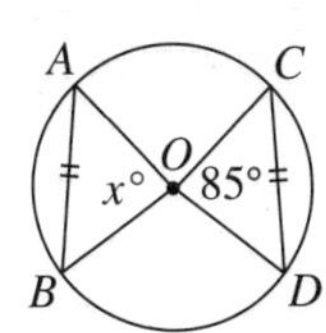

d

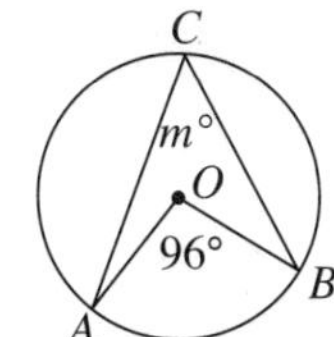

e

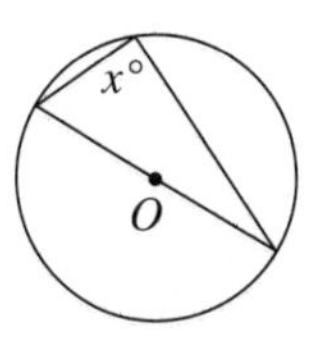

f

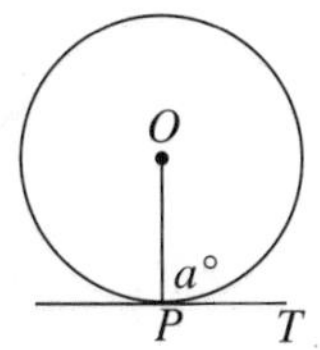

g

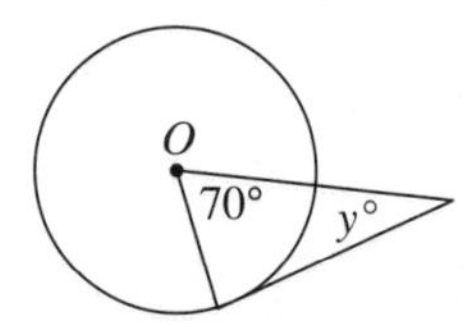

h

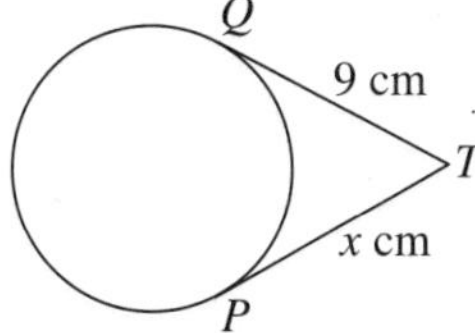

i

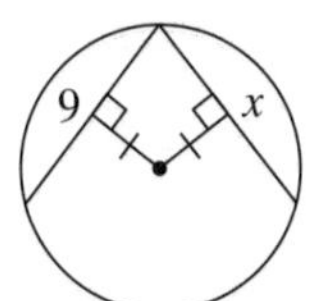

j

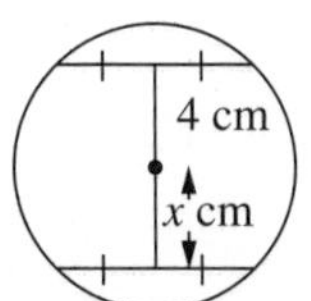

k

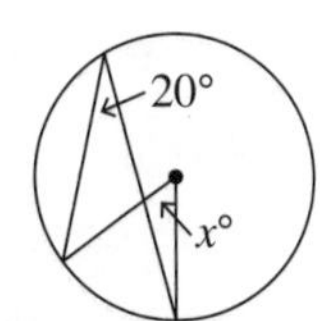

l

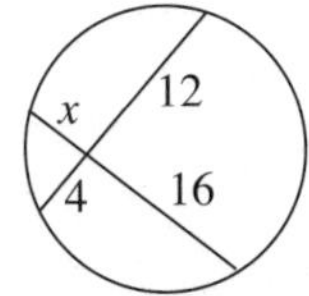

QUESTION 2 Find the value of each pronumeral.

a

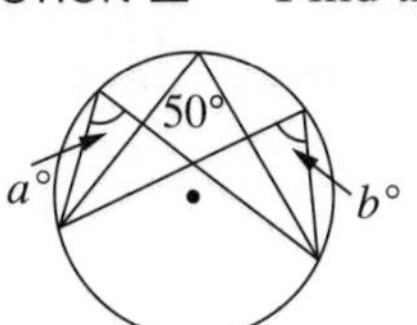

b

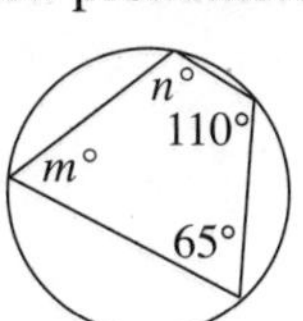

c

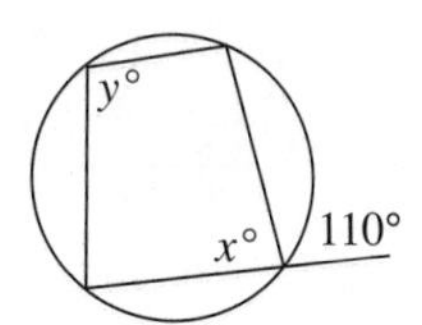

d

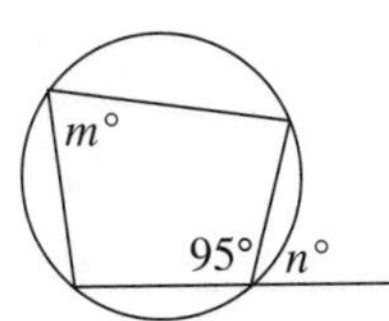

e

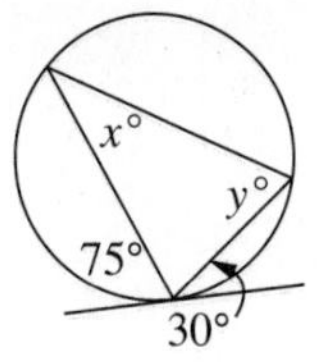

f

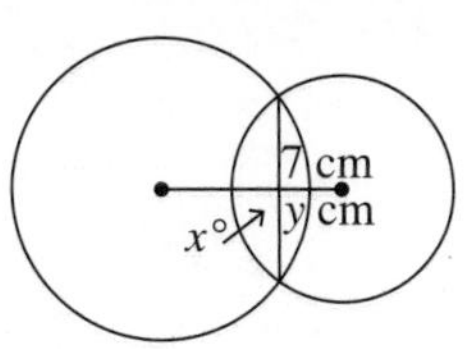

g

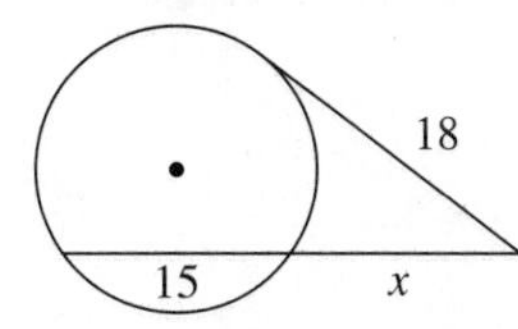

h

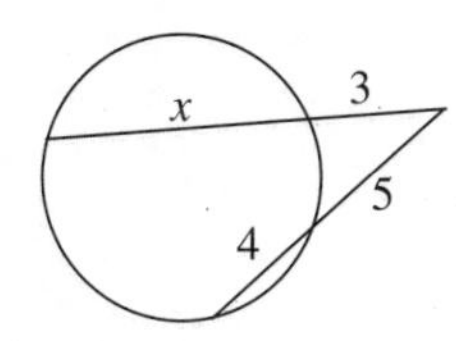

UNIT 9: Deductive exercises (1)

QUESTION 1 Find the value of the pronumeral.

a

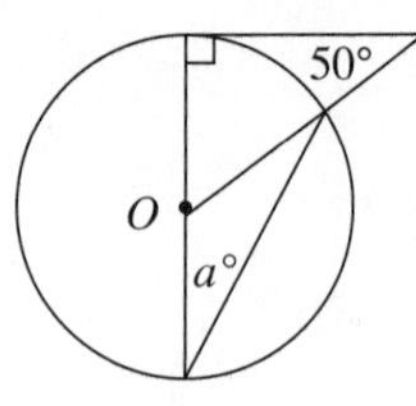

b

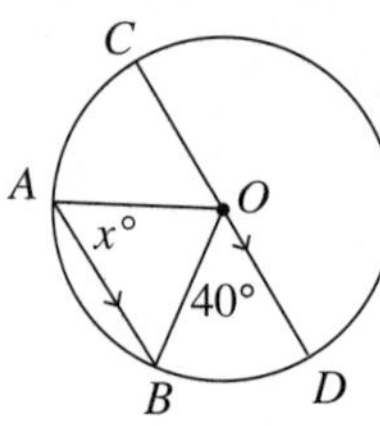

c

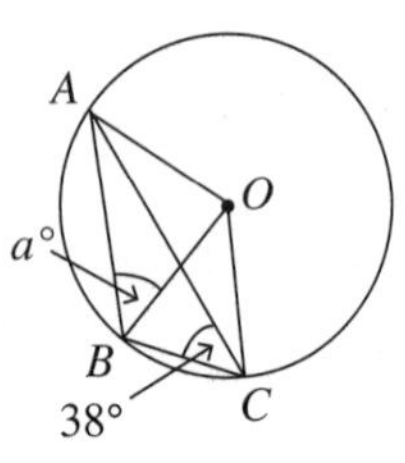

d

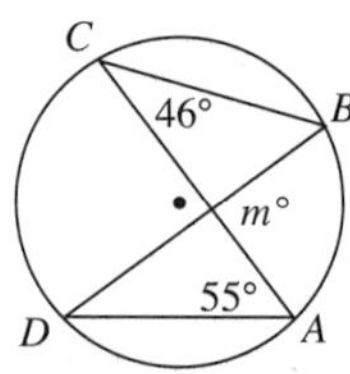

QUESTION 2 Find the value of each pronumeral.

a

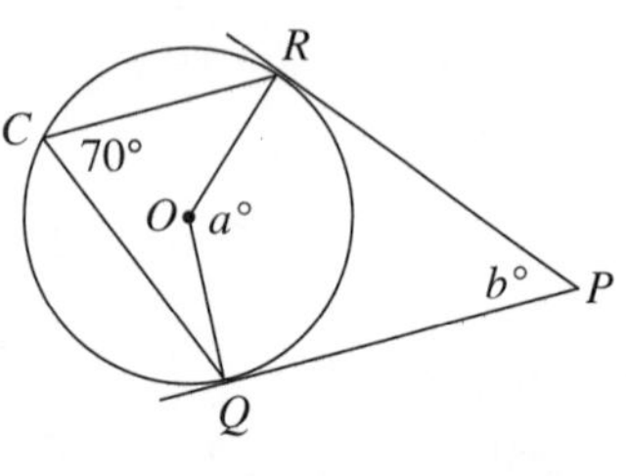

b

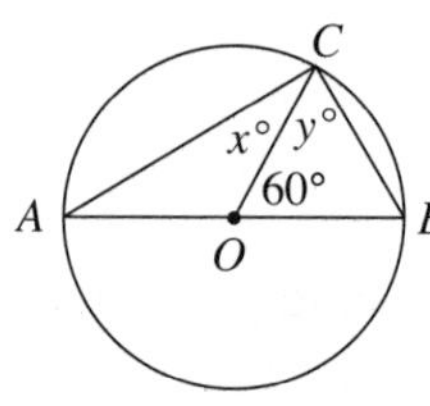

c

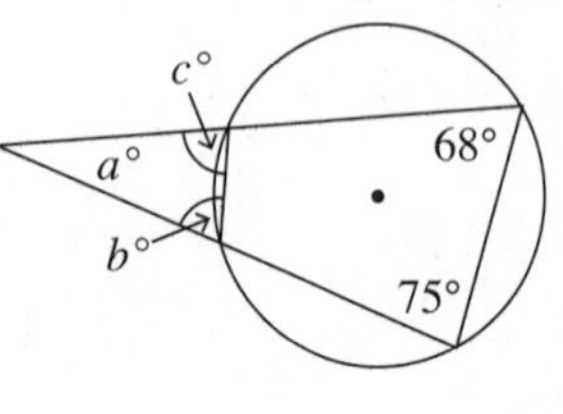

d

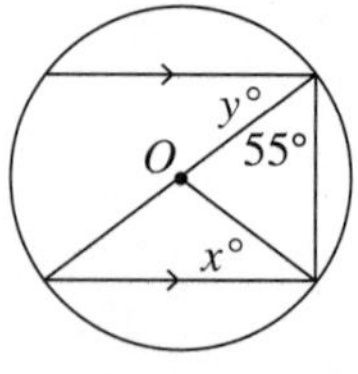

Circle geometry

Excel Advanced-level Mathematics Study Guide Years 9–10
Pages 263–286

UNIT 10: Deductive exercises (2)

QUESTION 1 Find the size of the required angle. Give reasons.

a O is the centre of the circle. $\angle AOC = 120°$.
Find the size of $\angle ABC$.

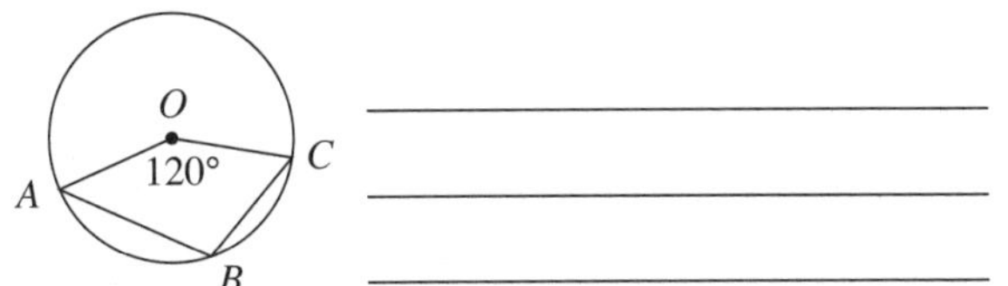

b Chords AB and DE are parallel.
AE and BD intersect at C. $\angle ACB = 110°$.
Find the size of $\angle CED$.

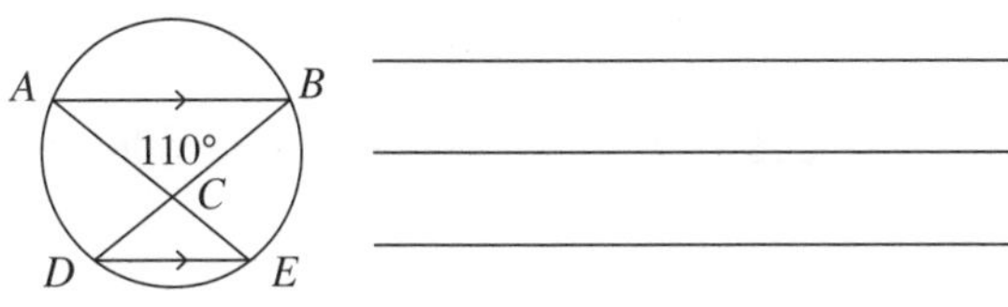

c $AB = BC$. $\angle BDC = 65°$.
Find the size of $\angle ABC$.

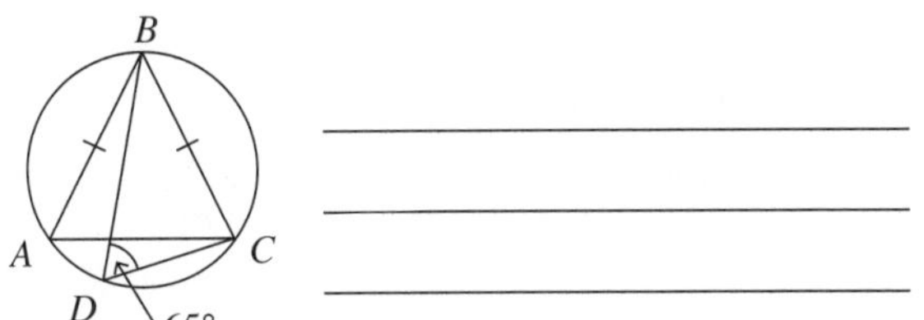

d AB is a diameter of a circle, centre O.
CD is parallel to AB. $\angle ODC = 72°$.
Find the size of $\angle ACD$.

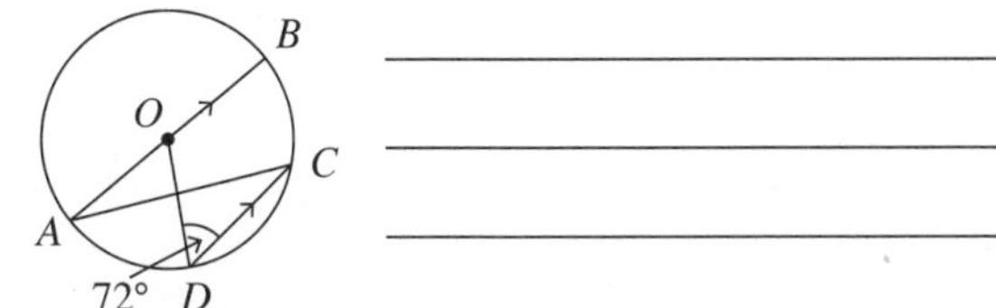

QUESTION 2 The diameter AB of the circle, when produced, meets the tangent at T. $PT = 8$ cm and $BT = 5$ cm. Find the radius of the circle.

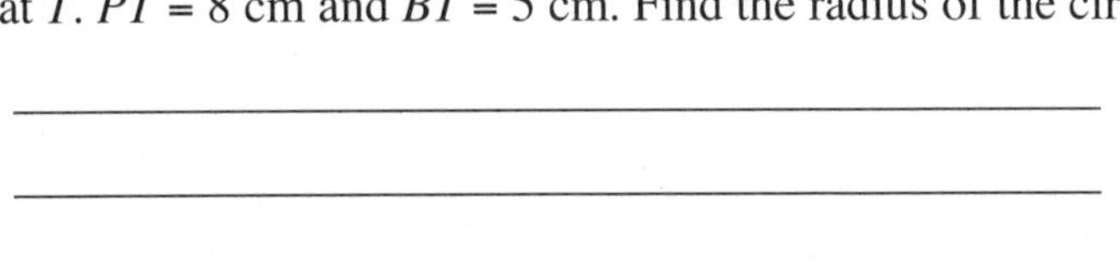

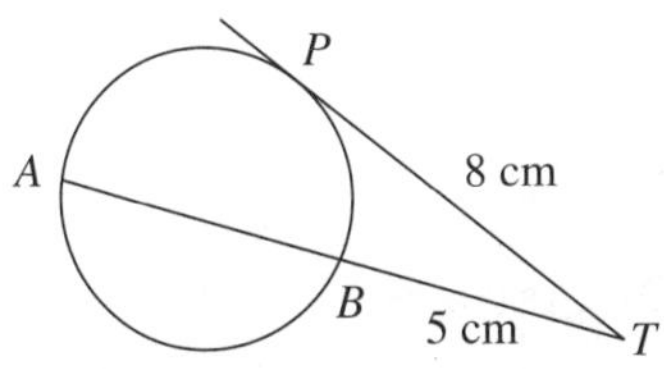

QUESTION 3 The chord BC of a circle is produced to meet the tangent at A at T. $AB = 15$ cm. $BC = 9$ cm and $CT = 16$ cm.

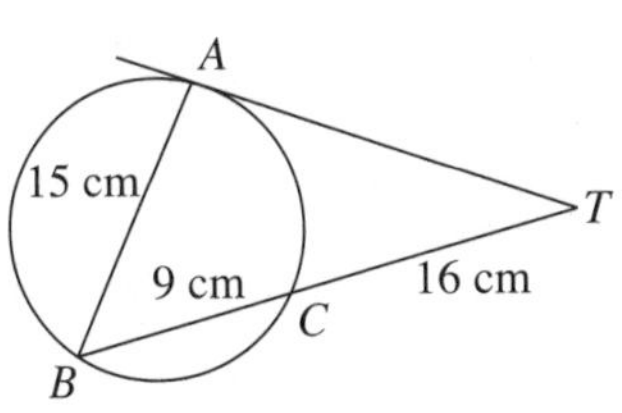

a Find the length of AT.

b Show that AB is a diameter of the circle.

c Find the length of AC.

QUESTION 4 Two circles intersect at B and at P. The smaller circle passes through O, the centre of the larger circle. APT is a tangent to the smaller circle. $\angle BPT = 120°$.

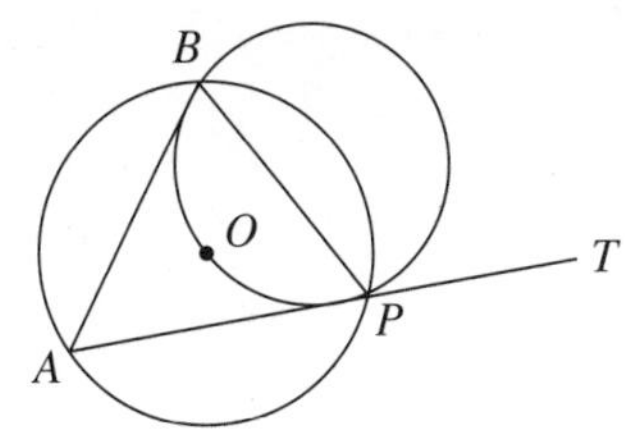

a Find the size of $\angle BOP$.

b Hence show that $\triangle ABP$ is equilateral.

Circle geometry

Excel Advanced-level Mathematics Study Guide Years 9–10
Pages 263–286

UNIT 11: Proofs involving circle geometry

QUESTION 1 The tangent to the circle at A is parallel to the chord BC. Show that $AB = AC$.

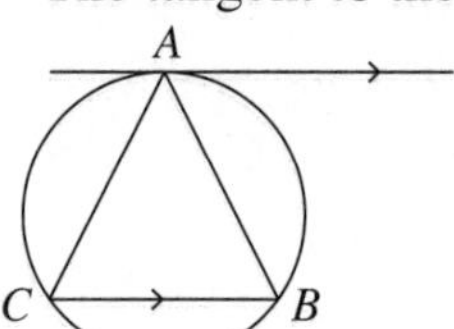

QUESTION 2 O is the centre of the circle. M is the midpoint of chord AB. Prove that OM is perpendicular to AB.

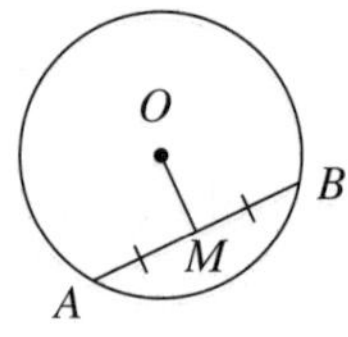

QUESTION 3 A, B and C are three points on a circle. The tangent at C meets AB produced at T.

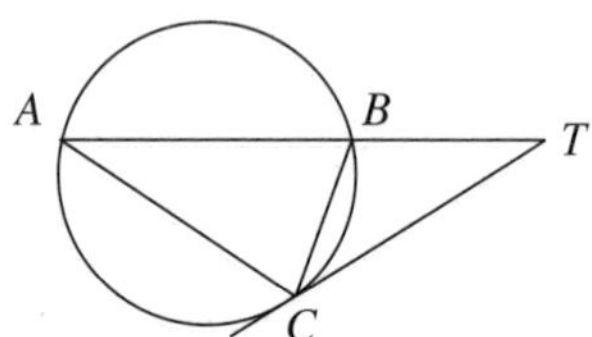

a Prove that triangles ACT and CBT are similar.

b Hence prove that $AT \times BT = CT^2$.

QUESTION 4 A, B and C lie on a circle, centre O. The tangent at A and the tangent at B meet at T. C lies on the line joining O to T. Prove that $AT = BT$.

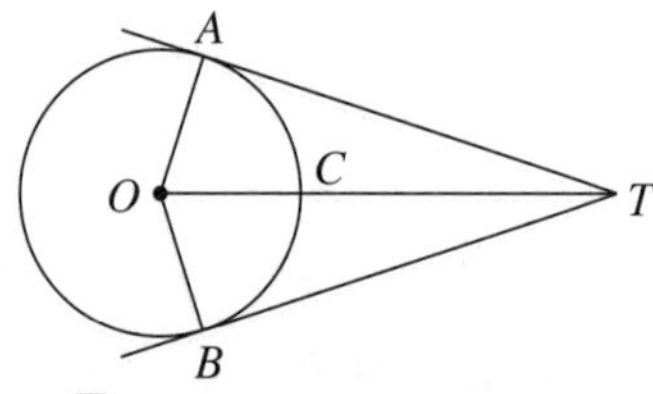

QUESTION 5 AC and BD are diameters of a circle. Show that AB is parallel to DC.

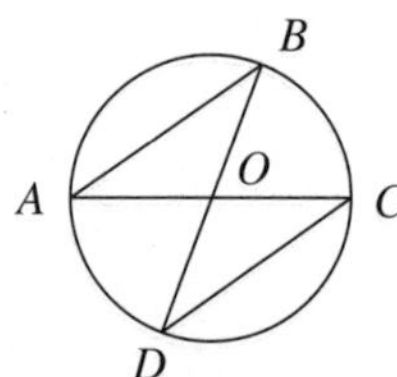

QUESTION 6 AC and BD are two chords of a circle intersecting at E. AB and DC produced meet at P.

$BD = BP$. Let $\angle APD = \alpha$.

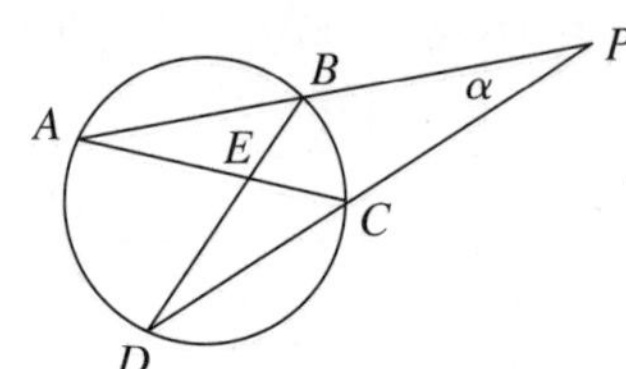

a Show that $AC = CP$.

b Find the size of $\angle AED$ in terms of α.

Circle geometry

TOPIC TEST — PART A

Instructions
- This part consists of 10 multiple-choice questions.
- Fill in only ONE CIRCLE for each question.
- Each question is worth 1 mark.

Time allowed: 10 minutes **Total marks: 10**

Marks

1 The angle between the tangent and the radius through the point of contact is

Ⓐ 45° Ⓑ 90° Ⓒ 135° Ⓓ 170° 1

2 A straight line that touches a circle at one point only is called a

Ⓐ chord. Ⓑ diameter. Ⓒ tangent. Ⓓ secant. 1

3 What is the value of x?

Ⓐ 80 Ⓑ 85

Ⓒ 95 Ⓓ 100 1

4 What is the value of a?

Ⓐ 6 Ⓑ 8 Ⓒ 10 Ⓓ 13.5 1

5 What is the value of n?

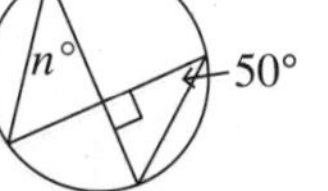

Ⓐ 40 Ⓑ 45

Ⓒ 50 Ⓓ none of these 1

6 What is the value of y?

Ⓐ 90 Ⓑ 100 Ⓒ 110 Ⓓ 120 1

7 The tangent at A meets the tangent at B at T. CB is a diameter. $\angle ABC = 25°$. What is the size of $\angle ATB$?

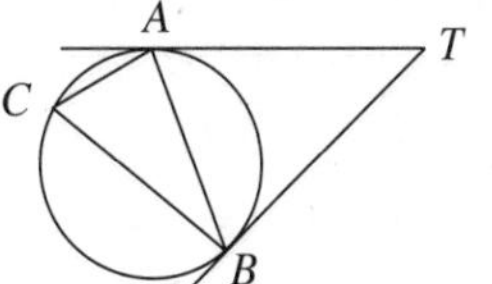

Ⓐ 25° Ⓑ 40°

Ⓒ 50° Ⓓ 65° 1

8 What is the size of $\angle OBC$?

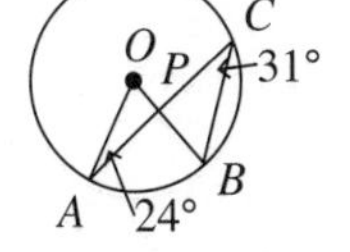

Ⓐ 24° Ⓑ 31° Ⓒ 48° Ⓓ 55° 1

9 What is the value of m?

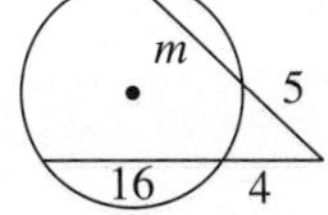

Ⓐ 11 Ⓑ 15

Ⓒ 18 Ⓓ 20 1

10 What is the value of p?

Ⓐ 6 Ⓑ 16

Ⓒ 18 Ⓓ 25 1

Total marks achieved for PART A

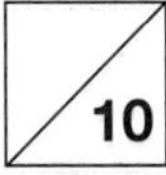

Circle geometry

TOPIC TEST — PART B

Instructions
- This part consists of 6 questions.
- Each question part is worth 1 mark.
- Show all working.

Time allowed: 20 minutes **Total marks: 15**

	Marks

1 A circle has centre O, COD is a straight line and $\angle BCA = 60°$.

a Find the size of the obtuse angle AOB. ____________

b What is the size of $\angle DAC$? ____________ **2**

2 O is the centre of the circle. Chord PQ is parallel to OR. S is the point of intersection of OQ and PR. $\angle SPQ = 23°$.

a What is the size of $\angle ROQ$? ____________

b What is the size of $\angle PSQ$? ____________ **2**

3 Two circles intersect at B. AD is a diameter of one circle. The second circle intersects BD at E. AB produced meets this circle at C.

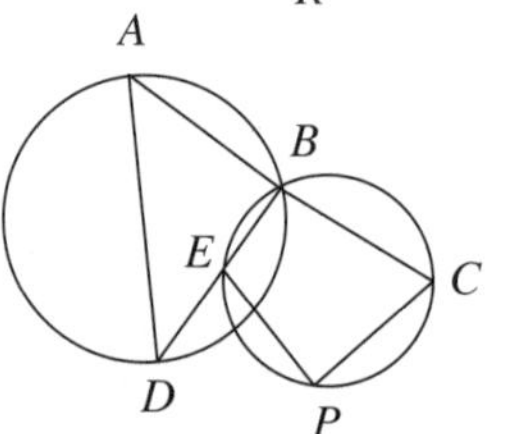

a Why does $\angle ABD = 90°$? **b** Why does $\angle EPC = 90°$?

____________ ____________

c Show that EC is a diameter of the second circle.

____

____________ **3**

4 AT and BT are tangents to a circle. C is a point on the circle such that CA is parallel to BT.

a Why does $\angle ABT = \angle ACB$?

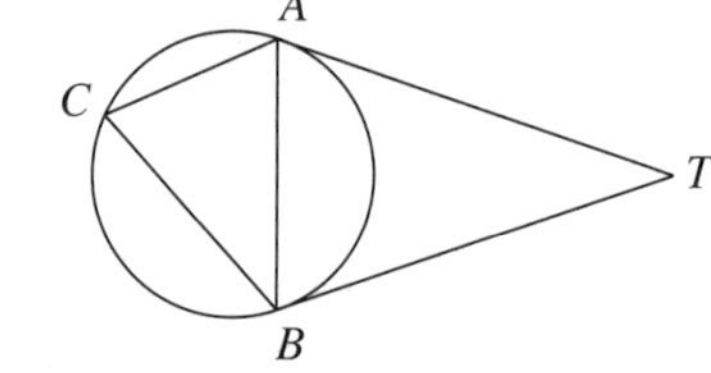

____

b Show that triangle ACB is isosceles.

____

____________ **2**

5 O is the centre of a circle. PQ is a chord. X is a point on PQ such that OX is perpendicular to PQ.

a Prove that $\triangle OPX$ is congruent to $\triangle OQX$. **b** Show that X is the midpoint of PQ.

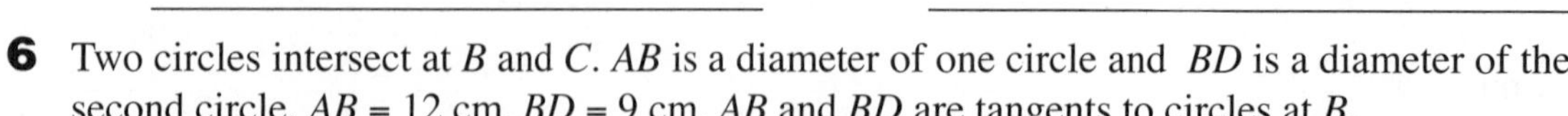

____________ ____________

____________ ____________ **2**

6 Two circles intersect at B and C. AB is a diameter of one circle and BD is a diameter of the second circle. $AB = 12$ cm. $BD = 9$ cm. AB and BD are tangents to circles at B.

a Show that AD passes through C. ____________ **b** Find the length of AD. ____________

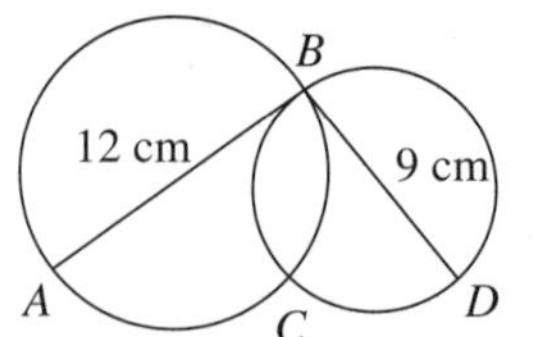

____________ ____________

c Show that $\triangle ABC$ and $\triangle ADB$ are similar. ____ **d** Find the length of BC. ____________

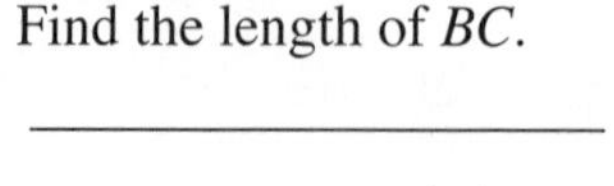

____________ ____________ **4**

Total marks achieved for PART B

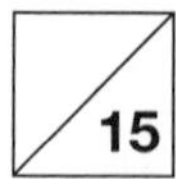

CHAPTER 9

Statistics and probability

Excel Advanced-level Mathematics Study Guide Years 9–10
Pages 187–228

UNIT 1: Review of probability

QUESTION 1 Diana has a box containing three red and two green marbles. She selects two marbles at random. Find the probability of two green marbles if, before the second marble is drawn, the first marble is:

a replaced. **b** not replaced.

QUESTION 2 Suppose we wish to throw a total of 6. Which is the better chance — rolling one die or two dice?

QUESTION 3 What is the probability of rolling two even numbers in one roll of a pair of dice?

QUESTION 4 A coin and a die are thrown simultaneously. Find the probability of throwing a head and an odd number.

QUESTION 5 In a class of 30 students who study either French or German or both, 18 study French and 22 study German.

a Draw a Venn diagram to represent this.

b How many study both French and German? ____________

c If a student is chosen at random from this class, what is the probability that the student studies only French?

QUESTION 6 There are 2 prizes in a raffle in which 100 tickets are sold. Michelle buys 2 tickets. What is the probability that Michelle will:

a win first prize? **b** win both prizes? **c** not win a prize?

QUESTION 7 A family consists of 4 children—Holly, Alex, James and George. A pair of children are chosen at random to go shopping.

a Draw a tree diagram to show all possible shopping pairs.

b Find the probability of choosing a pair that includes:

i Holly. **ii** George but not Holly.

Statistics and probability

Excel Advanced-level Mathematics Study Guide Years 9–10
Pages 187–228

UNIT 2: Review of statistics

QUESTION 1 For the scores 5, 6, 6, 7, 7, 7, 7, 7, 8, 8, 8, 8, 9, 9, 9, 10, 10, 11, 11, 12 find the:

a mean ____________ **b** mode ____________ **c** median ____________

d range ____________ **e** interquartile range ____________

QUESTION 2 The mean mass of 3 students is 53 kg. A fourth student of mass 61 kg joins the group. What is the mean mass of the 4 students?

____

____

QUESTION 3 The following back-to-back stem-and-leaf plot shows the amounts spent (in dollars) at a restaurant in Darling Harbour by a group of 50 people in one evening.

Amount spent ($)

Males		Females
6 6 6 5	4	5 6 6 7
7 7 7 5 3	5	0 1 1 1 1 2
2 2 2 1	6	0 2 3 5 5
7 4 3 3 3	7	2 2 4 4 4
2 1 1 1	8	1 3 6 7
1 0	9	2 3

a What was the median amount spent by the:

i males? ____________ **ii** females? ____________

b What is the range for the:

i males? ____________ **ii** females? ____________

c What is the mode for the males and females together? ____________

____

d Briefly comment on any similarities and differences between the two groups referring to both the location and spread of the amounts and the shape of the distribution.

____

____

QUESTION 4 The number of years that people have been employed by 2 different companies are shown in the following data.

Company A

3	3	4	4	4	5	5	6	6
6	7	7	7	7	8	9	9	9
9	9	9	10	10	14	14	17	18
20	22	25	28	30				

Company B

4	4	4	4	5	6	7	7	7
7	8	8	8	8	11	11	11	11
11	11	12	22	22	22	22	23	24
25	25	25	27	28				

a Determine the five-number summary for each data set.

____

____

b Draw box plots for the two sets of data on the same axis.

c Comment on any similarities and differences between the two data sets. ____________

____

Statistics and probability

UNIT 3: Standard deviation (1)

QUESTION 1 Consider the scores 5, 7, 11, 2, 8, 6, 3.

a What is the number (n) of scores? _______

b What is the mean ($\bar{x}$) of the scores? _______________________________

c Complete the table.

x	5	7	11	2	8	6	3
$x - \bar{x}$							
$(x - \bar{x})^2$							

d Find (the sum) $\Sigma(x - \bar{x})^2$ ____________________

e Find $\sqrt{\dfrac{\sum(x - \bar{x})^2}{n}}$ ____________________

QUESTION 2 Use the formula $s = \sqrt{\dfrac{\sum(x - \bar{x})^2}{n}}$ to find the standard deviation, to one decimal place for:

a 19, 25, 27, 29, 30

b 71, 77, 83, 89, 90, 94

c 42, 43, 43, 45, 48, 49

d 13, 17, 18, 21, 22, 24, 25

QUESTION 3 Use your calculator to find the mean ($\bar{x}$) and standard deviation (σ_n) correct to one decimal place for each set of scores.

a 3, 5, 6, 7, 8, 9, 10 $\bar{x}$ = ____________ σ_n = ____________

b 32, 43, 45, 37, 33, 38, 39, 34 $\bar{x}$ = ____________ σ_n = ____________

c 6, 9, 11, 16, 16, 11, 9, 10, 19, 21, 19, 16, 11, 16 $\bar{x}$ = ____________ σ_n = ____________

d $\bar{x}$ = ____________ σ_n = ____________

Score	5	10	15	20	25	30	35
Frequency	2	3	2	4	6	4	1

Statistics and probability

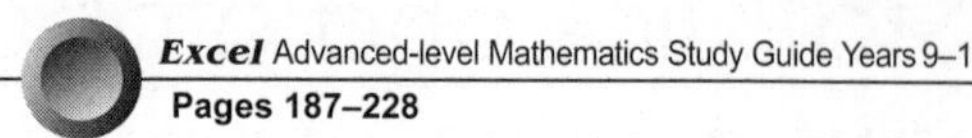

UNIT 4: Standard deviation (2)

QUESTION 1 Consider these scores: 12, 18, 22, 17, 19, 20, 18, 20, 16, 21, 18, 15. Use a calculator to find (to one decimal place where necessary) the:

a mean ____________________ **b** standard deviation (σ_n) ____________________

QUESTION 2 If another score of 18 was added to the scores in question 1:

a briefly explain what will happen to the mean. Justify your answer. ____________________

____

b briefly explain what will happen to the standard deviation. Justify your answer. ____________________

____

QUESTION 3 If a score of 10 was added to the scores in question 1:

a briefly explain what will happen to the mean. Justify your answer. ____________________

____

b briefly explain what will happen to the standard deviation. Justify your answer. ____________________

____

QUESTION 4 Consider these scores: 17, 23, 27, 22, 24, 25, 23, 25, 21, 26, 23, 20. Use a calculator to find (to one decimal place where necessary) the:

a mean ____________________ **b** standard deviation (σ_n) ____________________

QUESTION 5 Compare the scores in question 1 and question 4.

a Describe any relationship between the scores. ____________________

b Describe the relationship between the means of the two sets of scores and briefly explain why this is so.

____

c Describe the relationship between the standard deviations of the two sets of scores and briefly explain why this is so. ____________________

____

QUESTION 6 Consider these scores: 24, 36, 44, 34, 38, 40, 36, 40, 32, 42, 36, 30. Use a calculator to find (to one decimal place where necessary) the:

a mean ____________________ **b** standard deviation (σ_n) ____________________

QUESTION 7 Compare the scores in question 1 and question 6.

a Describe any relationship between the scores. ____________________

b Describe the relationship between the means of the two sets of scores and briefly explain why this is so.

____

c Describe the relationship between the standard deviations of the two sets of scores and briefly explain why this is so. ____________________

____

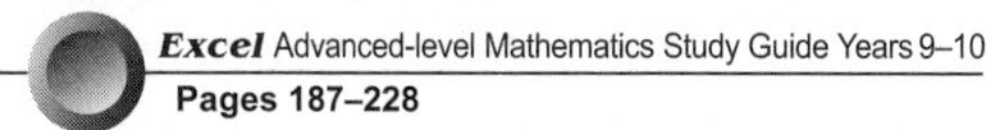

UNIT 5: Comparing data sets

QUESTION 1 12 students in a class sat for tests in three different areas of mathematics. These are the results.

Algebra	60, 84, 97, 75, 59, 80, 99, 65, 87, 78, 96, 56
Geometry	86, 69, 83, 93, 54, 68, 78, 84, 98, 75, 61, 87
Statistics	82, 79, 89, 86, 89, 72, 63, 82, 47, 83, 96, 68

a Find the mean and standard deviation of the marks in the algebra test.

____

b Find the mean and standard deviation of the marks in the geometry test.

____

c Find the mean and standard deviation of the marks in the statistics test.

____

d Briefly explain any similarities or differences in the means and standard deviations for the three tests.

____

____

e Which test had the greatest spread of marks? Briefly comment. ____________________

____

____

QUESTION 2 Five students sat for a mathematics test and a science test. Their marks are given below.

Science	56	60	69	59	65
Mathematics	70	75	86	82	80

a Find the mean and standard deviation for each set of scores. ____________________

____

b Michael scored 69 in science and 86 in mathematics. Which was the better mark?

____

c If Matthew scored 65 in science and 80 in mathematics, in which subject did Matthew perform better compared with the class average?____________________

____

QUESTION 3 The results for Tom and Theo in ten tests given during a term are found below.

Tom:	10	12	15	15	16	17	18	20	18	19
Theo:	5	13	17	17	11	12	9	18	18	21

a Find the mean and standard deviation of both Tom's and Theo's results. ____________________

____

____

b Which person had the most consistent results? Justify your answer. ____________________

____

____

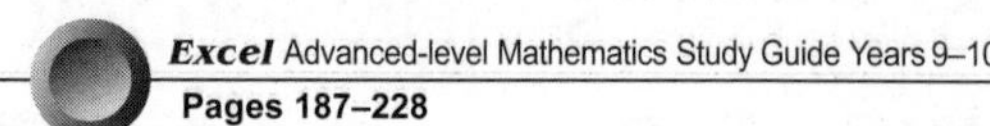

UNIT 6: Comparing measures of spread

QUESTION 1 Consider these scores: 9, 12, 7, 16, 8, 10, 11, 9, 15, 13, 12, 9, 15

a What is the range? __________

b What is the interquartile range? __________

c What is the standard deviation? __________

d Briefly comment on the advantages or disadvantages of each measure of spread. __________

QUESTION 2 A new score of 39 is included with the scores in question 1.

a What is the new range? __________

b What is the interquartile range? __________

c What is the standard deviation? __________

d Which measure of spread was most affected by the inclusion of the extra score? __________

e Which measure of spread was least affected? __________

f Briefly comment on the effect on the measures of spread of a score that is much larger or smaller than the existing scores. __________

QUESTION 3 The stem-and-leaf plot shows scores in the practical and theory parts of a competition.

a Find the range of:

(i) practical marks __________

(ii) theory marks __________

b Find the interquartile range for:

(i) practical marks __________

(ii) theory marks __________

c Find the standard deviation for:

(i) practical marks __________

(ii) theory marks __________

Practical		Theory
9	4	8
8 8 7 4 2 1	5	0 2 7
7 3 0	6	1 3 3 5 6 8
6 4 1	7	2 4 6 7 9 9
9 6 3 1 1 0	8	3 5 9
2	9	4

d Comment on similarities and differences between the two sets of marks. Refer to the measures of spread and to the shape of the distribution. __________

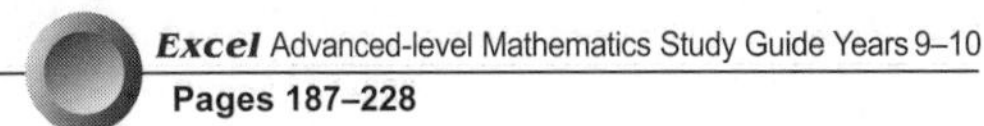

UNIT 7: Scatter plots

QUESTION 1 In a class, the number of hours each student spent studying for an examination and the mark each one was awarded were recorded as shown in the table below. Construct a scatter plot to show these data, and comment on any trends.

Student	Mark	Hours	Student	Mark	Hours
1	15	2	11	72	21
2	93	35	12	82	29
3	30	5	13	85	30
4	52	8	14	9	2
5	61	15	15	27	3
6	82	30	16	39	4
7	97	36	17	48	6
8	100	39	18	92	36
9	5	1	19	67	20
10	38	7	20	99	38

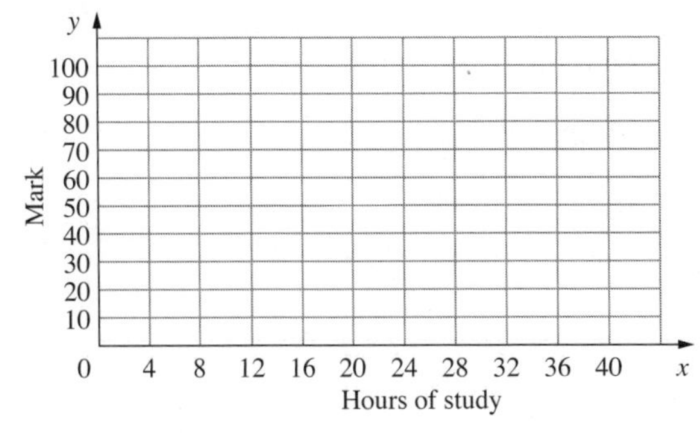

QUESTION 2 The following table shows the ages and advertised prices of a particular model car. Construct a scatter plot displaying the data and comment on any trends.

Age (years)	Price ($)	Age (years)	Price ($)
3	20 500	10	5 000
7	12 800	9	7 000
2	26 900	6	9 000
1	30 000	2	29 000
8	10 000	1	32 000
2	28 000	3	27 000
9	9 000	8	11 000
5	14 000	9	7 500
6	10 000	6	9 500

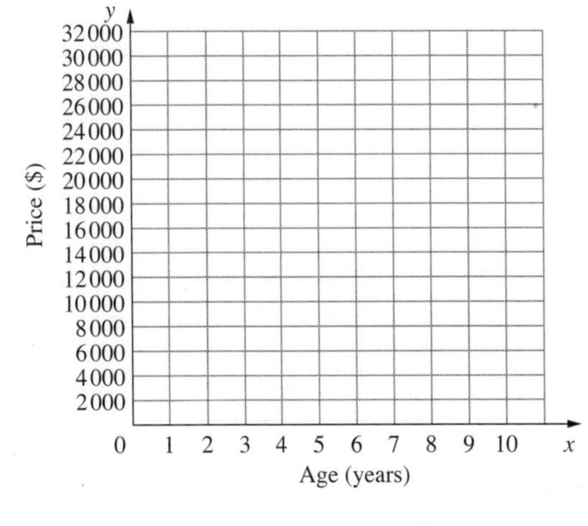

QUESTION 3 The assessment test results in Maths and Science of a class of 15 students are given in the table. Draw a scatter plot of these results and then draw in the line of best fit.

Student	Maths mark	Science mark	Student	Maths mark	Science mark
1	70	58	9	32	36
2	37	40	10	53	58
3	52	55	11	42	48
4	66	62	12	64	56
5	36	32	13	27	34
6	46	50	14	67	73
7	30	35	15	57	49
8	62	68			

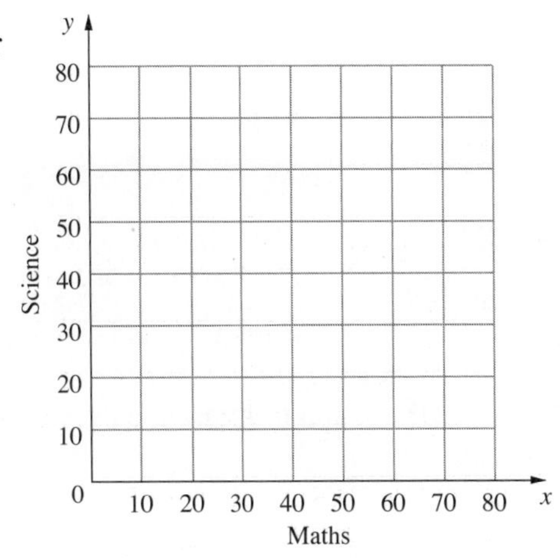

Excel Advanced-level Mathematics Study Guide Years 9–10
Pages 187–228

UNIT 8: Lines of best fit (1)

QUESTION 1 The data in the table to the right gives the marks out of 100 in a Maths and Biology test for a group of students.

Maths mark	Biology mark
84	74
88	82
51	47
62	55
100	90
68	59
79	73
74	64
95	90
48	42
40	34

a Plot this data on this graph.

b Draw a line of best fit.

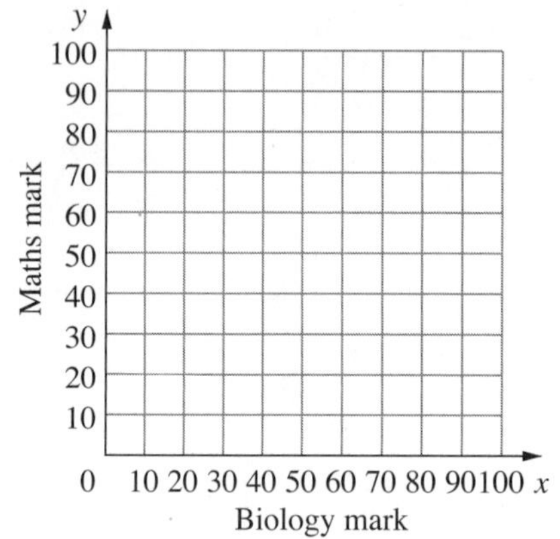

c Use your graph to find an equation for the line of best fit.

____

____

____

QUESTION 2 The table to the right shows the assessment results in Maths and Science of a class of 12 students.

Students	Maths	Science
1	75	63
2	42	47
3	57	62
4	71	69
5	41	39
6	51	57
7	35	42
8	67	75
9	37	43
10	58	65
11	47	55
12	69	63

a Draw a scatter plot of these results.

b Draw in the line of best fit.

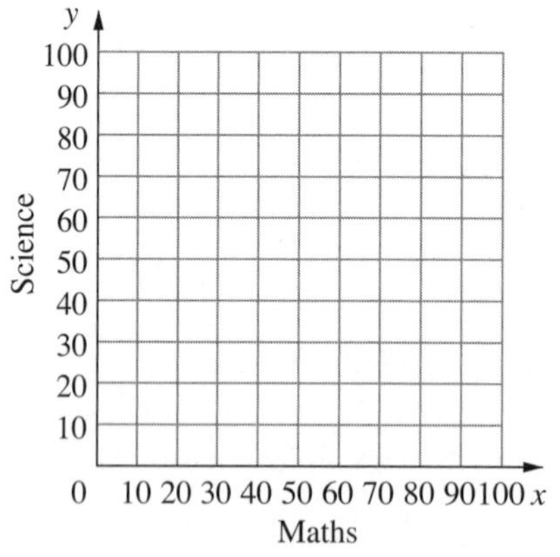

c Use this line of best fit to predict Science marks for students with the following Maths results.

i 95 **ii** 20

______________________ ______________________

d Use the graph to find the equation of the line of best fit.

____

____

UNIT 9: Lines of best fit (2)

QUESTION 1 A scatter plot shows the height (h cm) of some seedlings n weeks after they were planted. A line of best fit has been drawn on the scatter plot.

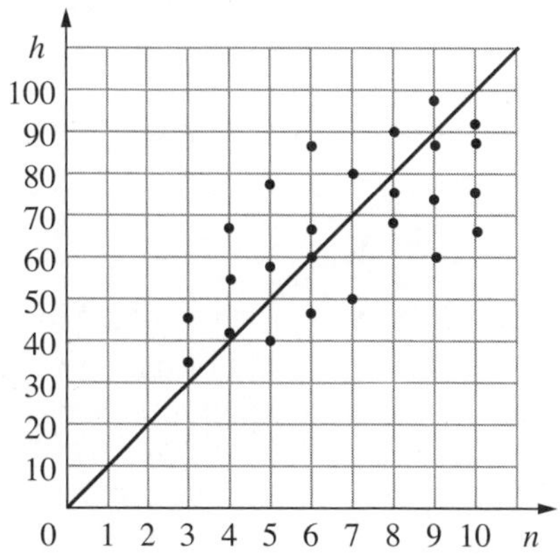

a What is the equation of the line of fit?

____

b Using this equation, what would be the height of a seedling planted for:

i $8\frac{1}{2}$ weeks? ______________________________

ii 13 weeks? ______________________________

c Using this equation, how many weeks would you predict that a seedling had been planted if it had a height of:

i 55 cm? ______________________________

ii 160 cm? ______________________________

QUESTION 2 A different line of fit was drawn on the same scatter plot from question 1.

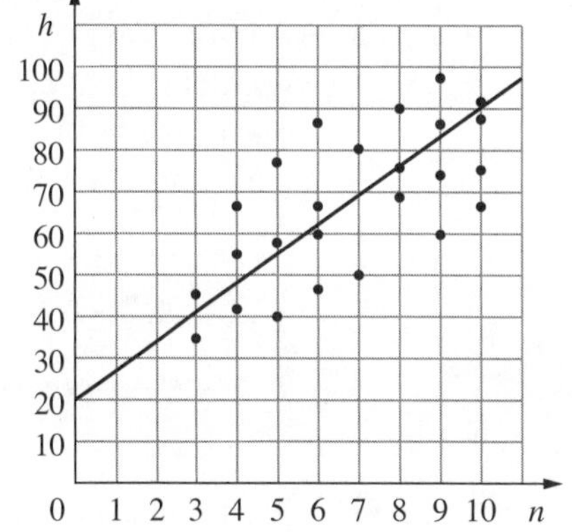

a What is the equation of the line of fit?

____

b Using this equation, what would be the height of a seedling planted for:

i $8\frac{1}{2}$ weeks? ______________________________

ii 13 weeks? ______________________________

c Using this equation, how many weeks ago would you predict that a seedling had been planted if it had a height of:

i 55 cm? ______________________________

ii 160 cm? ______________________________

d Briefly comment on the differences between predictions based on the different lines of fit.

____

____

QUESTION 3 A line of fit was drawn on this scatter plot that shows the age and value of some cars.

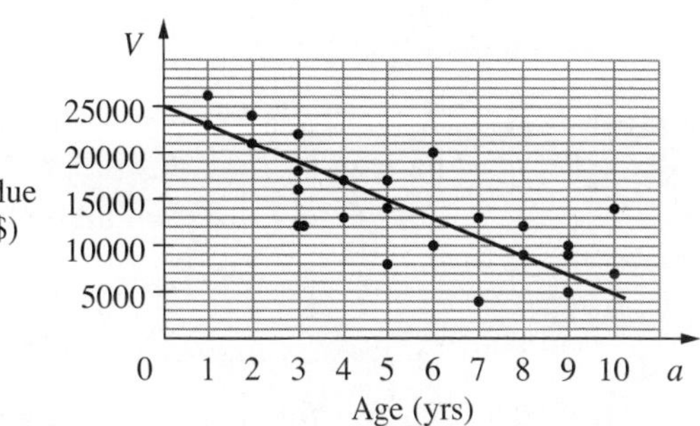

a What is the equation of the line of fit?

____

b What is the predicted value of a car that is 5 years old?

____

c What is the predicted age of a car that is valued at $5000?

____

d After what age will the equation be no longer valid? Justify your answer.

____

____

Statistics and probability

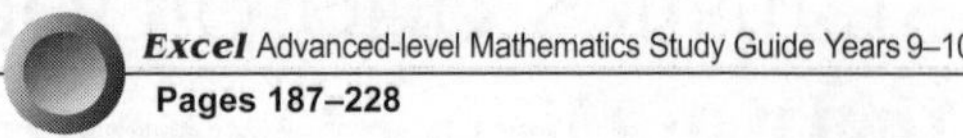

UNIT 10: Investigating reports

QUESTION 1 A survey of some year 10 students found that 35% of the students couldn't swim the length of a 50 m pool. A politician, appalled by this statistic, suggested that all school students should have swimming lessons. Do you think that is a good response? Justify your answer.

QUESTION 2 A particular vitamin supplement, (brand X), is being advertised with a claim that 'studies show that brand X is the best'. What information about these studies would be helpful in deciding whether or not the claim is correct?

QUESTION 3 A phone survey just before an election asked whether voters would still vote for a particular politician if she changed her stance on an environmental issue. The survey was conducted for one of the politician's opponents. The politician claimed that she had no intention of changing her stance. Why do you think the question was being asked in the survey? Comment.

QUESTION 4 A current affairs show on television showed a program about a particular crime and the leniency of the sentence given to the perpetrator. At the end of the show the viewers were asked to vote whether or not they believed sentences for crimes were long enough. What problems might there be with the results of this survey?

QUESTION 5 An insurance company uses probabilities when determining the size of premiums that must be paid by their customers. Two neighbours, with exactly the same type of car, have to pay completely different amounts for insurance with the same company. Why might this be?

QUESTION 6 The results of a census in a local area showed that the area had a high percentage of young families. What implications might this have for government decision makers?

Statistics and probability

TOPIC TEST **PART A**

Instructions
- This part consists of 10 multiple-choice questions.
- Fill in only ONE CIRCLE for each question.
- Each question is worth 1 mark.

Time allowed: 10 minutes **Total marks: 10**

Marks

1 The scores 3, 5, 6, 7, 8, 5, 3 and 5 have a mean 5.25 and standard deviation 1.64. How many of these scores are more than 1 standard deviation from the mean?
Ⓐ 2 Ⓑ 3 Ⓒ 4 Ⓓ 5 **1**

2 Six students were tested on their understanding of data analysis. The marks were 38, 48, 22, 38, 68 and 24. After a refresher course, they were tested again. Each student's mark had increased by exactly 12 compared with the first set of marks. The second set had a
Ⓐ larger mean and smaller standard deviation. Ⓑ larger mean and larger standard deviation.
Ⓒ same mean and larger standard deviation. Ⓓ larger mean and same standard deviation. **1**

3 A score of 10 is added to this sample.
Which of these measures will change?
Ⓐ range Ⓑ median
Ⓒ mode Ⓓ mean **1**

Score	Frequency
6	4
7	5
8	7
9	6
10	3

4 After 5 Maths tests, Amy's mean mark was 87.
In the next 3 tests she scored 90, 95 and 100. Calculate Amy's mean mark for all these tests.
Ⓐ 88 Ⓑ 89 Ⓒ 90 Ⓓ 92 **1**

5 A set of scores has mean 20 and standard deviation 1.5. The lowest score is 18 and the range is 4. What percentage of the scores lie within 2 standard deviations of the mean?
Ⓐ 100% Ⓑ 95% Ⓒ 99.7% Ⓓ 68% **1**

6 Find the standard deviation, correct to two decimal places, of the set of scores given in question 3.
Ⓐ 1.38 Ⓑ 1.25 Ⓒ 1.15 Ⓓ 1.83 **1**

7 A bag holds 2 white and 3 blue balls. Two balls are taken from the bag, one after the other, with replacement. What is the probability that both balls are blue?
Ⓐ 24% Ⓑ 30% Ⓒ 36% Ⓓ 60% **1**

8 Which of these is not a measure of spread?
Ⓐ median Ⓑ range Ⓒ interquartile range Ⓓ standard deviation **1**

9 Consider these 3 sets of scores. X: 6, 8, 10, 12, 14 Y: 7, 9, 11, 13, 15 Z: 12, 14, 16, 18, 20
Which statement is correct?
Ⓐ X has the highest standard deviation. Ⓑ Y has the highest standard deviation.
Ⓒ Z has the highest standard deviation. Ⓓ The standard deviation is the same for all three sets of scores. **1**

10 Consider these scores: 19, 15, 12, 18, 1, 14, 18, 16
Which of these will have the biggest percentage change in value if the score of 1 is omitted?
Ⓐ range Ⓑ interquartile range
Ⓒ standard deviation Ⓓ All will have the same percentage change. **1**

Total marks achieved for PART A /10

Statistics and probability

TOPIC TEST PART B

Instructions
- This part consists of 5 questions.
- Each question part is worth 1 mark.
- Show all working.

Time allowed: 20 minutes **Total marks: 15**

Marks

1 The contents of soft-drink cans are to have mean capacity 375 mL with standard deviation 3.5 mL. Five cans of soft drink were found to contain 380 mL, 370 mL, 378 mL, 368 mL and 381 mL. Which cans, if any, have contents that differ from the mean by more than 2 standard deviations? ______ 1

2 Ten students sat for examinations in Maths and English. They scored the following marks:

Maths

67	70	82	75	68
88	78	71	88	86

English

68	86	76	84	74
72	90	84	62	53

a Find the mean and standard deviation of the Maths marks. ______

b Find the mean and standard deviation of the English marks. ______

c Which subject had the greater spread of marks? Justify your answer. ______

d One student scored 68 in both English and Maths. Which was the better result? Justify your answer. ______ 4

3 The following results for an art competition are given below.

23	18	11	36	22	14	20
21	19	22	20	23	21	22
19	25	29	24	25	30	22
21	20	20	24	25	29	30

a Construct a dot plot of the data.

Find. **b** the range ______ **c** interquartile range ______ **d** standard deviation ______

e Briefly comment on the relative merits of each measure of spread. ______ 5

4 This scatter plot was drawn to show the profit ($\$p$) made when n toys have been sold at a market.

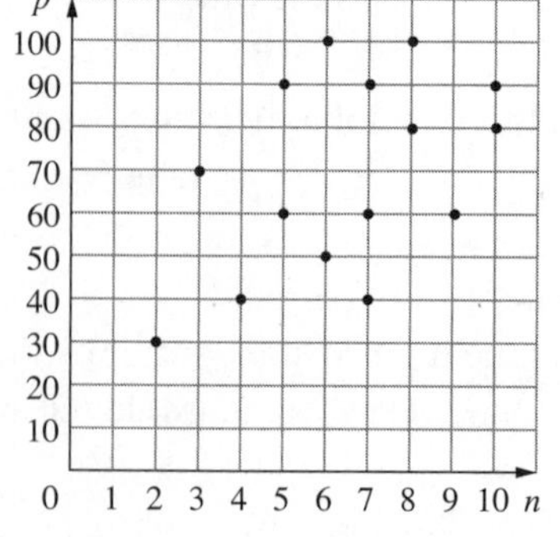

a Draw a line of best fit on the plot.

b What is the equation of your line of best fit? ______

c How much profit does the equation predict will be made if 17 toys are sold? ______

d About how many toys sell for a \$320 profit ______ 4

5 A survey of 15 year ten students found that only 20% eat the recommended amount of fruit each day. Is this something to be concerned about? Justify your answer. ______ 1

Total marks achieved for PART B

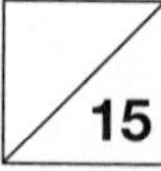

Exam Paper 1

Instructions for all parts
Time allowed: 1 hour

- Attempt all questions.
- Calculators are allowed.

Part A: Allow about 10 minutes for this part.
Part B: Allow about 20 minutes for this part.
Part C: Allow about 30 minutes for this part.

Total marks: 50

EXAM PAPER 1 PART A

Fill in only one circle for each question.

Marks

1 $\sqrt[3]{a^2}$ written in index notation is

Ⓐ a^6 Ⓑ $a^{\frac{3}{2}}$ Ⓒ $a^{\frac{2}{3}}$ Ⓓ a^2 **1**

2 $\sqrt{5} + \sqrt{5}$ equals

Ⓐ 5 Ⓑ 10 Ⓒ $\sqrt{10}$ Ⓓ $2\sqrt{5}$ **1**

3 What are the coordinates of the vertex of the parabola $y = x^2 - 4x + 9$?

Ⓐ (2, 5) Ⓑ (−2, 21) Ⓒ (−2, 13) Ⓓ (2, 9) **1**

4 Which is NOT a factor of $x^3 - x^2 - 14x + 24$?

Ⓐ $x - 2$ Ⓑ $x - 3$ Ⓒ $x + 4$ Ⓓ $x + 6$ **1**

5 The area of this triangle is closest to

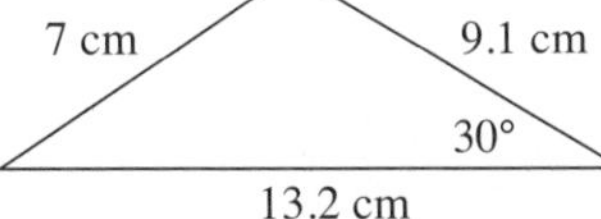

Ⓐ 21 cm^2 Ⓑ 23 cm^2
Ⓒ 30 cm^2 Ⓓ 52 cm^2 **1**

6 What is the value of x?

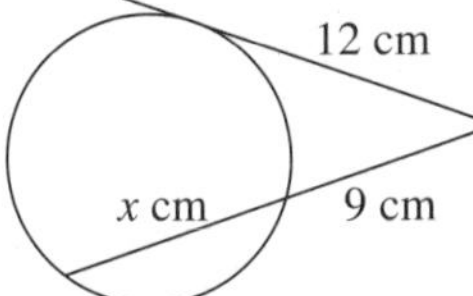

Ⓐ 7 Ⓑ 9
Ⓒ 15 Ⓓ 16 **1**

7 $\log_6 4 + \log_6 9 =$

Ⓐ $\log_6 13$ Ⓑ $\log_{12} 13$ Ⓒ 2 Ⓓ 1 **1**

8 What are the roots of the equation $x^2 - 6x - 5 = 0$?

Ⓐ $x = \dfrac{6 \pm \sqrt{41}}{2}$ Ⓑ $x = \dfrac{-6 \pm \sqrt{41}}{2}$ Ⓒ $x = 3 \pm \sqrt{14}$ Ⓓ $x = -3 \pm \sqrt{14}$ **1**

9 Which is closest to the surface area of a sphere of diameter 9 cm?

Ⓐ 254 cm^2 Ⓑ 339 cm^2 Ⓒ 382 cm^2 Ⓓ 1018 cm^2 **1**

10 A set of scores has mean 18 and standard deviation of 3.2. Another score of 18 is added to the set. What will happen to the standard deviation?

Ⓐ It will increase. Ⓑ It will decrease.
Ⓒ It will stay the same. Ⓓ There is not enough information. **1**

Total marks achieved for PART A /10

EXAM PAPER 1 PART B

Write only the answer in the answer column. For any working use the question column.

Questions	Answers	Marks
1 Evaluate $64^{\frac{2}{3}}$		1
2 What is the degree of the polynomial $5 - 2x - 3x^2 - 6x^3$?		1
3 What are the coordinates of the centre of the circle $(x - 2)^2 + (y + 3)^2 = 16$?		1
4 Evaluate $\log_a \sqrt{a}$		1
5 A pentagonal pyramid has base area 15 m^2 and perpendicular height 6 m. What is the volume of the pyramid?		1
6 What is the exact value of cos 150°?		1
7 What is the value of x?		1
8 Find, to one decimal place, the standard deviation (σ_n) of these scores. 3, 5, 5, 6, 8, 12, 13		1

Continued on the next page

Write only the answer in the answer column. For any working use the question column.

Questions	Answers	Marks
9 How many solutions does the equation $2x^2 - 3x + 7 = 0$ have?		1
10 Simplify $\sqrt{12} + 3\sqrt{5} - \sqrt{3}$		1
11 P is due north of Q. The bearing of O from Q is 292°. $\angle POQ = 90°$. What is the bearing of P from O? 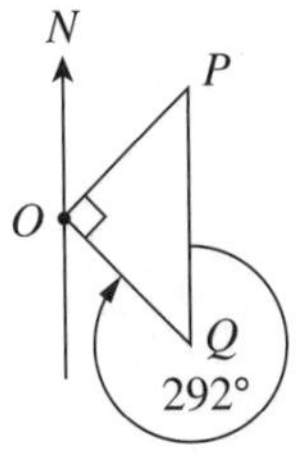		1
12 What is the remainder when $x^4 - 5x^2 + 7x + 2$ is divided by $(x - 3)$?		1
13 If $\sin\theta = 0.6$ and $\cos\theta = 0.8$, what is the value of $\tan\theta$?		1
14 What are the equations of the asymptotes of $y = \frac{2}{x}$?		1
15 The probability of a serious accident occurring on a particular road on a given day has been calculated to be 5%. What is the probability, as a percentage to one decimal place, of no serious accident over the next three days?		1

Total marks achieved for PART B /15

EXAM PAPER 1 PART C

Show all working for each question.

Marks

1 Express in simplest surd form (with a rational denominator).

a $\sqrt{12}+\sqrt{48}$ **b** $(2\sqrt{3}-3)^2$ **c** $\dfrac{3\sqrt{7}}{2\sqrt{3}}$

3

2 Solve.

a $x^2=16x$ **b** $x^2-2x-15=0$ **c** $4^x=\dfrac{1}{\sqrt{2}}$

3

3 Consider the parabola $y=2x^2-x-6$

a Find the equation of the axis of symmetry.

b Find the coordinates of the vertex.

c What are the coordinates of the intercept with the y-axis?

d Find the coordinates of any intercepts with the x-axis.

e Draw a neat sketch of the curve on the axes at right.

5

4 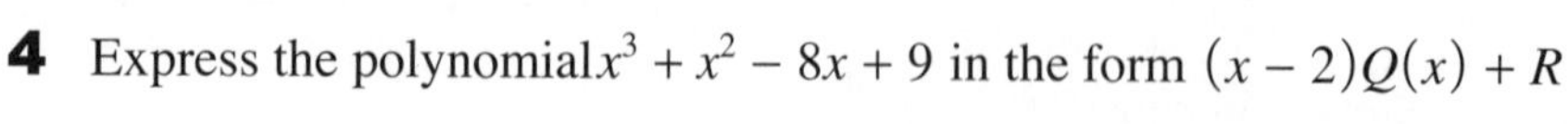
Express the polynomial x^3+x^2-8x+9 in the form $(x-2)Q(x)+R$

Continued on the next page

EXAM PAPER 1 PART C

Show all working for each question.

Marks

5 O is the centre of each circle below. Find the size of angle θ in each diagram.

a

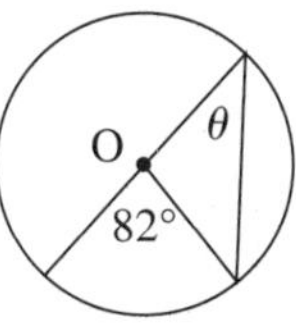

b

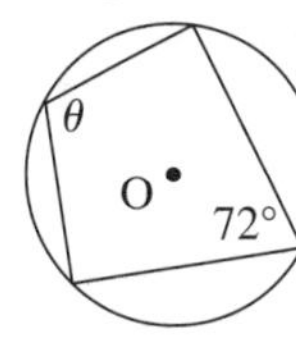

2

6 This figure consists of a cone on top of a cylinder. Find, to two decimal places:

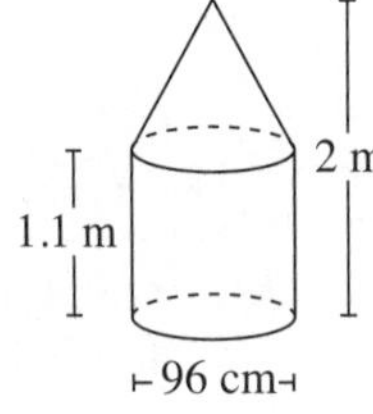

a the slant height of the cone in metres.

b the volume of the figure in m^3

c the surface area of the figure in m^2

3

7 On the axes provided sketch $y = \log_6 x$ showing essential features.

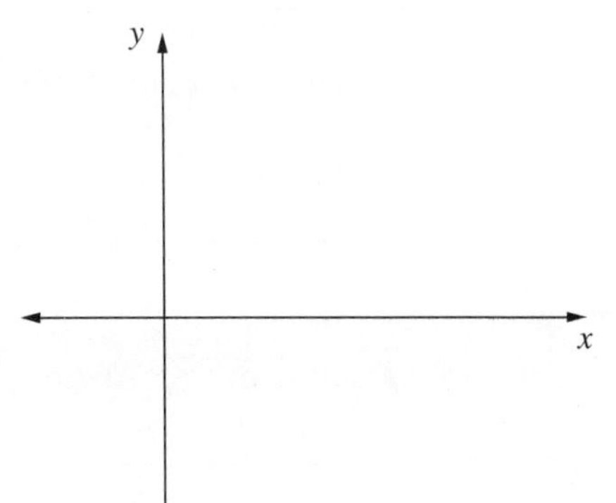

1

Continued on the next page

EXAM PAPER 1 PART C

Show all working for each question.

Marks

8 Find.

a x to one decimal place

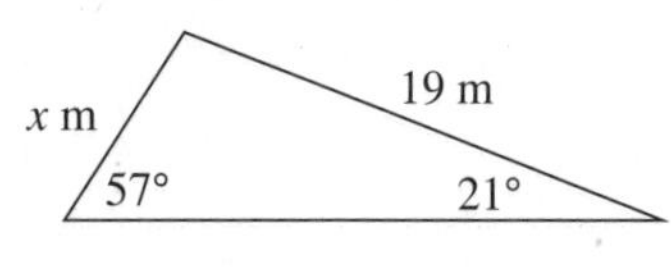

b θ to the nearest degree

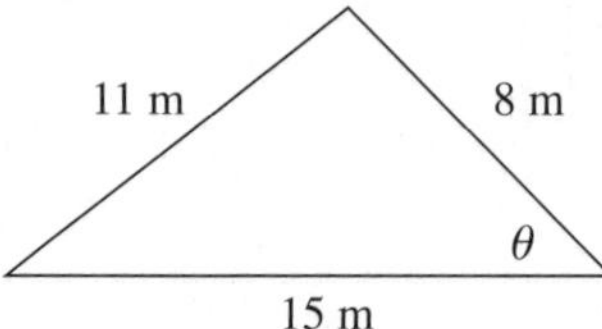

2

9 Some students were asked to look at a group of objects for different numbers of minutes. This scatter plot was drawn to show the number (n) of objects that could be remembered after t minutes.

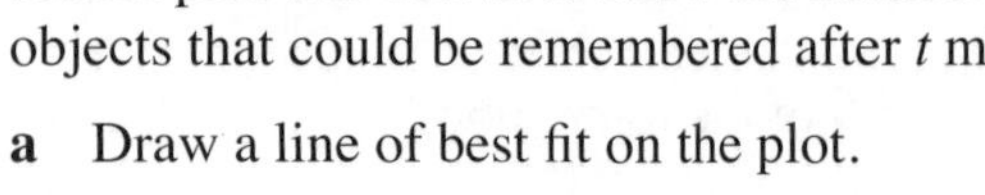

a Draw a line of best fit on the plot.

b Find the equation of the line of fit.

c Use the equation to predict the number of objects that would be remembered after 10 minutes.

3

10 **a** Find the length of BD to one decimal place.

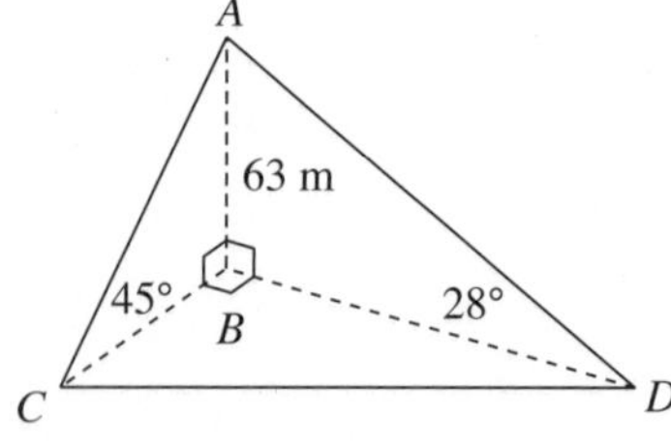

b Find the length of CD to the nearest metre.

2

Total marks achieved for PART C

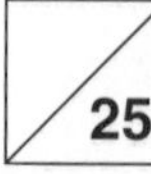

Exam Paper 2

Instructions for all parts
Time allowed: 1 hour

- Attempt all questions.
- Calculators are allowed.

Part A: Allow about 10 minutes for this part.
Part B: Allow about 20 minutes for this part.
Part C: Allow about 30 minutes for this part.

Total marks: 50

EXAM PAPER 2 — PART A

Fill in only one circle for each question.

Marks

1 $\frac{25}{\sqrt{5}}$ equals

Ⓐ 1 Ⓑ $\sqrt{5}$ Ⓒ 5 Ⓓ $5\sqrt{5}$ — 1

2 If $(2x + 7)(x - 2) = 0$ then x equals

Ⓐ –7 or 2 Ⓑ 7 or –2 Ⓒ $3\frac{1}{2}$ or –2 Ⓓ $-3\frac{1}{2}$ or 2 — 1

3 Evaluate $27^{\frac{1}{3}} + \left(\frac{1}{2}\right)^0$

Ⓐ 3 Ⓑ 4 Ⓒ 9 Ⓓ 10 — 1

4 $\sqrt{9^6}$ equals

Ⓐ 3^3 Ⓑ $\frac{1}{3^6}$ Ⓒ 9^3 Ⓓ 81^2 — 1

5 What is the exact value of $\cos\theta$?

Ⓐ $\frac{11}{16}$ Ⓑ $\frac{3}{4}$

Ⓒ $\frac{13}{16}$ Ⓓ $\frac{7}{8}$ — 1

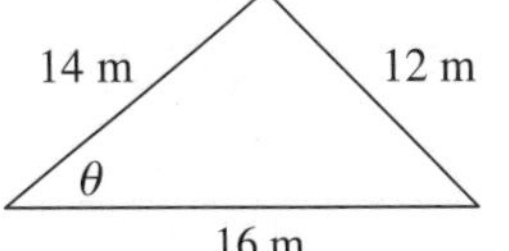

6 What type of graph has equation $y^2 = 9 - x^2$?

Ⓐ parabola Ⓑ hyperbola

Ⓒ exponential curve Ⓓ circle — 1

7 $\log_4 8 =$

Ⓐ $\frac{2}{3}$ Ⓑ $\frac{3}{2}$ Ⓒ 2 Ⓓ 1.68 — 1

8 What is the degree of the polynomial $3 + 5x - 2x^2 + 5x^5 - 4x^6$?

Ⓐ 3 Ⓑ 4 Ⓒ 5 Ⓓ 6

9 What is the value of x?

Ⓐ 54 Ⓑ 58

Ⓒ 61 Ⓓ 68

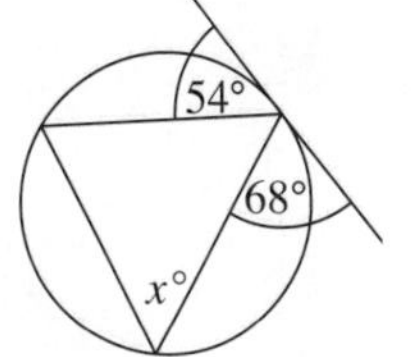

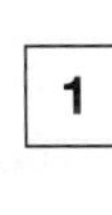

10 If the surface area of a sphere is 15.2 m^2, which is closest to the volume?

Ⓐ 5.6 m^3 Ⓑ 19.4 m^3 Ⓒ 54.9 m^3 Ⓓ 59.3 m^3 — 1

Total marks achieved for PART A /10

EXAM PAPER 2 PART B

Write only the answer in the answer column. For any working use the question column.

Questions	Answers	Marks
1 Find the maximum value of $9 - x^2$		1
2 Let $P(x) = 8x^5 + 6x^4 - 5x^3 + 4x^2 + 9x + 7$. Write the value of the leading coefficient.		1
3 Find x correct to two decimal places. 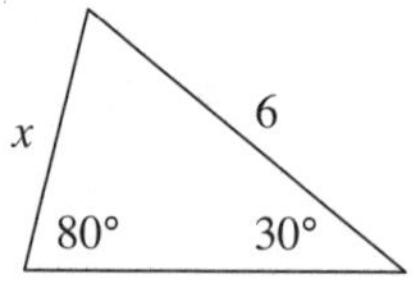		1
4 Write $0.3\dot{4}\dot{5}$ as a fraction in simplest form.		1
5 Simplify $\log_a a^3$		1
6 Write down the exact value of sin 60°.		1
7 In this diagram $\angle CDE = 84°$. Which other angle must equal 84°? 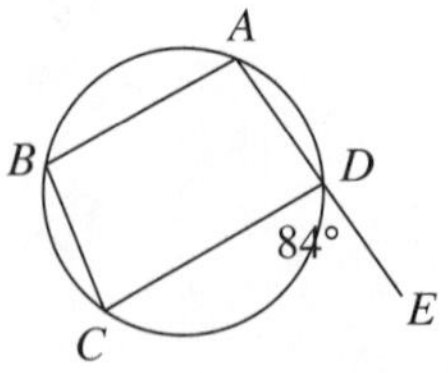		1
8 Solve $\log_2 x = 8$		1

Continued on the next page

EXAM PAPER 2 — PART B

Write only the answer in the answer column. For any working use the question column.

Questions	Answers	Marks
9 What is the remainder when $3x^4 - 2x^2 + 7x - 5$ is divided by $(x + 4)$?		1
10 Simplify $2\sqrt{3}(3 - \sqrt{3})$		1
11 Sketch $y = -\frac{6}{x}$ 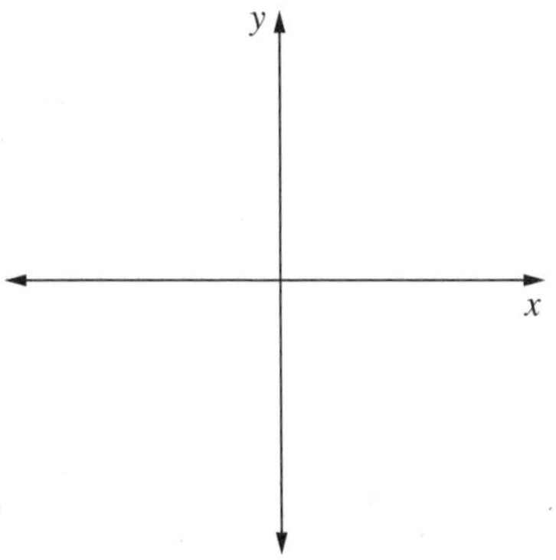		1
12 A cone has diameter 4 m and perpendicular height 6 m. Find its volume.		1
13 Solve $3^n = 3\sqrt{3}$		1
14 Find the value of x correct to one decimal place. 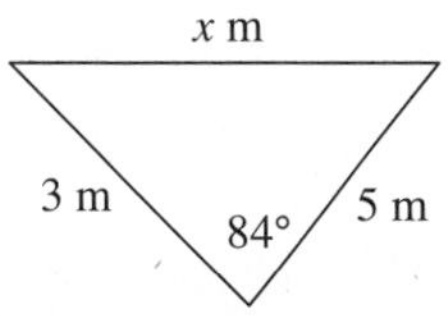		1
15 What are the permissible x-values for $y = \log_7 x$?		1

Total marks achieved for PART B ___ / 15

EXAM PAPER 2 PART C

Show all working for each question.

Marks

1 a Expand $(2x-3)(x^4-5x^3+6x^2-8x-2)$ b Factorise $6x^2+11x-10$

2

2 Solve.

a $2x^2+x-6=0$ b $x^2+3x-7=0$ (to three decimal places)

2

3 a Show that $(x-3)$ is a factor of x^3-4x^2+x+6

b Hence express x^3-4x^2+x+6 as a product of its factors.

c Sketch $y=x^3-4x^2+x+6$

4 If $\log_a 2 = 0.356$ and $\log_a 3 = 0.565$ find the value of:

a $\log_a \sqrt{2}$ b $\log_a 18$

2

Continued on the next page

EXAM PAPER 2 PART C

Show all working for each question.

Marks

5 Consider the circle $x^2 + 6x + y^2 - 8y = 144$

a What is the radius of the circle?

b What are the coordinates of the centre?

c Does the point $(9, -1)$ lie inside, on, or outside the circle?

3

6 In this rectangular pyramid, $AB = 18$ cm, $BC = 10$ cm, $OP = 12$ cm. M is the midpoint of AB and N is the midpoint of BC.

a Find the volume of the pyramid.

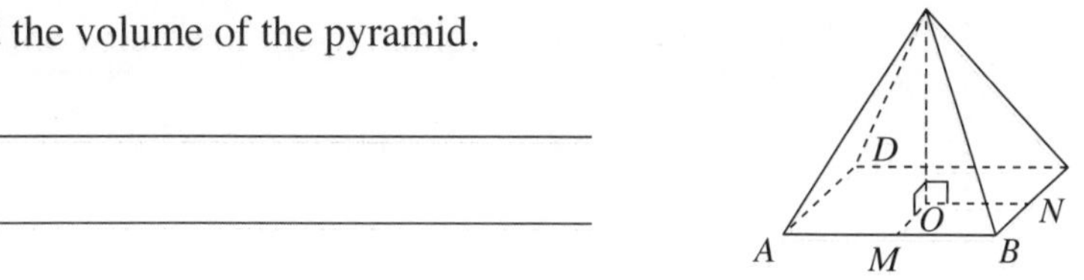

b Find the slant height PN.

c Find the slant height PM.

d Find the total surface area of the pyramid.

4

7 Find, to the nearest degree, the size of the obtuse angle ABC.

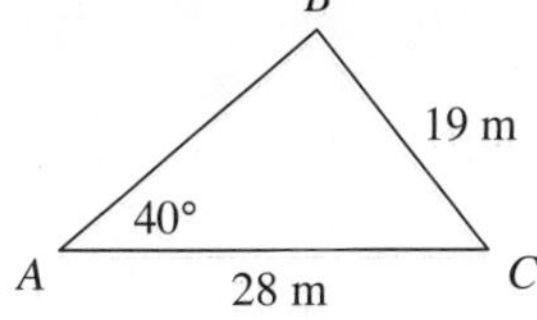

1

Continued on the next page

EXAM PAPER 2 PART C

Show all working for each question.

Marks

8 Find the obtuse angle θ for which $\sqrt{2}\cos\theta + 1 = 0$.

1

9 The following marks were achieved by students in a class in an exam.

57 59 60 63 63 66 68 70 70 71 74 75 77 79 79 79 80 83 85 88 89 91

a Find the mean.

b Find the standard deviation.

c What is the range?

d What is the interquartile range?

e Simone is in the class, She scored 83 in the exam. In another exam Simone scored 79. The mean mark in this second exam was 73 and the standard deviation was 5.25. Which was the better result for Simone? Justify your answer.

5

10 Two circles intersect at Q and S. PQ is a diameter of one circle and SR is the diameter of the second circle. PQ and SR are tangents to circles at Q and S respectively.

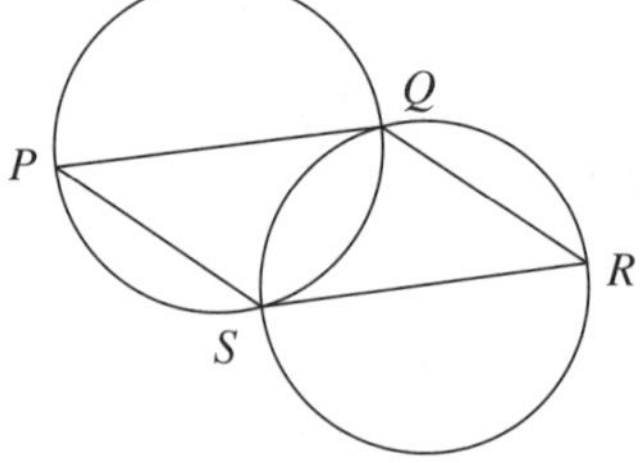

a Show that PS is parallel to QR.

b If $PS = QR$ show that the circles have equal radii.

2

Total marks achieved for PART C

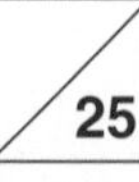

25

Exam Paper 3

Instructions for all parts
Time allowed: 1 hour

- Attempt all questions.
- Calculators are allowed.

Part A: Allow about 10 minutes for this part.
Part B: Allow about 20 minutes for this part.
Part C: Allow about 30 minutes for this part.

Total marks: 50

EXAM PAPER 3 — PART A

Fill in only one circle for each question.

Marks

1 $(x-5)^2$ equals

Ⓐ x^2-25 Ⓑ x^2+25 Ⓒ $x^2-10x+25$ Ⓓ $x^2-10x-25$ **1**

2 $\sqrt{16x^{16}}$

Ⓐ $4x^4$ Ⓑ $4x^8$ Ⓒ $8x^4$ Ⓓ $8x^8$ **1**

3 Which could be the graph of $y=-2^x$?

Ⓐ
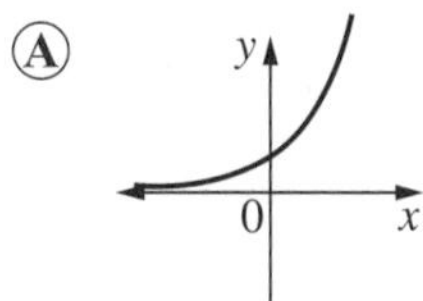

Ⓑ

Ⓒ
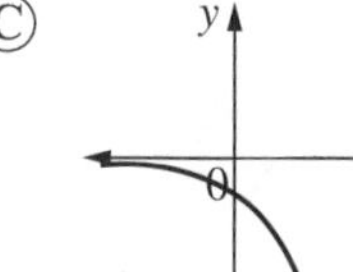

Ⓓ
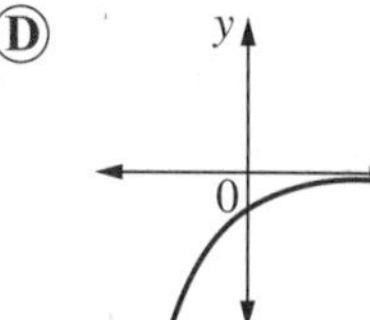

1

4 $\log_2 1=$

Ⓐ 0 Ⓑ $\frac{1}{2}$ Ⓒ 1 Ⓓ 2 **1**

5 Which is NOT a factor of $x^4+6x^3+5x^2-24x-36$?

Ⓐ $x-2$ Ⓑ $x+2$ Ⓒ $x-3$ Ⓓ $x+3$ **1**

6 Which is an asymptote of the curve $y=\frac{3}{x-2}$?

Ⓐ $x=2$ Ⓑ $x=-2$ Ⓒ $y=2$ Ⓓ $y=-2$ **1**

7 A circle has its centre at $(-3, 4)$ and radius 5 units. What is the equation of the circle?

Ⓐ $x^2+6x+y^2-8y=0$ Ⓑ $x^2-6x+y^2+8y=0$

Ⓒ $x^2+6x+y^2-8y=50$ Ⓓ $x^2-6x+y^2+8y=50$ **1**

8 What are the solutions of $x^2-17x-60=0$

Ⓐ $x=5, x=-12$ Ⓑ $x=-5, x=12$ Ⓒ $x=3, x=-20$ Ⓓ $x=-3, x=20$ **1**

9 What is the exact value of x?

Ⓐ $\sqrt{7}$ Ⓑ $\sqrt{8}$ Ⓒ 3 Ⓓ $\sqrt{10}$ **1**

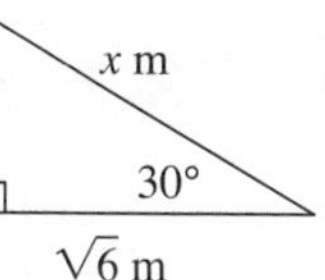

10 A cone has a diameter twice that of a cylinder of the same height as the cone. What fraction of the volume of the cylinder is the volume of the cone?

Ⓐ $\frac{1}{3}$ Ⓑ $\frac{2}{3}$ Ⓒ $\frac{3}{4}$ Ⓓ $\frac{4}{3}$ **1**

Total marks achieved for PART A /10

EXAM PAPER 3 PART B

Write only the answer in the answer column. For any working use the question column.

Questions	Answers	Marks
1 If $7\sqrt{3} = \sqrt{x}$, find x.		1
2 Find A if $\sin A = \cos 30°18'$ (A is acute).		1
3 If AT is a tangent, $BC \parallel AT$ and $\angle TAC = 58°$, find $\angle BAC$. 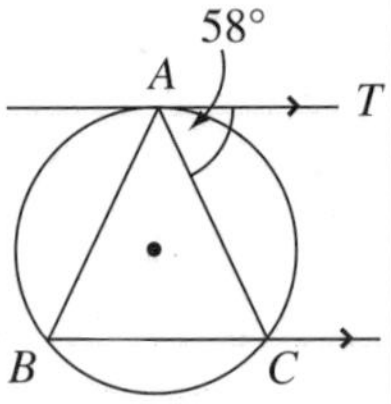		1
4 AD is a diameter. Find the value of x. 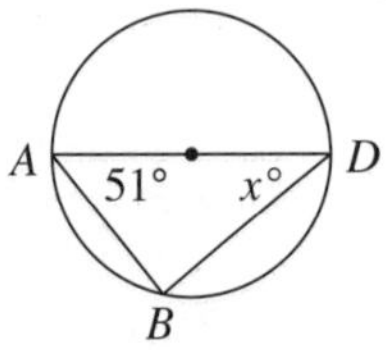		1
5 Evaluate $\log_5 125$		1
6 Simplify $\log_5 80 - \log_5 16$.		1
7 Find the standard deviation of these scores.		1
8 Factorise $3x^2 - 2x - 1$		1
9 $\sqrt{40} + \sqrt{90} =$		1
10 Find the coordinates of the vertex of the parabola $y = x^2 - 8x + 5$		1

Question 3 diagram labels: 58°, A, T, B, C

Question 4 diagram labels: A, D, B, 51°, x°

Question 7 table:

x	0	2	3	4	7	8
f	3	2	4	7	8	2

Continued on the next page

EXAM PAPER 3 PART B

Write only the answer in the answer column. For any working use the question column.

Questions	Answers	Marks
11 Find the radius, to one decimal place, of a sphere that has volume 1150 m^3.		1
12 When $x^3 + px^2 - 4x + 5$ is divided by $(x + 1)$ the remainder is 3. Find the value of p.		1
13 Find the curved surface area of a cone with radius 3 cm and slant height 8 cm. Give the answer to the nearest square centimetre.		1
14 Give a reason why the following question is not a good survey question. "Do you think the current drab uniform should be replaced with something more colourful and vibrant?"		1
15 Sketch the graph of $y = \sin x°$ for $0° \le x° \le 360°$. 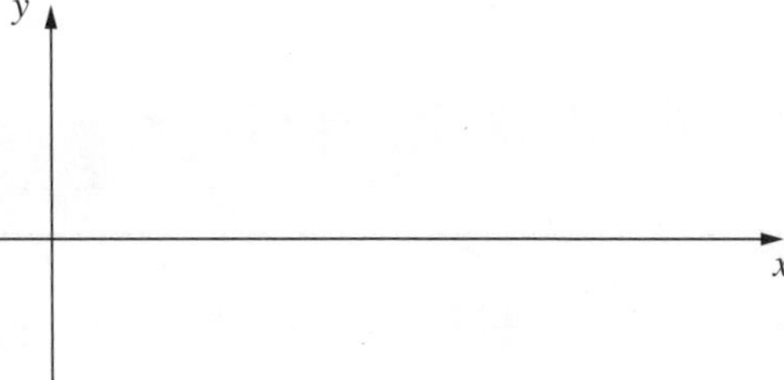		1

Total marks achieved for PART B /15

EXAM PAPER 3 PART C

Show all working for each question.

Marks

1 a Express $\frac{2\sqrt{3}}{\sqrt{3}-1}$ with a rational denominator.

b Express $x^{-\frac{2}{3}}$ in surd form

2

2 Solve.

a $7x^2 - 5x - 1 = 0$

b $x^4 - 13x^2 + 36 = 0$

2

3 a If $\log_a 5 = 1.8$ and $\log_a 3 = 1.23$ find $\log_a 45$

b Find the value of m if $\log_x 4m - \log_x 3 = \log_x (m + 4)$

2

4 From a point P due south of the base, B, of a tower, the angle of elevation of the top of the tower, A, is 23°. From a point Q due East of B the angle of elevation of A is 19°. The tower is 125 m tall.

a Show this information on a diagram.

b Find the distance from P to B.

c Find the distance from Q to B.

d Find the distance from P to Q.

4

Continued on the next page

EXAM PAPER 3 PART C

Show all working for each question.

Marks

5 **a** Find the size of $\angle ABC$ to the nearest minute.

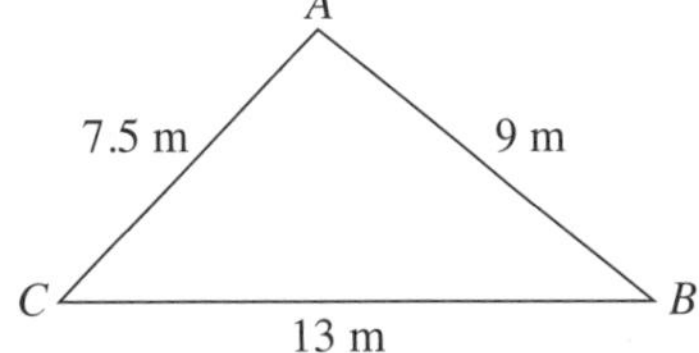

b Find the area of the triangle to one decimal place.

2

6 Consider the polynomial $P(x) = x^3 + 6x^2 - 32$

a Divide $P(x)$ by $(x - 2)$

b Factorise $P(x)$ fully.

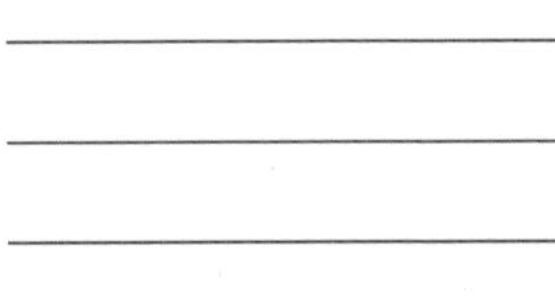

c Sketch $y = P(x)$.

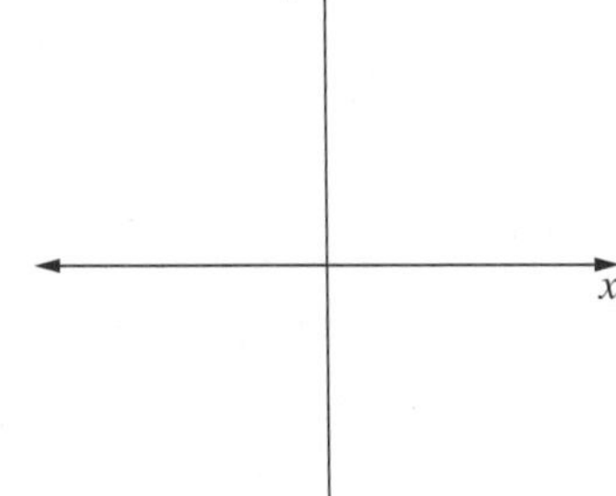

3

7 A solid consists of a pyramid on top of a cube of side 28 cm.

a Find the perpendicular height of the pyramid.

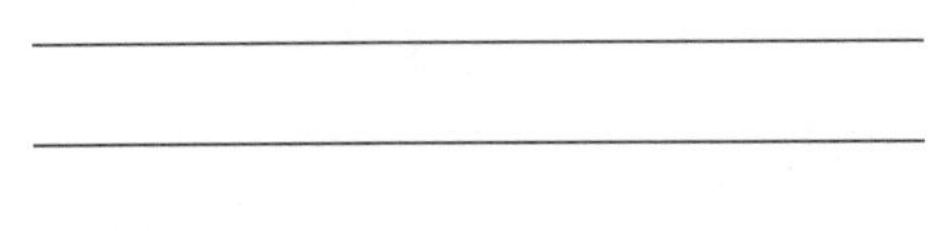

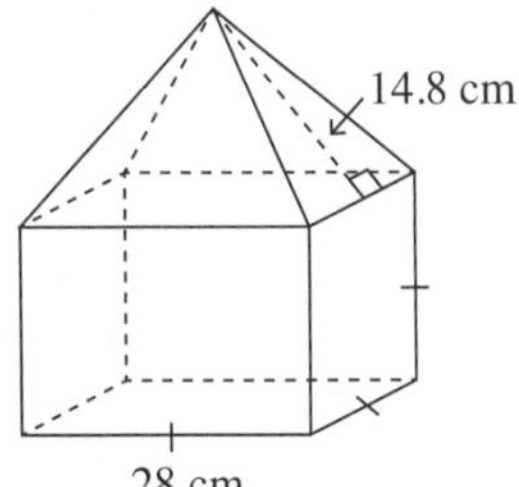

b Find the volume of the solid.

c Find the total surface area.

3

Continued on the next page

EXAM PAPER 3 PART C

Show all working for each question.

Marks

8 This scatter plot has been drawn to show the number of minutes (N) to achieve a task after the number of weeks (w) employed at that task. A line of best fit has been drawn on this scatter plot.

a What is the equation of the line of best fit?

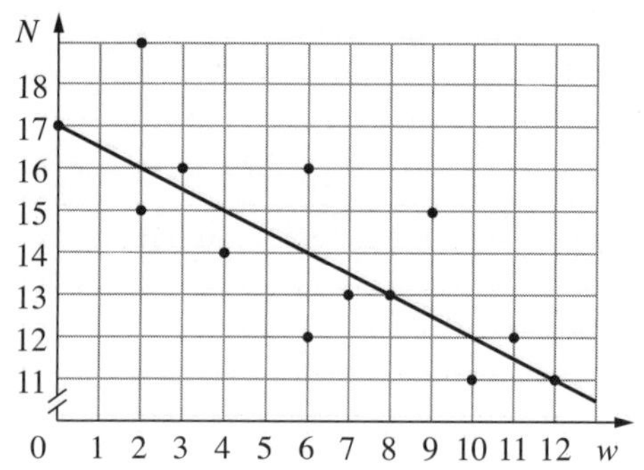

b After how many weeks employed does the line of best fit predict a person will complete the task in 8 minutes?

c Briefly explain why this equation cannot be used to predict the number of minutes to complete the task after a person has been employed for 40 weeks.

3

9 **a** On the same diagram sketch $y = x$ and $y = \frac{4}{x}$

b Write down the coordinates of the two points of intersection.

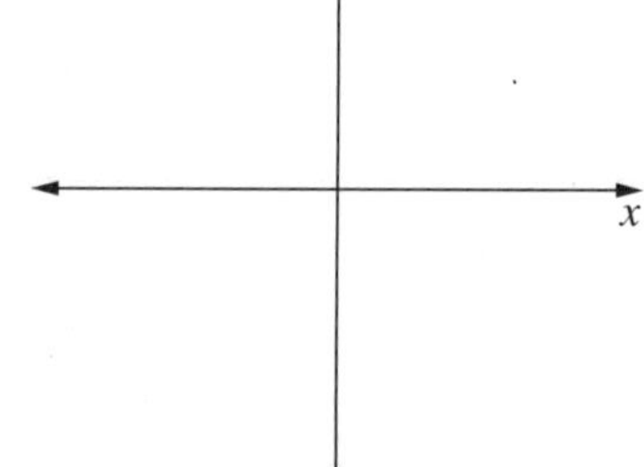

2

10 The tangent from T meets a circle at P. $PT = 12$ cm. The line from A to T cuts the circle again at B. $AB = 10$ cm. $BT = x$ cm.

a Find the value of x.

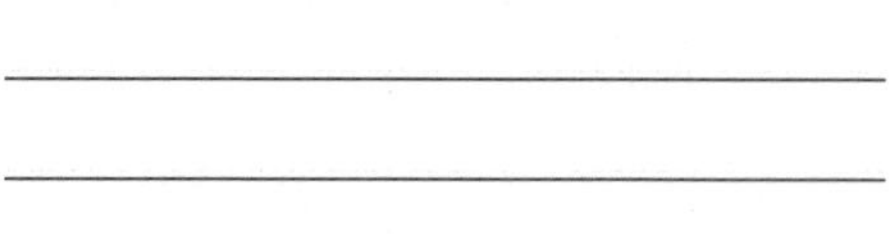

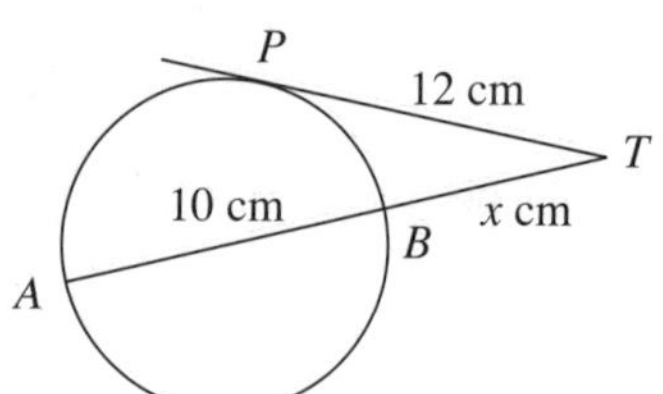

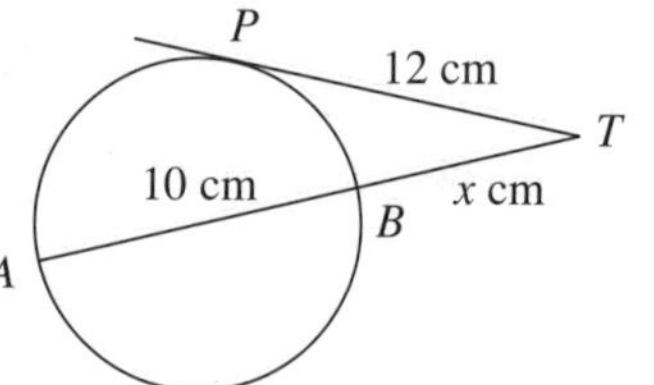

b Is AB a diameter of the circle? Justify your answer.

2

Total marks achieved for PART C

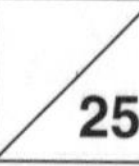

Exam Paper 4

Instructions for all parts • Attempt all questions.
Time allowed: 1 hour • Calculators are allowed.

Part A: Allow about 10 minutes for this part.
Part B: Allow about 20 minutes for this part.
Part C: Allow about 30 minutes for this part. **Total marks: 50**

EXAM PAPER 4 PART A

Fill in only one circle for each question.

	Marks
1 $25^{\frac{1}{2}}$ equals Ⓐ $\frac{1}{5}$ Ⓑ $\frac{1}{25}$ Ⓒ 5 Ⓓ 12.5	1
2 A die is thrown twice. What is the probability of throwing a two both times? Ⓐ $\frac{1}{6}$ Ⓑ $\frac{1}{36}$ Ⓒ $\frac{1}{4}$ Ⓓ $\frac{3}{4}$	1
3 O is the centre of the circle. The value of x is Ⓐ 29 Ⓑ 32 Ⓒ 58 Ⓓ 116 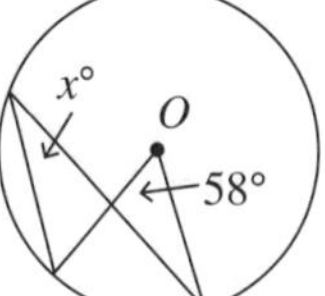 	1
4 $\frac{3+\sqrt{3}}{\sqrt{3}}$ equals Ⓐ $3\sqrt{3}$ Ⓑ $\sqrt{3}+1$ Ⓒ 4 Ⓓ 3	1
5 The remainder when $x^5 - 2x^3 + 4x^2 - 9x + 2$ is divided by $(x + 1)$ is Ⓐ −4 Ⓑ −2 Ⓒ 12 Ⓓ 16	1
6 $\cos(180° - \theta) =$ Ⓐ $\cos\theta$ Ⓑ $-\cos\theta$ Ⓒ $\sin\theta$ Ⓓ $-\sin\theta$	1
7 Which of these is NOT a function? Ⓐ $y = 9x^2$ Ⓑ $y = \frac{9}{x}$ Ⓒ $x^2 + y^2 = 9$ Ⓓ $y = \frac{x}{9}$	1
8 Which is the correct factorisation of $2x^2 + 7x - 15$? Ⓐ $(2x+3)(x-5)$ Ⓑ $(2x-3)(x+5)$ Ⓒ $(2x+5)(x-3)$ Ⓓ $(2x-5)(x+3)$	1
9 What is the volume of a triangular pyramid with base area 18 m^2 and perpendicular height 3 m? Ⓐ 6 m^3 Ⓑ 9 m^3 Ⓒ 18 m^3 Ⓓ 54 m^3	1
10 $3\log_a 6 - 2\log_a 3 =$ Ⓐ $\log_a 2$ Ⓑ $\log_a 3$ Ⓒ $\log_a 24$ Ⓓ $\log_a 27$	1

Total marks achieved for PART A 

EXAM PAPER 4 — PART B

Write only the answer in the answer column. For any working use the question column.

Questions	Answers	Marks
1 Simplify $\sqrt{162} + \sqrt{32}$		1
2 Solve $x^2 = 25$		1
3 Find the exact value of $3^{-2} \times 8^{\frac{1}{3}}$		1
4 Solve $(4x - 1)(x + 2) = 0$		1
5 Find the equation of the circle. 		1
6 What is the total surface area, to two decimal places, of this solid hemisphere of radius 3 m? 		1
7 Solve $3^{x+1} = 27$		1

Continued on the next page

EXAM PAPER 4 PART B

Write only the answer in the answer column. For any working use the question column.

Questions	Answers	Marks
8 PT is a tangent. C is the centre of the circle. Find the value of a 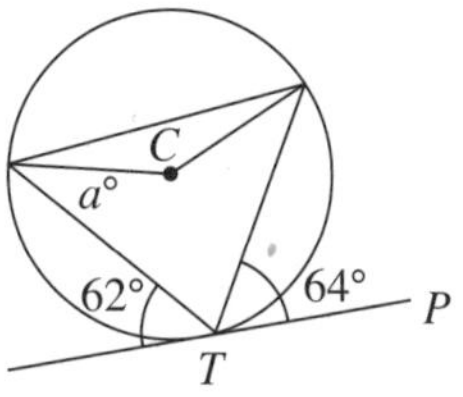		1
9 If $f(x) = -x^3$, find $f(-1)$		1
10 Find the exact value of x. 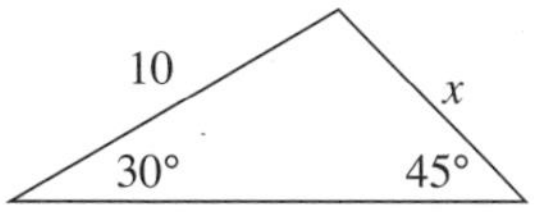		1
11 What is the area of this triangle to the nearest square metre? 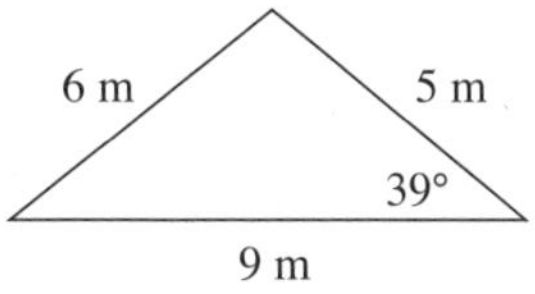		1
12 Factorise $4x^2 + 12x - 27$		1
13 Solve $\log_9 x = -3$		1
14 Expand $(2\sqrt{3} - \sqrt{5})^2$		1
15 Write down the equation of the horizontal asymptote of the hyperbola $y = 1 - \dfrac{3}{x+2}$		1

Total marks achieved for PART B /15

EXAM PAPER 4 PART C

Show all working for each question.

Marks

1 Solve.

a $y^2 - 8y = -5$ (by completing the square)

b $5x^2 - 7x + 1 = 0$ (by the quadratic formula, leaving your answer in simplest surd form)

c $(x - 3)(x - 5) = 24$ (by completing the square)

d $x - y = 1$ (simultaneously)
$xy = 6$

4

2 Consider these scores.

4	8
5	4 9
6	3 6 7 8
7	0 2 3 5 7 9 9 9
8	1 2 4 6 6 8
9	0 3 7

a What is the range?

b Find the interquartile range.

c What is the mean?

d What is the standard deviation (to 1 decimal place)?

4

3 Consider the parabola $y = x^2 - 6x + 13$

a Find the equation of the axis of symmetry.

b Will the parabola cut the x-axis? Justify your answer.

2

Continued on the next page

EXAM PAPER 4 PART C

Show all working for each question.

Marks

4 This figure is made up of a cylinder and hemisphere. Find the:

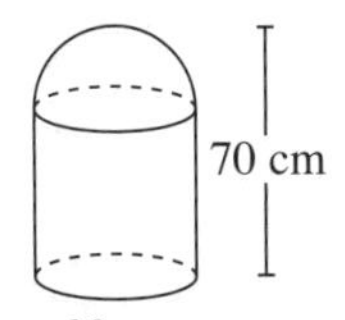

a height of the cylinder.

b figure's volume to one decimal place.

c figure's capacity to the nearest litre.

3

5 Find the total surface area of this cone.

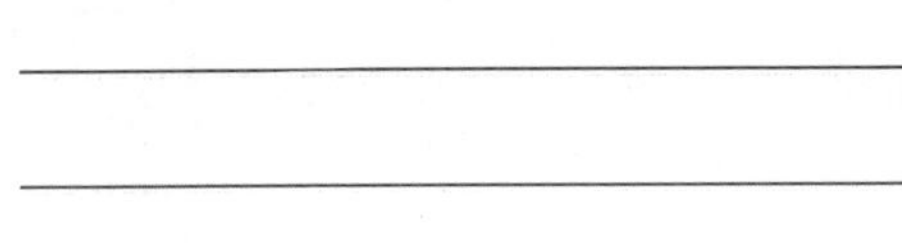

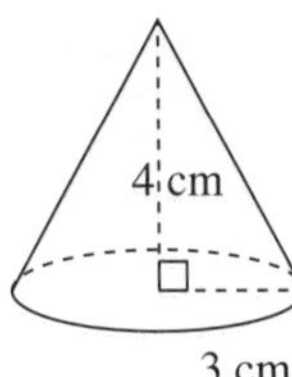

1

6 $(x - 2)$ and $(x + 3)$ are factors of $P(x) = 2x^3 + px^2 + qx - 6$.

a Find the values of p and q.

b Factorise $P(x)$ fully.

c Sketch $y = P(x)$.

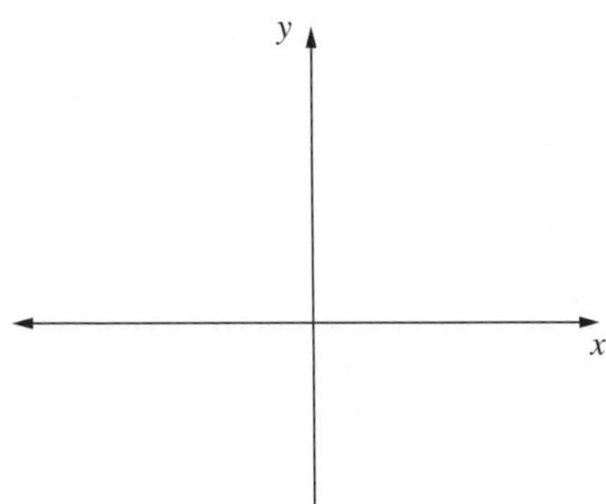

3

Continued on the next page

EXAM PAPER 4 PART C

Show all working for each question.

Marks

7 Q is 23 km from P on a bearing of 104°. The bearing of R from P is 169°. R is 36 km from Q.

a What is the size of $\angle QPR$?

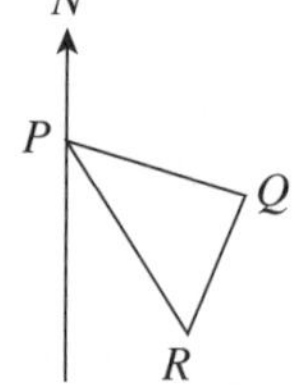

b Find the size of $\angle PRQ$ to the nearest degree.

c What is the bearing of Q from R?

d Find the distance from P to R.

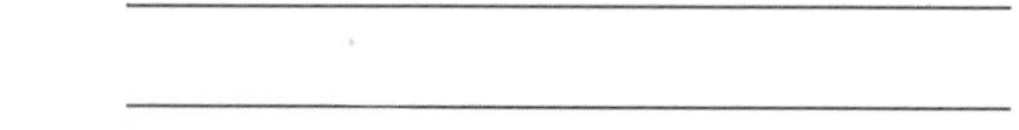

4

8 **a** On the same diagram sketch $y = 4^x$ and $y = \log_4 x$.

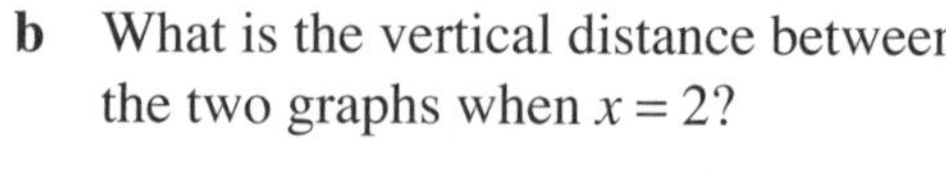

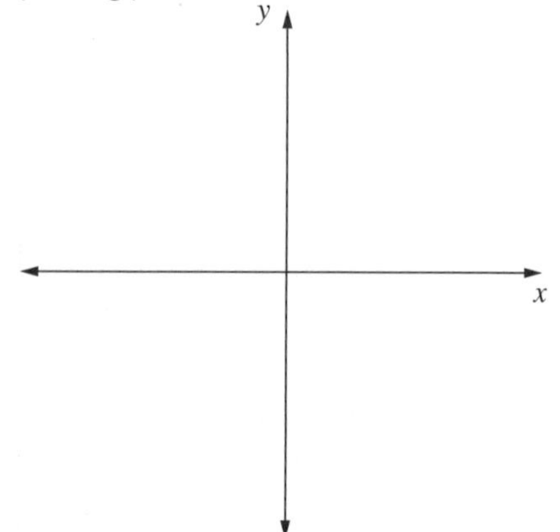

b What is the vertical distance between the two graphs when $x = 2$?

2

9 Two circles intersect at P and at Q. The smaller circle passes through O, the centre of the larger circle. AP is a chord of the larger circle. It cuts the smaller circle at B. Let $\angle BAQ = \alpha$.

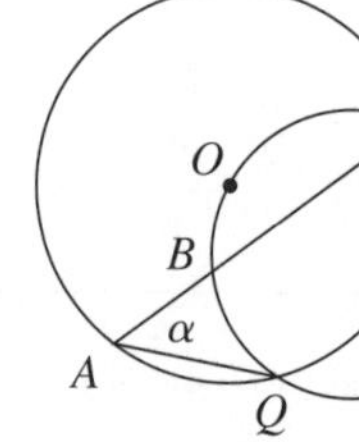

a Show that $\angle PBQ = 2\alpha$.

b Show that $AB = QB$.

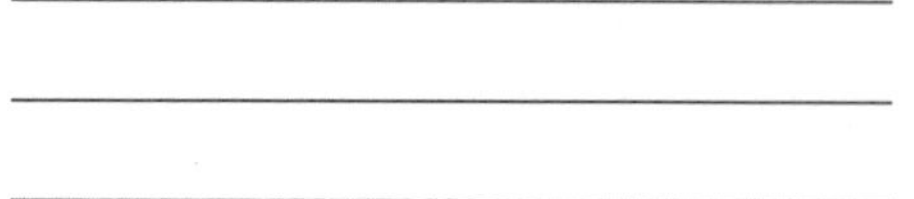

2

Total marks achieved for PART C /25

Answers

Chapter 1 – Rational and irrational numbers

Page 1 **1 a** number line **b** integers **c** irrational **d** rational **2 a** $0.\dot{7}$ **b** $0.\dot{3}\dot{6}$ **c** $0.\overline{285714}$ **d** $0.41\dot{6}$ **e** $0.\dot{3}\dot{9}$ **f** $0.\dot{1}2\dot{3}$ **3 a** $\frac{4}{9}$ **b** $\frac{5}{11}$ **c** $\frac{41}{90}$ **d** $\frac{152}{333}$ **e** $\frac{226}{495}$ **f** $\frac{137}{300}$ **4 a** $\frac{5}{9}$ **b** $\frac{1}{6}$ **c** $\frac{29}{33}$ **d** $\frac{7}{12}$ **e** $\frac{28}{37}$ **f** $\frac{68}{165}$

Page 2 **1 a** Q **b** Q **c** Q′ **d** Q **e** Q′ **f** Q′ **g** Q **h** Q **i** Q **j** Q′ **k** Q **l** Q′ **2 a** 1.41 **b** 1.73 **c** 2.45 **d** 2.65 **e** 3.32 **f** 4.12 **g** 5.39 **h** 6.40 **3 a** 2 and 3 **b** 4 and 5 **c** 6 and 7 **d** 7 and 8 **e** 9 and 10 **f** 5 and 6 **g** 8 and 9 **h** 14 and 15 **4 a** $\sqrt{2}, \sqrt{3}, 2, 5$ **b** $\sqrt{8}, 3, \sqrt{17}, 6$ **c** $\sqrt{3}, \sqrt{8}, 3, \sqrt{15}$ **d** $7, \sqrt{60}, \sqrt{80}, 9$ **5 a i** 36 **ii** 36 **b** 6 and –6 **6 a** 7 **b** 9 **c** –10 **d** –13 **e** ±5 **7 a iii** $\sqrt{-51}$ and **iv** $-\sqrt{-27}$
b For a real square root, the number under the square root should always be greater than or equal to zero ($x \geq 0$). **c** $x < 0$ **8**

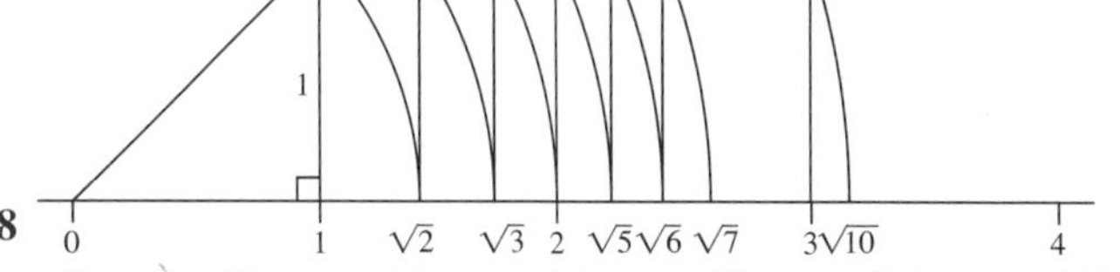

Page 3 **1 a** 6 **b** 10 **c** 2 **d** 3 **e** 5 **f** 8 **2 a** $2\sqrt{2}$ **b** $2\sqrt{3}$ **c** $3\sqrt{2}$ **d** $2\sqrt{6}$ **e** $2\sqrt{7}$ **f** $5\sqrt{2}$ **g** $2\sqrt{10}$ **h** $2\sqrt{11}$ **i** $5\sqrt{3}$ **j** $4\sqrt{3}$ **k** $3\sqrt{3}$ **l** $4\sqrt{2}$ **m** $6\sqrt{2}$ **n** $7\sqrt{2}$ **o** $11\sqrt{2}$ **p** $10\sqrt{2}$ **3 a** $3\sqrt{6}$ **b** $5\sqrt{6}$ **c** $4\sqrt{6}$ **d** $4\sqrt{10}$ **e** $15\sqrt{2}$ **f** $5\sqrt{5}$ **g** $3\sqrt{7}$ **h** $6\sqrt{3}$ **i** $8\sqrt{3}$ **j** $12\sqrt{2}$ **k** $9\sqrt{2}$ **l** $9\sqrt{3}$ **4 a** $6\sqrt{3}$ **b** $12\sqrt{2}$ **c** $15\sqrt{3}$ **d** $9\sqrt{2}$ **e** $16\sqrt{3}$ **f** $6\sqrt{6}$ **g** $36\sqrt{2}$ **h** $14\sqrt{14}$ **i** $20\sqrt{2}$ **j** $27\sqrt{2}$ **k** $-2\sqrt{3}$ **l** $-50\sqrt{3}$

Page 4 **1 a** true **b** false **c** false **d** false **e** false **f** true **2 a** $\sqrt{6}$ **b** $\sqrt{15}$ **c** $\sqrt{30}$ **d** 6 **e** 8 **f** $\sqrt{21}$ **g** $\sqrt{6}$ **h** 2 **i** $\sqrt{10}$ **j** 20 **k** $\sqrt{3}$ **l** 2 **3 a** 3 **b** 7 **c** 9 **d** 5 **e** 10 **f** 8 **g** 45 **h** 24 **i** 90 **j** 176 **k** 175 **l** 32 **4 a** $2\sqrt{5}$ **b** $3\sqrt{5}$ **c** $3\sqrt{10}$ **d** $4\sqrt{7}$ **e** $8\sqrt{2}$ **f** $7\sqrt{3}$ **g** $10\sqrt{5}$ **h** $11\sqrt{3}$ **i** $3\sqrt{15}$ **5 a** $6\sqrt{2}$ **b** $6\sqrt{2}$ **c** $25\sqrt{5}$ **d** $12\sqrt{6}$ **e** $30\sqrt{2}$ **f** $40\sqrt{3}$ **g** $9\sqrt{6}$ **h** $16\sqrt{6}$ **i** $28\sqrt{2}$ **j** $24\sqrt{2}$ **k** $24\sqrt{2}$ **l** $21\sqrt{5}$ **m** $15\sqrt{2}$ **n** $30\sqrt{3}$ **o** $27\sqrt{7}$ **6 a** $\sqrt{28}$ **b** $\sqrt{45}$ **c** $\sqrt{75}$ **d** $\sqrt{250}$ **e** $\sqrt{96}$ **f** $\sqrt{640}$

Page 5 **1 a** $5\sqrt{2}$ **b** $\sqrt{34}$ **c** $6\sqrt{2}$ **d** $\sqrt{7}$ **e** $2\sqrt{13}$ **f** $4\sqrt{5}$ **g** $2\sqrt{66}$ **h** $6\sqrt{2}$ **i** $5\sqrt{15}$ **j** $3\sqrt{17}$ **k** $6\sqrt{5}$ **l** $3\sqrt{21}$ **2** $\sqrt{73}$ **3 a** 7 units^2 **b** $\sqrt{14}$ units

Page 6 **1 a** $11\sqrt{2}$ **b** $15\sqrt{5}$ **c** $15\sqrt{3}$ **d** $4\sqrt{5}$ **e** $11\sqrt{3}$ **f** $13\sqrt{2}$ **g** $5\sqrt{x}$ **h** $8\sqrt{y}$ **i** $16\sqrt{m}$ **j** $11\sqrt{2}$ **k** $10\sqrt{3}$ **l** $5a\sqrt{x}$ **2 a** $10\sqrt{3} + 5\sqrt{2}$ **b** $8\sqrt{7} - 2\sqrt{11}$ **c** $2\sqrt{2} - 4\sqrt{3}$ **d** $3\sqrt{5} + 6\sqrt{3}$ **e** $8\sqrt{2}$ **f** $5\sqrt{3}$ **g** 0 **h** $15\sqrt{3}$ **i** $2\sqrt{10} + \sqrt{3}$ **j** $3\sqrt{5}$ **k** $17\sqrt{2} - 6\sqrt{3}$ **l** $7\sqrt{7}$ **m** $12\sqrt{2}$ **n** $19\sqrt{3} - 3\sqrt{5}$ **3 a** $5\sqrt{2}$ **b** $7\sqrt{10}$ **c** $8\sqrt{3}$ **d** $\sqrt{6}$ **e** $\sqrt{3}$ **f** $9 - 4\sqrt{6}$ **g** $8\sqrt{2}$ **h** $-2\sqrt{6}$ **i** $2\sqrt{3} + 18\sqrt{2}$ **j** $4\sqrt{5}$ **k** $18\sqrt{3}$ **l** $3\sqrt{2}$ **m** $5\sqrt{5} - 5\sqrt{3}$ **n** $5\sqrt{2}$

Page 7 **1 a** $\sqrt{6}$ **b** $\sqrt{30}$ **c** $\sqrt{14}$ **d** $\sqrt{33}$ **e** $\sqrt{42}$ **f** $\sqrt{70}$ **g** $30\sqrt{6}$ **h** $6\sqrt{6}$ **i** $20\sqrt{35}$ **j** $24\sqrt{15}$ **k** $15\sqrt{77}$ **l** $30\sqrt{30}$ **2 a** 10 **b** 4 **c** 6 **d** $2\sqrt{15}$ **e** $3\sqrt{10}$ **f** 40 **g** $280\sqrt{2}$ **h** 24 **i** 42 **j** $40\sqrt{3}$ **3 a** $\sqrt{3}$ **b** $\sqrt{5}$ **c** $\sqrt{7}$ **d** $6\sqrt{7}$ **e** $2\sqrt{5}$ **f** $5\sqrt{2}$ **4 a** 2 **b** 2 **c** 4 **d** 3 **e** 6 **f** $7\sqrt{7}$ **g** $4\sqrt{10}$ **h** $3\sqrt{5}$ **i** $\sqrt{2}$ **j** $\sqrt{7}$ **k** $\sqrt{2}$ **l** $2\sqrt{2}$

Page 8 **1 a** $2\sqrt{3}+4\sqrt{2}$ **b** $15\sqrt{6}+25$ **c** $\sqrt{14}+\sqrt{6}$ **d** $\sqrt{10}+\sqrt{15}$ **e** $5\sqrt{5}-\sqrt{15}$ **f** $2\sqrt{15}-\sqrt{21}$ **g** $\sqrt{22}+2\sqrt{14}$ **h** $2\sqrt{30} + 3\sqrt{42}$ **i** $3\sqrt{22} + 3\sqrt{10}$ **j** $2\sqrt{6} - 6\sqrt{10}$ **k** $6\sqrt{14} + 9\sqrt{21}$ **l** $20\sqrt{6} - 12\sqrt{15} + 4\sqrt{3}$ **2 a** $\sqrt{15} + 3$ **b** $5 + \sqrt{15}$ **c** $21 - 2\sqrt{7}$ **d** 4 **e** $5\sqrt{6} - 3$ **f** $2\sqrt{33} - 66$ **g** $2\sqrt{3} + 3\sqrt{2}$ **h** $3\sqrt{5} - 3\sqrt{2}$ **i** $30 - 2\sqrt{6}$ **j** $64\sqrt{5} - 48$ **k** $50\sqrt{2} - 25\sqrt{3}$ **l** $36\sqrt{2} + 90\sqrt{7}$ **3 a** $7\sqrt{3} - 14\sqrt{2}$ **b** $\sqrt{21} - 2\sqrt{3} - \sqrt{10} - 2$ **c** $6\sqrt{15} - 18\sqrt{10} + 3\sqrt{6}$ **d** $6\sqrt{14} - 7\sqrt{35} - 8\sqrt{5}$

Page 9 **1 a** $3 + \sqrt{6} + 4\sqrt{3} + 4\sqrt{2}$ **b** $\sqrt{15} + 2\sqrt{3} + \sqrt{35} + 2\sqrt{7}$ **c** $8 - 6\sqrt{3}$ **d** $7 - 18\sqrt{3}$ **e** $9 + 6\sqrt{2}$ **f** $\sqrt{35} - 3\sqrt{5} + 2\sqrt{7} - 6$ **g** $10\sqrt{5} + 2\sqrt{10} - 10 - 2\sqrt{2}$ **h** 20 **i** $\sqrt{6} + 2 + \sqrt{3} + \sqrt{2}$ **j** $70 + 5\sqrt{6} + 14\sqrt{2} + 2\sqrt{3}$ **k** $6 + 3\sqrt{6} + 2\sqrt{3} + 3\sqrt{2}$ **l** $-\sqrt{2} - 11$ **2 a** $3 + 2\sqrt{2}$ **b** $21 + 12\sqrt{3}$ **c** $12 + 2\sqrt{35}$ **d** $8 - 2\sqrt{15}$ **e** $4 - 2\sqrt{3}$ **f** $9 - 2\sqrt{14}$ **g** $39 - 12\sqrt{3}$ **h** $37 + 20\sqrt{3}$ **i** $46 - 6\sqrt{5}$ **j** $5 + 2\sqrt{6}$ **k** $5 - 2\sqrt{6}$ **l** $38 + 12\sqrt{10}$ **m** $21 + 4\sqrt{5}$ **n** $29 - 4\sqrt{7}$ **3 a** 1 **b** 5 **c** 53 **d** 2 **e** 4 **f** 9 **g** 3 **h** 20 **i** 56 **j** 19 **k** 54 **l** 49 **m** 5 **n** 167

Page 10 **1 a** $\frac{\sqrt{6}}{6}$ **b** $\frac{3\sqrt{7}}{7}$ **c** $\sqrt{2}$ **d** $\frac{5\sqrt{11}}{11}$ **e** $\frac{7\sqrt{5}}{5}$ **f** $\frac{2\sqrt{15}}{5}$ **g** $\frac{\sqrt{15}}{5}$ **h** $\frac{\sqrt{14}}{7}$ **i** $\frac{\sqrt{55}}{11}$ **j** $\frac{3\sqrt{10}}{5}$ **k** $\frac{2\sqrt{21}}{3}$ **l** $\frac{7\sqrt{10}}{5}$ **2 a** $\frac{\sqrt{3}}{5}$ **b** $\frac{5\sqrt{2}}{4}$ **c** $\frac{7\sqrt{3}}{9}$ **d** $\frac{9\sqrt{5}}{25}$ **e** $\frac{4\sqrt{14}}{21}$ **f** $\frac{5\sqrt{6}}{4}$ **g** $\frac{3\sqrt{10}}{20}$ **h** $\frac{8\sqrt{30}}{27}$ **i** $\frac{\sqrt{42}}{9}$ **j** $\frac{\sqrt{35} - \sqrt{21}}{7}$ **k** $\frac{2 - \sqrt{2}}{2}$ **l** $\frac{3 + 2\sqrt{3}}{3}$

Page 11 **1 a** $\sqrt{2} - 1$ **b** $\frac{\sqrt{3} + 1}{2}$ **c** $\frac{\sqrt{5} + 1}{4}$ **d** $3\sqrt{2} + 3$ **e** $5\sqrt{5} - 10$ **f** $\sqrt{7} + \sqrt{5}$ **g** $\sqrt{5} + \sqrt{2}$ **h** $2\sqrt{7} + 2\sqrt{3}$ **i** $\frac{4 - \sqrt{3}}{26}$ **j** $\frac{5 + \sqrt{3}}{11}$ **k** $\frac{5 - \sqrt{7}}{6}$ **l** $-\sqrt{3} + \sqrt{5}$ **m** $\frac{\sqrt{30} - 2\sqrt{5}}{2}$ **n** $\sqrt{15} + 2\sqrt{2}$ **2 a** $3 + 2\sqrt{2}$ **b** $5 + 2\sqrt{6}$ **c** $-25 - 8\sqrt{10}$ **d** $25\sqrt{7} + 66$

Page 12 **1 a i** 25 **ii** 25 **b i** 49 **ii** 49 **c i** 6.25 **ii** 6.25 **d** positive **e** positive **2 a i** 125 **ii** –125 **b i** 125 **ii** –125 **c i** 15.625 **ii** –15.625 **d** positive **e** negative **3 a** $7^{\frac{3}{2}}$ **b** 5^2 **c** $10^{\frac{5}{3}}$ **d** $22^{\frac{2}{3}}$ **e** $18^{\frac{2}{5}}$ **f** $16^{\frac{3}{7}}$ **4 a** $\sqrt{17}$ **b** $\sqrt[3]{-10}$ **c** $\sqrt[5]{18}$ **d** $\sqrt[3]{12}$ **e** $\sqrt[4]{24}$ **f** $\sqrt[6]{28}$ **5 a** 14 **b** 12 **c** 17 **d** 5 **e** 2.6 **f** 2.9 **6 a** 5.27 **b** 3.93 **c** 2.52 **d** 3.09 **e** 22.9 **f** 3.97 **7 a** x **b** $a^{\frac{4}{3}}$ **c** y **d** m^4 **e** 1 **f** $2ab$ **8 a** $\frac{1}{3}$ **b** $\frac{1}{9}$ **c** $\frac{1}{4}$ **d** $\frac{1}{5}$ **e** $\frac{1}{8}$ **f** $\frac{16}{25}$

Answers

Page 13 **1** a $\sqrt{5}$ b $\sqrt{7}$ c $\sqrt{m}$ d $\sqrt[3]{2}$ e $\sqrt[3]{6}$ f $\sqrt[4]{11}$ g $\sqrt[6]{x}$ h $\sqrt[10]{n}$ i $\sqrt[3]{a^2}$ j $\sqrt[4]{p^3}$ k $\sqrt[3]{q^4}$ l $\sqrt[5]{k^2}$ **2** a $5^{\frac{1}{2}}$ b $3^{\frac{1}{2}}$ c $15^{\frac{1}{2}}$ d $43^{\frac{1}{2}}$ e $7^{\frac{1}{2}}$ f $6^{\frac{1}{3}}$ g $28^{\frac{1}{3}}$ h $240^{\frac{1}{3}}$ i $10^{\frac{1}{4}}$ j $37^{\frac{1}{5}}$ k $23^{\frac{1}{8}}$ l $14^{\frac{1}{n}}$ **3** a 2 b 5 c 7 d 9 e 3 f 4 g 8 h 3 i 6 j 2 k 4 l 5 **4** a 2 b 4 c $9a^{\frac{1}{4}}$ d 81 e 27 f $81x^3$ g 25 h 25 i x^3 j 125 k 8 l 1 **5** a 72 b 100 c 7 d $\frac{4}{15}$ e 10 f 9 g 64 h 10 i $5\frac{2}{5}$ j $\frac{16}{15}$ k 629 l 3 **6** a x b $y^{\frac{3}{2}}$ c $m^{\frac{4}{3}}$ d $24x^{\frac{3}{4}}$ e $p^{\frac{7}{3}}$ f $x^{\frac{1}{2}}$ g 128 h $16y^{\frac{5}{3}}$ i a^4b^2 j $8a^8$ k $7a^4b^6$ l y^3

Page 14 **1** a $\sqrt{25}$ b $\sqrt[3]{64}$ c $\sqrt[6]{a}$ d $\sqrt[3]{64}$ e $\sqrt[3]{5x}$ f $\sqrt[3]{125m}$ **2** a $11^{\frac{1}{2}}$ b $x^{\frac{1}{2}}$ c $(216a)^{\frac{1}{3}}$ d $65^{\frac{1}{7}}$ e $20^{\frac{1}{n}}$ f $21^{-\frac{1}{2}}$ **3** a 16 b 16 c 6 d $7a^3$ e a^2b^2 f $9y^8$ **4** a $a^{\frac{13}{9}}$ b $30m^{\frac{5}{4}}$ c $x^{\frac{2}{15}}$ d y e a^7b^5 f $3a^4$ **5** a $x^{\frac{14}{3}}$ b $y^{\frac{7}{6}}$ c $a^{\frac{1}{6}}$ d $10a+5a^{\frac{1}{2}}$ e $m^{\frac{-15}{4}}$ f $\frac{P}{\sqrt{3}}$ **6** a a^2 b $a^{-1}b^2$ c $a+\frac{1}{a}$ d $a^{\frac{4}{3}}$ e $a^{-\frac{9}{20}}$ f $x^{12}y^{-9}$ **7** a 6 b 32 c 20 d $\frac{1}{4}$ e 8 f $\frac{1}{4}$

Page 15 **1** a $\sqrt{64}$ b $\sqrt{100}$ c $\sqrt{121}$ d $\sqrt{144}$ e $\sqrt{169}$ f $\sqrt{\frac{1}{4}}$ g $\sqrt{\frac{1}{9}}$ h $\sqrt{\frac{1}{16}}$ i $\sqrt{\frac{1}{25}}$ j $\sqrt{\frac{1}{36}}$ k $\sqrt[3]{8}$ l $\sqrt[3]{27}$ m $\sqrt[6]{64}$ n $\sqrt[4]{16}$ o $\sqrt[5]{x}$ p $\sqrt[6]{y}$ q $\sqrt[4]{3p}$ r $\sqrt[3]{9x}$ s $\sqrt{5x}$ t $\sqrt[7]{8m}$ **2** a $5^{\frac{1}{2}}$ b $7^{\frac{1}{2}}$ c $14^{\frac{1}{2}}$ d $38^{\frac{1}{2}}$ e $46^{\frac{1}{2}}$ f $26^{\frac{1}{3}}$ g $115^{\frac{1}{3}}$ h $34^{\frac{1}{3}}$ i $61^{\frac{1}{3}}$ i $41^{\frac{1}{3}}$ k $11^{\frac{1}{4}}$ l $37^{\frac{1}{5}}$ m $82^{\frac{1}{6}}$ n $19^{\frac{1}{7}}$ o $123^{\frac{1}{9}}$ p $x^{\frac{1}{n}}$ q $p^{\frac{1}{y}}$ r $k^{\frac{1}{m}}$ s $15^{-\frac{1}{2}}$ t $48^{-\frac{1}{4}}$ **3** a 3 b 3 c 8 d 4 e 3 f 2 g 5 h 9 i 6 j 2 k 7 l $\frac{1}{2}$ m 1 n 5 o $\frac{1}{4}$ p $\frac{1}{4}$ q 2 r 2 s 5 t 3 **4** a 16 b 243 c 8 d 4 e 1000 f 125 g 512 h 216 i 2 j 9 k 1 l 3 m 7 n 9 o 81 p 32 q 81 r 2 s 3 t 729 **5** a 8.07 b 17.89 c 4.47 d 2.80 e 5.95 f 22.58 g 4.41 h 18.71 i 4.55 j 9.90 k 7.46 l 9.38 m 7.44 n 3.87 o 7.68 p 3.19 q 5.57 r 4.48 s 1.84 t 3.00 **6** a 216 b 16 c 4 d $\frac{3}{8}$ e $\frac{3}{4}$ f 49 g 256 h 64 i 8 j 32 k 9 l 9 m 16 n 25 o 8 p 49 q 9 r 16 s 27 t 25

Page 16 **1** B **2** B **3** C **4** A **5** A **6** C **7** D **8** D **9** A **10** A

Page 17 **1** 12 **2** $7\sqrt{2}$ **3** $2\sqrt{3}-4\sqrt{2}$ **4** $x^{\frac{2}{3}}$ **5** 243 **6** $4x^8$ **7** 14 **8** $6+5\sqrt{21}$ **9** $32-10\sqrt{7}$ **10** $\frac{7\sqrt{15}}{6}$ **11** $10\sqrt{3}-4\sqrt{7}$ **12** 6 **13** $5\sqrt{5}$ **14** $\sqrt{21}+7$ **15** 41

Chapter 2 – Quadratic expressions and equations

Page 18 **1** a $x(x+9)$ b $7(x^2+2yz)$ c $(x+4)(x-4)$ d $2xy^2(2xy-3)$ e $4abc^2(2a^3+3)$ f $9xy(2y-1)$ **2** a $(x+3)(x+4)$ b $(x-7)(x+2)$ c $(x+5)(x-2)$ d $(x-4)(x-5)$ e $(x+5)(x+6)$ f $(x-6)(x-4)$ g $(x-12)(x+2)$ h $(x-4)(x+3)$ i $(x-5)(x-1)$ j $(x-5)^2$ k $(5-x)(3+x)$ l $(6-x)(7+x)$ **3** a $(p+9)(p+q)$ b $(2a+b)(a-3)$ c $(n-4)(3m-2)$ d $(x+3)(2x+1)$ e $(2a+3)(5a-2)$ f $(y+1)(2x+1)$ **4** a $2(x+6)(x-2)$ b $3(x+2)(x+3)$ c $7(x+2)^2$ d $5(x-3)(x+1)$ e $6(x-5)(x+4)$ f $4(x-6)(x-9)$ g $3(x+3)(x-3)$ h $4(x+2)(x+4)$ i $2(x-3)(x-5)$

Page 19 **1** a $(2x+3)(x+5)$ b $(5t-1)(t+1)$ c $(2x+5)(x+1)$ d $(2y+3)(3y-2)$ e $(7x-3)(x+2)$ f $(3x+2)(4x-1)$ g $(3a+1)(a+3)$ h $(2a+1)(a+2)$ i $(3y+1)(y-1)$ j $(2x+1)(3x+4)$ k $(2a+1)(a+3)$ l $(2x-1)(x-2)$ m $(4x+3)(x+1)$ n $(2y-3)(y-1)$ o $(2m+1)(m-2)$ p $(2y-1)(y+3)$ q $(4x-7)(x-5)$ r $(3t-1)(t+2)$ s $(2x-3)(x-2)$ t $(3m-2)(m+3)$ u $(8a+3)(a-2)$ v $(3y-10)(y+2)$ w $(2n+7)(n-6)$ x $(4x+5)(x-1)$

Page 20 **1** a $x=\pm3$ b $x=\pm4$ c $x=\pm5$ d $x=\pm1$ e $x=\pm2$ f $x=\pm8$ g $x=\pm6$ h $x=\pm7$ i $x=\pm11$ j $x=\pm12$ k $x=\pm10$ l $x=\pm9$ m $x=\pm13$ n $x=\pm20$ o $x=\pm0.5$ **2** a $x=\pm\sqrt{7}$ b $x=\pm\sqrt{6}$ c $x=\pm\sqrt{3}$ d $x=\pm\sqrt{87}$ e $x=\pm\sqrt{93}$ f $x=\pm2\sqrt{3}$ g $x=\pm6\sqrt{2}$ h $x=\pm\sqrt{5}$ i $x=\pm3\sqrt{2}$ j $x=\pm10\sqrt{7}$ **3** a $x=\pm4.80$ b $m=\pm7.28$ c $y=\pm5.39$ d $n=\pm8.19$ e $c=\pm12.25$ f $k=\pm4.36$ g $y=\pm2.41$ h $d=\pm1.46$ i $p=\pm2.68$ j $x=\pm3.16$ **4** a $x=\pm\sqrt{\frac{10}{3}}$ b $x=\pm2\sqrt{2}$ c $x=\pm\frac{5}{4}$ d $x=\pm\sqrt{6}$ e $x=\pm\sqrt{\frac{11}{2}}$ f $x=\pm3\sqrt{2}$ g $x=\pm6$ h $h=\pm\sqrt{5}$

Page 21 **1** a $x=2, x=5$ b $x=1, x=-9$ c $x=2, x=10$ d $x=0, x=-7$ e $x=0, x=3$ f $x=0, x=\frac{5}{4}$ g $x=4, x=6$ h $x=-4, x=9$ i $x=-7, x=9$ j $x=-2, x=7$ k $x=6, x=-8$ l $x=3, x=-8$ m $x=-8, x=\frac{3}{2}$ n $x=-3, x=\frac{4}{5}$ o $x=2, x=7$ **2** a $x=6, x=11$ b $x=-3, x=7$ c $x=\frac{3}{2}, x=-5$ d $x=0, x=-8$ e $x=0, x=\frac{3}{2}$ f $x=0, x=4$ g $x=-3, x=-9$ h $x=0, x=\frac{5}{4}$ i $x=0, x=9$ j $x=-\frac{1}{2}, x=0$ k $x=0, x=8$ l $x=0, x=4$ **3** a $x=7, x=10$ b $x=8, x=-10$ c $x=0, x=5$ d $x=\frac{1}{4}, x=-5$ e $x=-\frac{1}{2}, x=\frac{5}{3}$ f $x=0, x=1$ g $x=3, x=7$ h $x=-2, x=\frac{7}{3}$ i $x=-6, x=8$

Answers

Page 22 **1 a** $x=3, x=-3$ **b** $x=5, x=-5$ **c** $x=11, x=-11$ **d** $x=4, x=-4$ **e** $x=8,\ x=-8$ **f** $x=12, x=-12$ **g** $x=6, x=-6$ **h** $x=9, x=-9$ **i** $x=13, x=-13$ **j** $x=7, x=-7$ **k** $x=10, x=-10$ **l** $x=15, x=-15$ **2 a** $x=-\frac{7}{2}, x=\frac{7}{2}$ **b** $x=-\frac{9}{2}, x=\frac{9}{2}$ **c** $x=-\frac{5}{4}, x=\frac{5}{4}$ **d** $x=-\frac{5}{2}, x=\frac{5}{2}$ **e** $x=-\frac{1}{4}, x=\frac{1}{4}$ **f** $x=-\frac{8}{5}, x=\frac{8}{5}$ **g** $x=-6, x=6$ **h** $x=-3, x=3$ **i** $x=-\frac{5}{6}, x=\frac{5}{6}$ **j** $x=\frac{5}{3}, x=-\frac{5}{3}$ **k** $x=-2, x=2$ **l** $x=-6, x=0$ **m** $x=-\sqrt{7}, x=\sqrt{7}$ **n** $x=-\sqrt{11}, x=\sqrt{11}$ **o** $x=-2\sqrt{2}, x=2\sqrt{2}$ **d** $x=\frac{1}{4}, x=-5$ **e** $x=-\frac{1}{2}, x=\frac{5}{3}$ **f** $x=0, x=1$ **g** $x=3, x=7$ **h** $x=-2, x=\frac{7}{3}$ **i** $x=6, x=8$

Page 23 **1 a** $x=0, x=2$ **b** $x=0, x=3$ **c** $x=0, x=6$ **d** $x=0, x=-5$ **e** $x=0,\ x=-8$ **f** $x=0, x=-3$ **g** $x=0, x=9$ **h** $x=0, x=-12$ **i** $x=0, x=2$ **j** $x=0, x=7$ **k** $x=0, x=2$ **l** $x=0, x=4$ **m** $x=0, x=5$ **n** $x=0, x=-2$ **o** $x=0, x=-2$ **2 a** $x=0, x=-13$ **b** $x=0, x=3$ **c** $x=0, x=9$ **d** $x=0, x=-3$ **e** $x=0, x=4$ **f** $x=0, x=7$ **g** $x=0, x=3$ **h** $x=0, x=-7$ **i** $x=0, x=5$ **j** $x=0, x=-2$ **k** $x=0, x=2.5$ **l** $x=0, x=3$

Page 24 **1 a** $a=-9, a=1$ **b** $y=3, y=7$ **c** $p=3, p=5$ **d** $m=-8, m=3$ **e** $y=6,\ y=7$ **f** $x=-5, x=2$ **g** $a=1, a=10$ **h** $m=-7, m=5$ **i** $y=4, y=8$ **j** $y=-6, y=2$ **k** $m=-4, m=-9$ **l** $y=5, y=8$ **2 a** $x=6, x=-3$ **b** $x=6, x=9$ **c** $m=-9, m=4$ **d** $a=-7, a=4$ **e** $x=3, x=8$ **f** $x=3, x=4$ **g** $x=8, x=9$ **h** $x=8, x=-5$ **i** $x=-7, x=3$ **j** $x=6, x=3$ **k** $x=8, x=11$ **l** $x=-6, x=2$

Page 25 **1 a** $x=1, x=-\frac{3}{2}$ **b** $x=-2, x=\frac{1}{3}$ **c** $m=-4, m=-\frac{1}{2}$ **d** $p=-5, p=\frac{1}{2}$ **e** $a=-1, a=-\frac{1}{2}$ **f** $t=4, t=-\frac{3}{2}$ **g** $x=-5, x=\frac{2}{3}$ **h** $a=-2, a=-\frac{3}{2}$ **i** $x=-\frac{5}{3}, x=1$ **j** $x=-\frac{2}{3}, x=-3$ **k** $x=\frac{1}{5}, x=1$ **l** $x=-\frac{3}{4}, x=-2$ **2 a** $y=-\frac{7}{3}, y=1$ **b** $m=-\frac{5}{2}, m=2$ **c** $p=2, p=\frac{10}{3}$ **d** $y=-\frac{3}{2}, y=7$ **e** $t=-2, t=\frac{4}{3}$ **f** $x=\frac{5}{2}, x=2$ **g** $x=-2, x=-\frac{3}{2}$ **h** $y=3, y=-\frac{3}{2}$ **i** $y=2, y=1$ **j** $x=-\frac{5}{2}, x=\frac{1}{2}$ **k** $x=\frac{7}{2}, x=-6$ **l** $x=-\frac{3}{2}, x=\frac{2}{3}$

Page 26 **1 a** 4 **b** 9 **c** 121 **d** $2x$ **e** 4, 2 **f** $1, 4p+1$ **2 a** 36 **b** 100 **c** 64 **d** 144 **e** 1 **f** $6\frac{1}{4}$ **g** 9 **h** 25 **i** 49 **j** 121 **k** 1 **l** 81 **3 a** $x=2$ or -6 **b** $x=3$ or 9 **c** $x=3\pm\sqrt{13}$ **d** $x=4\pm\sqrt{21}$ **e** $x=5\pm\sqrt{22}$ **f** $x=-7\pm2\sqrt{14}$ **g** $x=-10\pm5\sqrt{5}$ **h** $x=\frac{-5\pm\sqrt{3}}{2}$ **i** $x=\frac{4\pm\sqrt{7}}{3}$ **j** $x=\frac{3\pm2\sqrt{2}}{10}$ **k** $x=\frac{1}{5}$ or $-1\frac{2}{5}$ **l** $x=3.5\pm\sqrt{3}$

Page 27 **1 a** $x=-17$ or -23 **b** $x=-2$ or $\frac{1}{4}$ **c** $x=\frac{4}{3}$ or $-4\frac{1}{2}$ **2 a** $x=\frac{-5\pm\sqrt{33}}{2}$ **b** $x=\frac{-7\pm\sqrt{37}}{2}$ **c** $x=\frac{3\pm\sqrt{41}}{4}$ **d** $x=\frac{1\pm\sqrt{5}}{2}$ **e** $x=\frac{7\pm\sqrt{37}}{6}$ **f** $x=\frac{11\pm\sqrt{21}}{10}$ **g** $x=-2\pm\sqrt{11}$ **h** $x=5\pm2\sqrt{11}$ **i** $x=\frac{-1\pm\sqrt{22}}{3}$

Page 28 **1 a** $x=\frac{-3\pm\sqrt{19}}{2}$ **b** $x=\frac{5\pm\sqrt{19}}{6}$ **c** $x=\frac{-9\pm3\sqrt{41}}{8}$ **2 a** $x=0.451$ or 2.215 **b** $x=-0.149$ or -3.351 **c** $x=10.525$ or 0.475 **d** $x=0.312$ or -4.812 **e** $x=2.120$ or -0.786 **f** $x=1.156$ or -1.556 **g** $x=0.518$ or -13.518 **h** $x=0.803$ or -2.803 **i** $x=2.573$ or -0.907

Page 29 **1 a** $m=8, m=-3$ **b** $x=\frac{2}{3}, x=-5$ **c** $x=4, x=9$ **d** $x=5, x=-1$ **e** $y=0, y=7$ **f** $n=0, n=9$ **g** $a=\frac{-3\pm\sqrt{29}}{10}$ **h** $x=\frac{7\pm\sqrt{61}}{2}$ **i** $x=\frac{-7\pm\sqrt{61}}{6}$ **j** $x=1, x=-8$ **k** $x=2, x=-4$ **l** $x=1\pm\sqrt{2}$ **2 a** $x=0.73, x=-0.23$ **b** $x=8, x=-8$ **c** $x=5.85, x=-0.85$ **d** $x=\frac{1}{2}, x=1\frac{1}{2}$ **e** $x=8.47, x=-0.47$ **f** $x=3, x=-3$ **g** $x=2, x=-10$ **h** $x=0.29, x=5.21$ **i** $t=2\frac{1}{2}, t=-1$

Page 30 **1 a** $x=6$ or $x=8$ **b** $x=6$ **2** (No real square root of a negative number) **3 a** 0, 1, 2 **b** 2, 0, 1 **4 a** 2 **b** 0 **c** 1 **d** 2 **e** 0 **f** 0 **g** 1 **h** 2 **i** 0 **j** 2 **k** 2 **l** 2

Page 31 **1 a** $t=-2$ or 8 **b** $t=8$, time cannot be negative **2 a** $x=4$ or -16 **b** If the shorter side is x, the longer side is $(x+12)$ and if the area is 64 then $x(x+12)=64$. **c** The length of a side cannot be negative **d** 16 cm long and 4 cm wide **3 a** $100\times(36-2x)\times x=16000$ **b** 100 cm by 16 cm by 10 cm or 100 cm by 20 cm by 8 cm **4** 9 cm, 12 cm and 15 cm or 4.5 cm. 6 cm and 7.5 cm

Page 32 **1 a** $x=6$ **b** $x=12$ **2 a** $x=10$ **b** 8 and 9 **3 a** 8 cm, 20 cm **b** 11 and 13

Page 33 **1 a** $x=1, y=2$ or $x=2, y=1$ **b** $x=1, y=4$ or $x=4, y=1$ **c** $x=1, y=11$ or $x=-3, y=7$ **d** $x=-1, y=-2$ or $x=-13, y=-14$ **e** $x=9, y=-2$ or $x=-2, y=9$ **f** $x=1, y=3$ or $x=3, y=1$ **g** $x=0, y=0$ or $x=1, y=1$ **h** $x=-1, y=2$ or $x=3, y=6$ **i** $x=-5, y=24$ or $x=3, y=0$ **j** $x=1, y=6$ or $x=2, y=4$ **k** $x=0, y=0$ or $x=2, y=4$ **l** $x=1, y=1$

Page 34 **1 a** $x=\pm1$ or ±3 **b** $x=\pm2$ or ±5 **c** $x=\pm2$ or ±3 **2 a** $x=\pm2$ **b** $x=\pm\sqrt{3}$ or $\pm\sqrt{5}$ **c** $x=\pm\sqrt{2}$ or ±3 **3 a** $x=1$ or -2 **b** $x=2$ **c** $x=\pm1$ or -3 **d** $x=1, 2, 3$ or 4

Page 35 **1** D **2** C **3** C **4** B **5** C **6** A **7** A **8** A **9** C **10** C

Page 36 **1 a** $(x-9)(x+5)$ **b** $3(x-7)(x-3)$ **c** $(2x+5)(3x-2)$ **2 a** $x=\frac{1}{2}$ **b** $x=4$ or -4 **c** $x=0$ or 15

Answers

3 a $x = 2$ or -2 **b** $x = 3$ or 9 **c** $x = -5$ or $\frac{1}{2}$ **4 a** $x = -6$ or 2 **b** $x = 1 \pm \sqrt{6}$ **c** $x = 0.908$ or -4.408 **5 a** $x = -1$ or -7 **b** -6 or 5 **6** $x = 7, y = -3$ or $x = -3, y = 7$

Chapter 3 – Non-linear relationships

Page 37

1

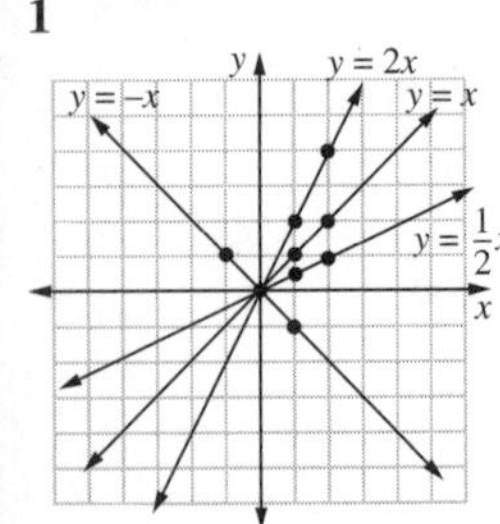

2

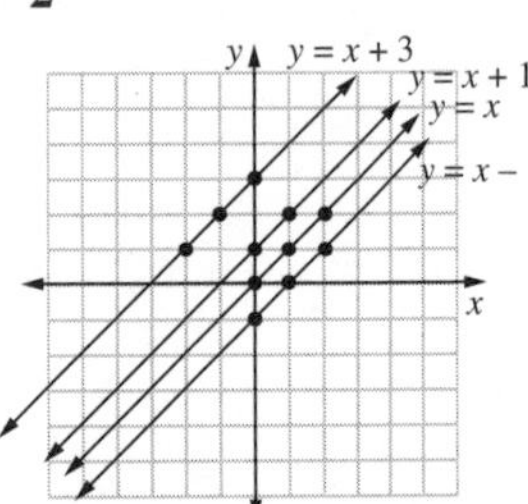

3

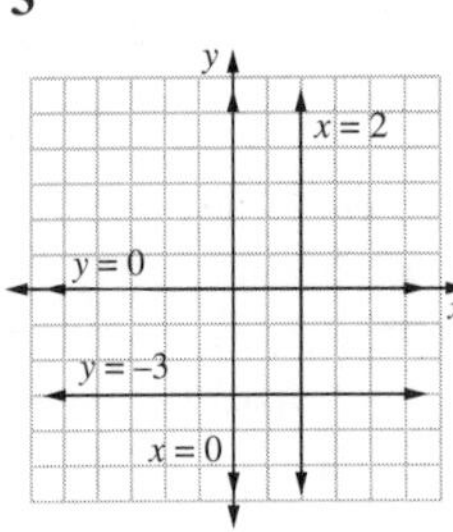

4

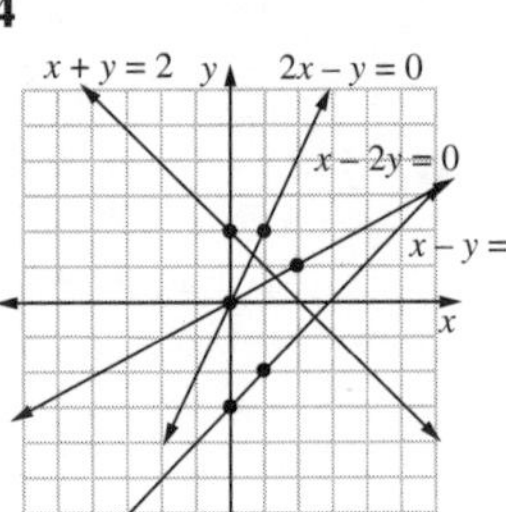

5

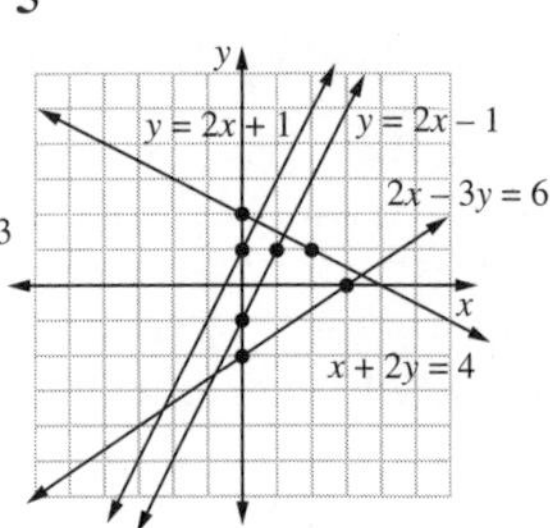

Page 38 **1 a** $-\frac{2}{3}$ **b** $\frac{3}{2}$ **2 a** $(1, -3)$ **b** $\frac{1}{2}$ **c** $4\sqrt{5}$ units **3 a** $\frac{3}{4}$ **b** -2 **c** $3x - 4y - 8 = 0$ **4 a** $y = 2x + \frac{3}{2}$ **b** 2 **c** $\frac{3}{2}$ **5 a** $y = -x + 5$ **b** $y = \frac{1}{2}x + 4$

Page 39 **1 a**

x	-3	-2	-1	0	1	2	3
$y = x^2$	9	4	1	0	1	4	9
$y = 3x^2$	27	12	3	0	3	12	27
$y = \frac{1}{3}x^2$	3	$\frac{4}{3}$	$\frac{1}{3}$	0	$\frac{1}{3}$	$\frac{4}{3}$	3

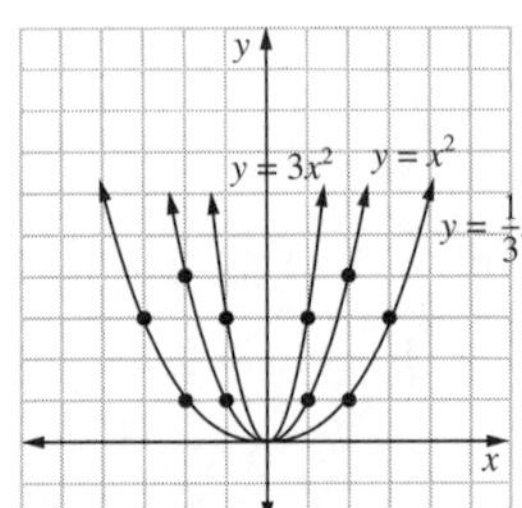

b the higher the (positive) coefficient the narrower the graph.

2 a

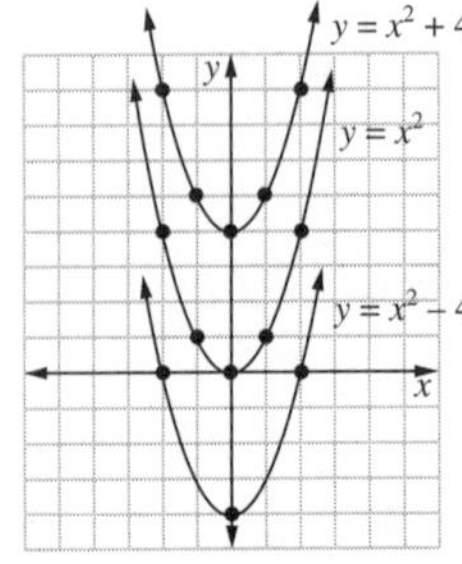

b All points that lie on curves above and below $y = x^2$ are 4 units along y-axis above or below those lying on $y = x^2$.

3 a

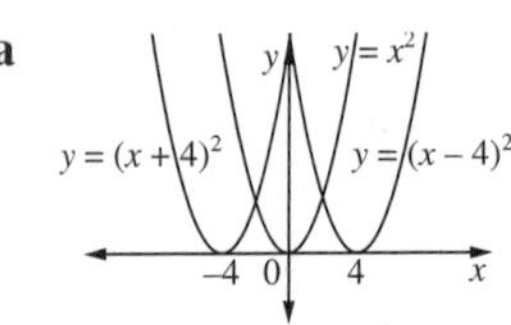

b The graph has its vertex at $(p, 0)$

4

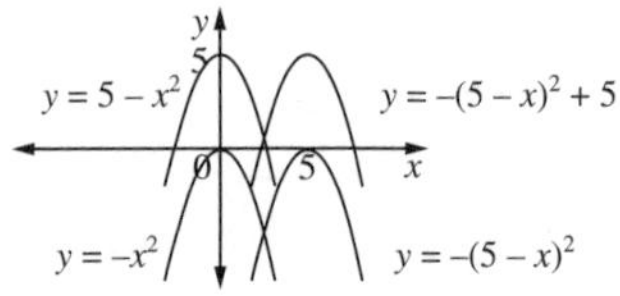

Page 40 **1 a** $x = 2$ and $x = -6$ **b** $x = 3$ and $x = 9$ **c** $x = 1\frac{1}{2}$ and $x = 5$ **2 a** $x = -2$ **b** $x = 6$ **c** $x = 3\frac{1}{4}$

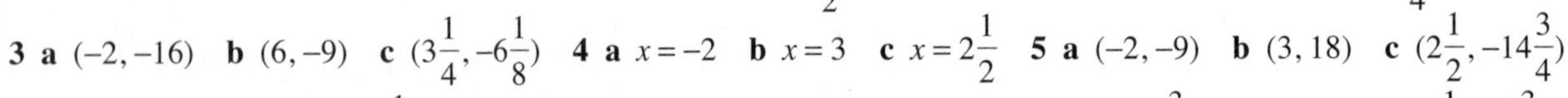

3 a $(-2, -16)$ **b** $(6, -9)$ **c** $(3\frac{1}{4}, -6\frac{1}{8})$ **4 a** $x = -2$ **b** $x = 3$ **c** $x = 2\frac{1}{2}$ **5 a** $(-2, -9)$ **b** $(3, 18)$ **c** $(2\frac{1}{2}, -14\frac{3}{4})$

6 a $(1, 5)$ **b** $(-3, -2)$ **c** $(\frac{1}{2}, -3)$ **7 a** $y = (x - 2)^2 + 7, (2, 7)$ **b** $y = (2x + 3)^2 - 4, (-\frac{3}{2}, -4)$ **c** $y = (x - 2\frac{1}{2})^2 + \frac{3}{4}, (2\frac{1}{2}, \frac{3}{4})$

Page 41

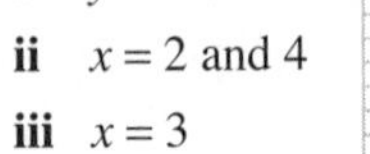

1 a i $y = 8$

ii $x = 2$ and 4

iii $x = 3$

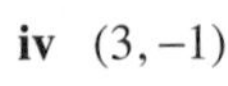

iv $(3, -1)$

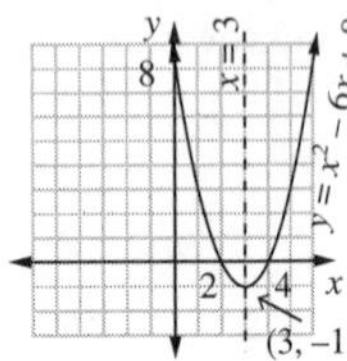

b i $y = 5$

ii $x = 1$ and 5

iii $x = 3$

iv $(3, -4)$

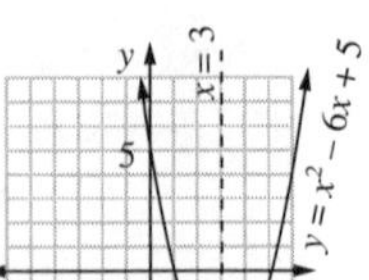

c i $y = -3$

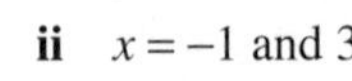

ii $x = -1$ and 3

iii $x = 1$

iv $(1, -4)$

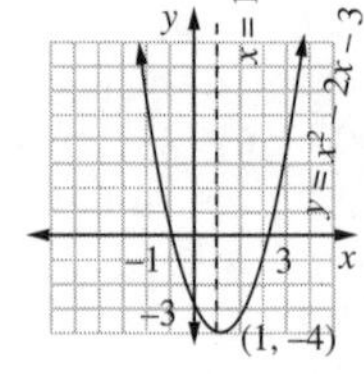

d i $y = -8$ **ii** $x = -2$ and 4 **iii** $x = 1$ **iv** $(1, -9)$

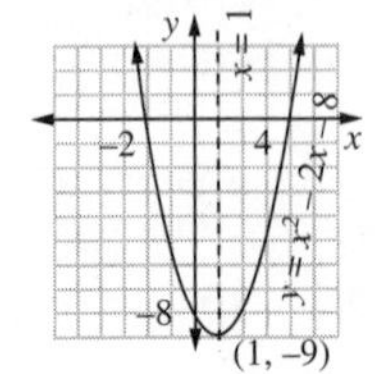

2 a

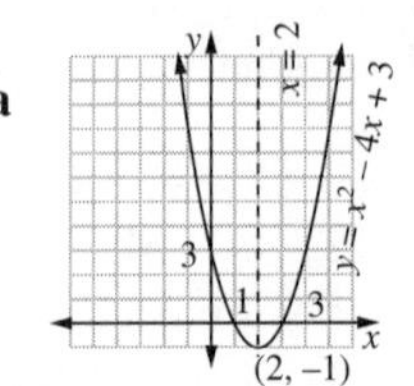

b

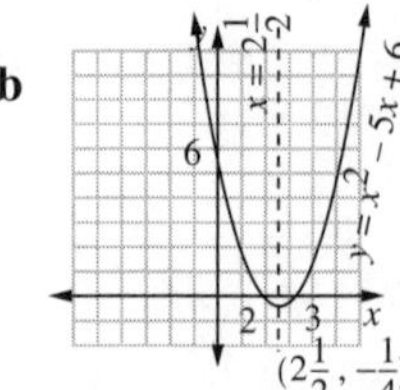

Answers

PAGE 42 1 **a** concave up **b** concave up **c** concave down **d** concave down **e** concave down **f** concave up

2 **a** **b** **c** **d** **e** **f**

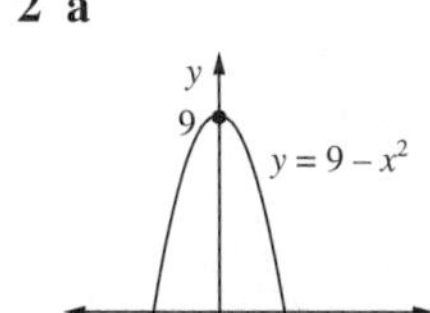

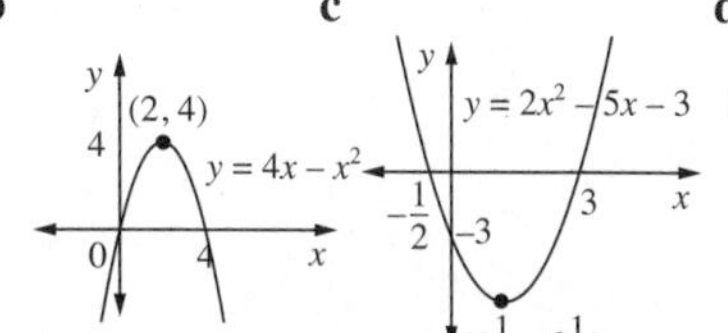

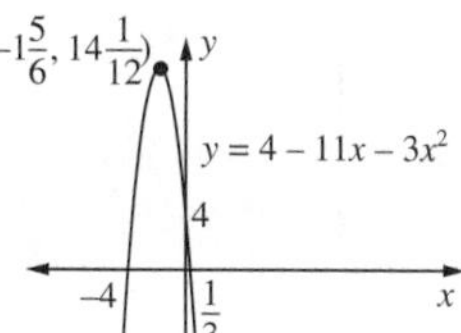

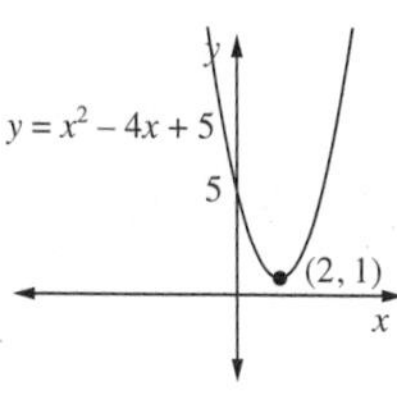

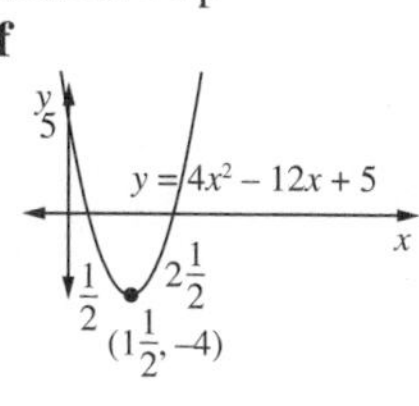

PAGE 43 1 **a** −2 **b** $\frac{1}{2}$ 2 **a** $y = x^2 - x - 2$ **b** $y = 3x^2 - 4x - 7$ **c** $y = 7 + 6x - x^2$ **d** $y = 2x^2 - 4x$

3 **a** $y = x^2 + 3$ **b** $y = x^2 - x$

PAGE 44 1 **a**

x	−6	−5	−4	−3	−2	−1	$-\frac{1}{2}$	$-\frac{1}{3}$	$-\frac{1}{4}$	$-\frac{1}{5}$	$-\frac{1}{6}$	0	$\frac{1}{6}$	$\frac{1}{5}$	$\frac{1}{4}$	$\frac{1}{3}$	$\frac{1}{2}$	1	2	3	4	5	6
y	$-\frac{1}{6}$	$-\frac{1}{5}$	$-\frac{1}{4}$	$-\frac{1}{3}$	$-\frac{1}{2}$	−1	−2	−3	−4	−5	−6	✖	6	5	4	3	2	1	$\frac{1}{2}$	$\frac{1}{3}$	$\frac{1}{4}$	$\frac{1}{5}$	$\frac{1}{6}$

b Division by zero is undefined **c** There will be a gap. **d**

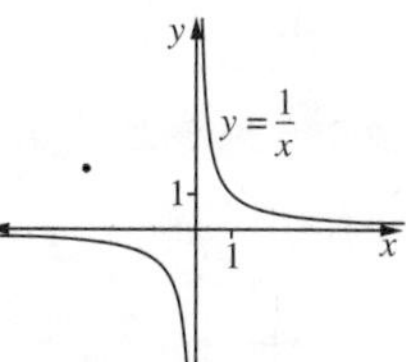

e They get closer to 0.

f They get larger and larger. **g** $x = 0, y = 0$

2 **a**

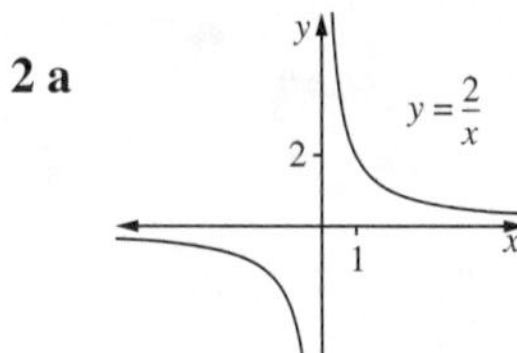

b

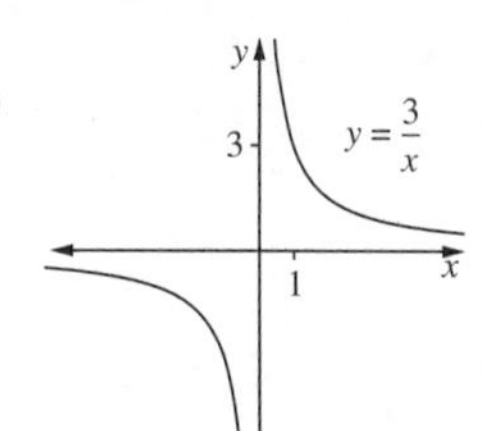

c

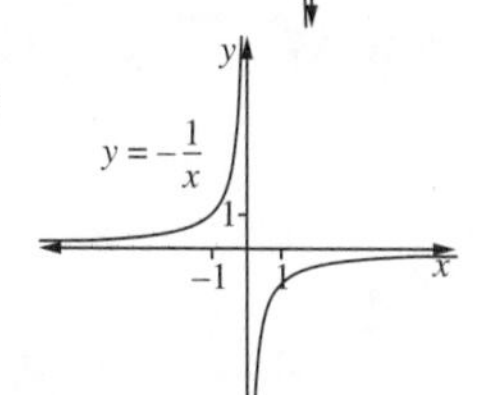

d

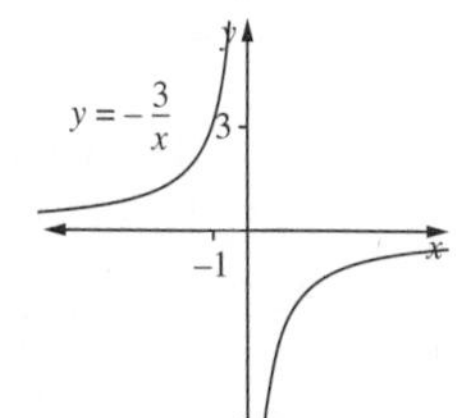

e

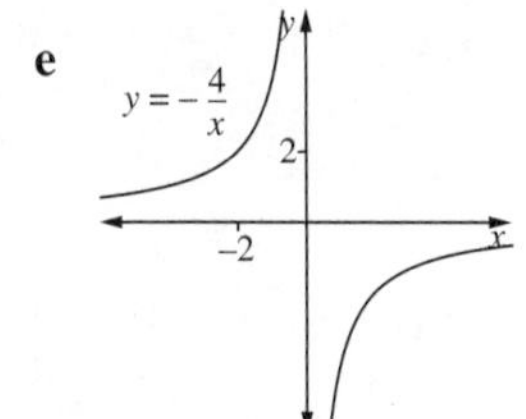

f

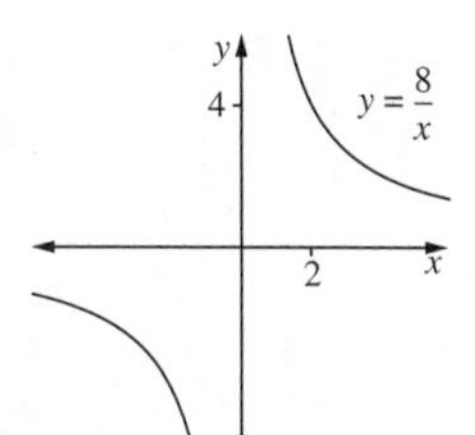

3 **a** The curve moves further from the asymptotes as the constant gets larger. **b** The curve is found in the opposite quadrants.

PAGE 45 1 **a** **i**

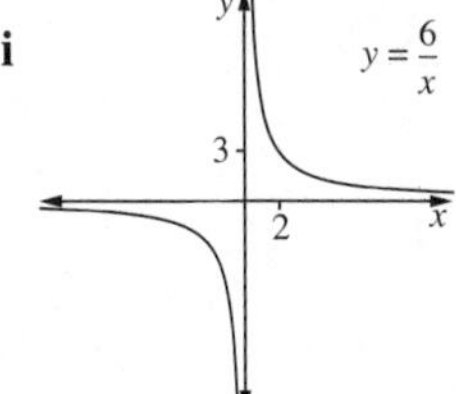

ii

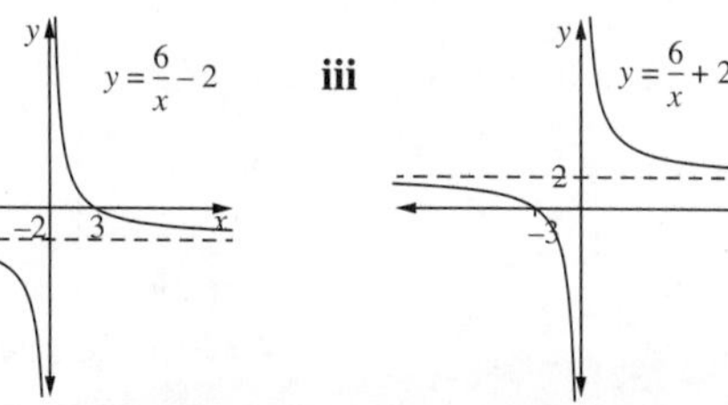

iii

b **i** $x = 0, y = 0$ **ii** $x = 0, y = -2$ **iii** $x = 0, y = 2$ **c** It moves the hyperbola, and horizontal asymptote, up or down.

2 **a** **i**

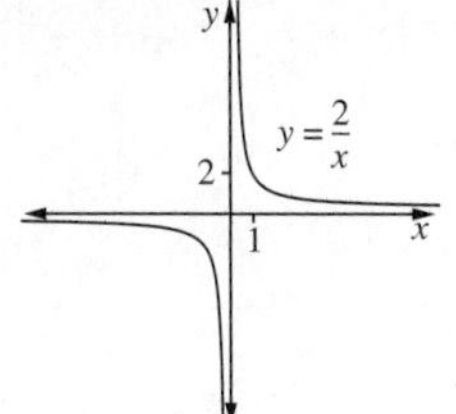

ii

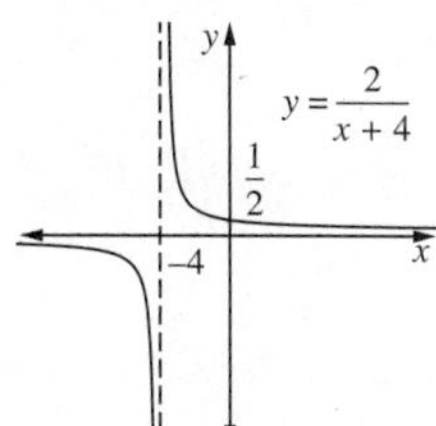

iii

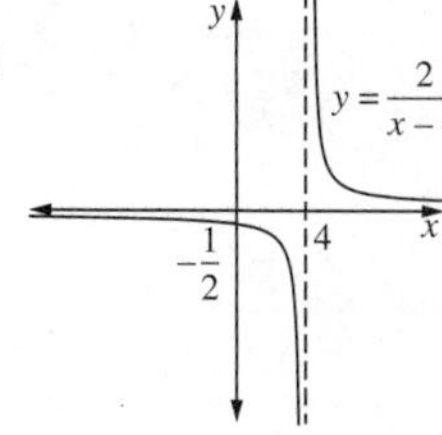

b **i** $x = 0, y = 0$ **ii** $x = -4, y = 0$ **iii** $x = 4, y = 0$ **c** It moves the curve, and the vertical asymptote right or left.

Answers

PAGE 46 **1 a** $x = 1, y = 0$ **b** no x-intercept, y-intercept $= -5$ **c**

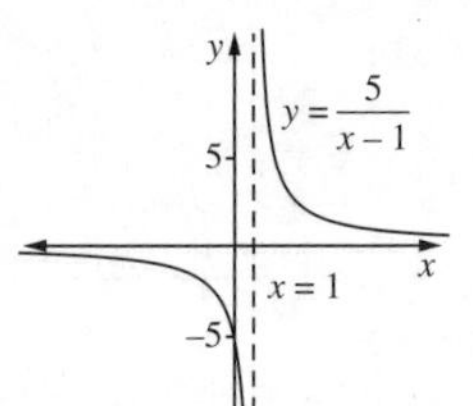

2 a $x = 0, y = 3$
b x-intercept $= \frac{1}{3}$, no y-intercept

c

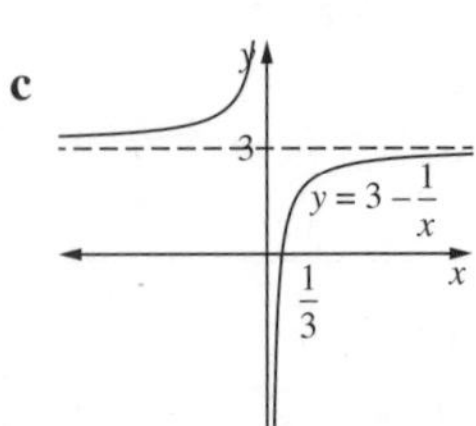

3 a **b**

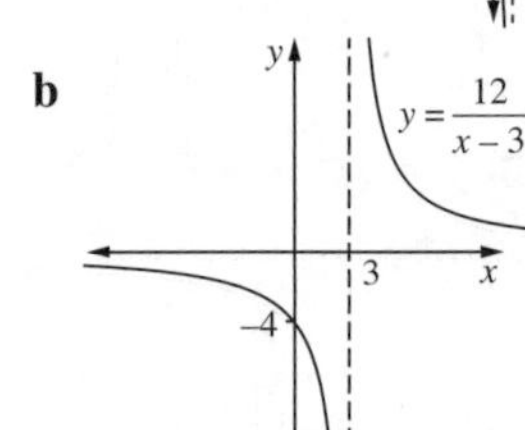

c

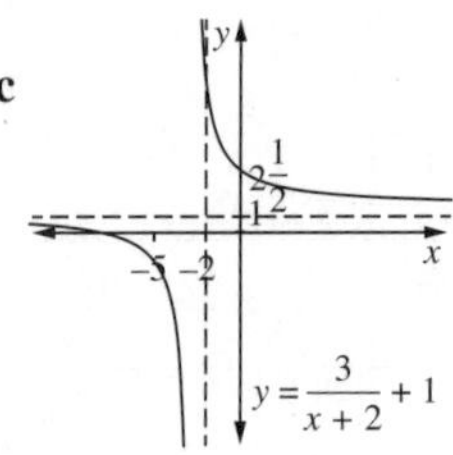

4 a The lengths cannot be negative. **b** 20 cm **c** 40 cm **d** 2 m **e** 98 m

PAGE 47 **1 a** $(0, 0)$, 20 units **b** $(0, 0)$, 11 units **c** $(0, 0)$, $\frac{2}{3}$ units **d** $(0, 0)$, 9 units **2 a** $x^2 + y^2 = 144$ **b** $x^2 + y^2 = 49$ **c** $x^2 + y^2 = 6\frac{1}{4}$ **d** $x^2 + y^2 = 100$ **3 a** $x^2 + y^2 = 4$ **b** $x^2 + y^2 = 25$ **c** $x^2 + y^2 = 64$

4 a $r = 4$, centre $(0, 0)$ **b** $r = 1$, centre $(0, 0)$ **c** $r = 3$, centre $(0, 0)$ **d** $r = 6$, centre $(0, 0)$

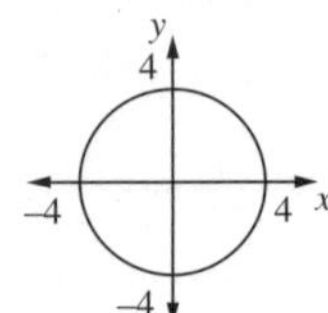

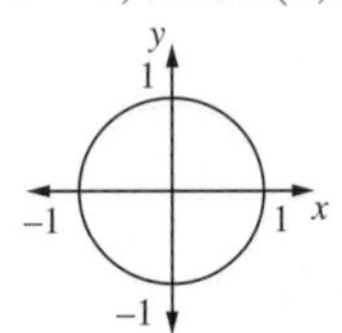

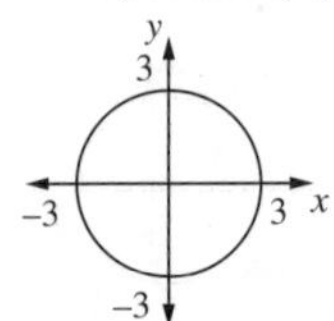

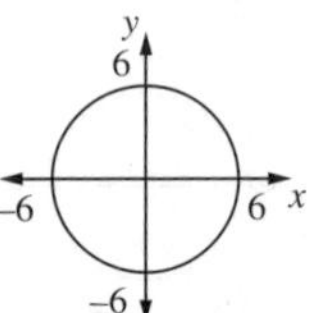

PAGE 48 **1 a** $(2, 3)$, 4 units **b** $(4, -1)$, 5 units **c** $(-3, -2)$, 7 units **d** $(0, 4)$, 1 unit **e** $(-5, 0)$, 9 units **f** $(-8, -1)$, 8 units **2 a** $(x-3)^2 + (y-5)^2 = 4$ **b** $(x-2)^2 + (y-6)^2 = 9$ **c** $(x+3)^2 + (y-2)^2 = 100$ **d** $(x-4)^2 + (y+1)^2 = 36$ **e** $(x+7)^2 + (y+2)^2 = 144$ **f** $(x-5)^2 + y^2 = 225$ **3 a** $(x-3)^2 + (y-3)^2 = 4$ **b** $(x+2)^2 + (y-1)^2 = 9$ **c** $(x+3)^2 + (y+2)^2 = 1$ **d** $(x-2)^2 + (y+1)^2 = 16$ **e** $(x-2)^2 + y^2 = 25$ **f** $(x-3)^2 + (y+1)^2 = 6\frac{1}{4}$

4 a **b**

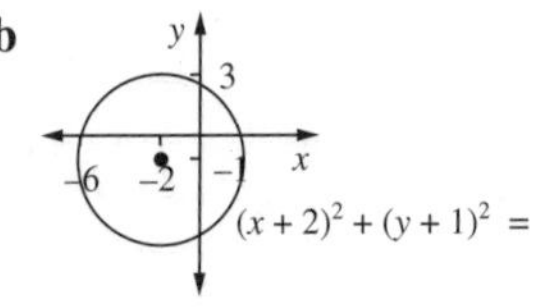

c

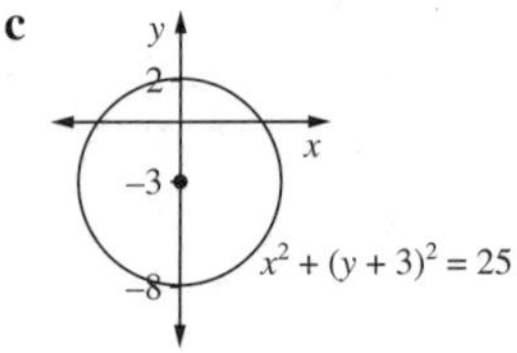

PAGE 49 **1 a** $(x-3)^2 + (y-2)^2 = 25$ **b** $(x+3)^2 + (y-1)^2 = 100$ **c** $(x+2)^2 + (y+1)^2 = 45$ **d** $(x-5)^2 + (y+2)^2 = 13$
2 a inside **b** outside **c** inside **d** on
3 a $(4, -4)$, 6 units **b** $(0, -6)$, 8 units **c** $(-3, 1)$, 4 units **d** $(2, 5)$, 3 units

PAGE 50 **1 a** $(0, -3)$, $(4, 5)$ **b** $(-3, -4)$, $(6, 2)$

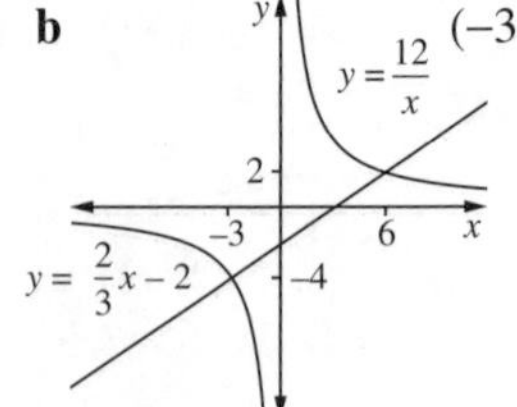

c $(-1, 5)$, $(6, 4)$

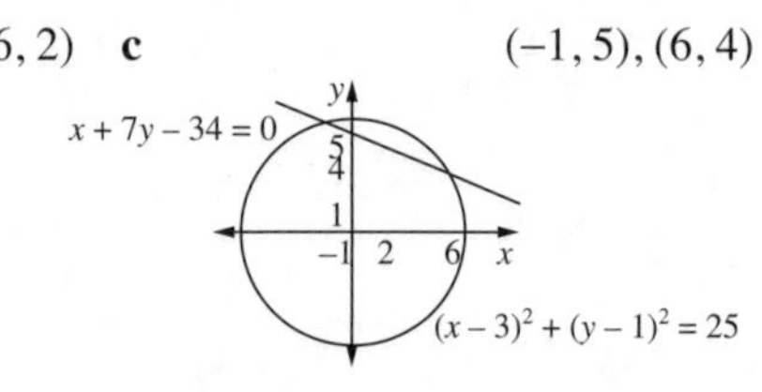

2 a$(-1, -16)$ and $(6, 5)$ **b** $(-6, 2)$ and $(-1, 1)$ **c** $(3, 3)$ and $(5, 1)$

PAGE 51 **1 a** **b**

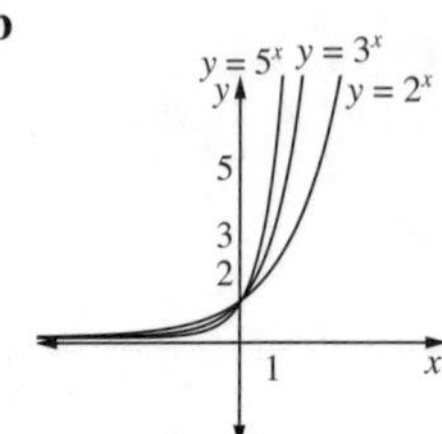

c

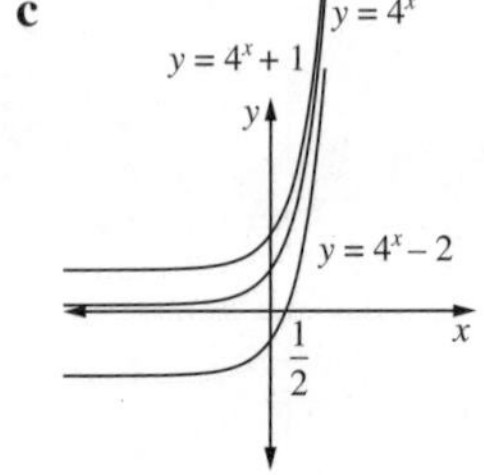

d

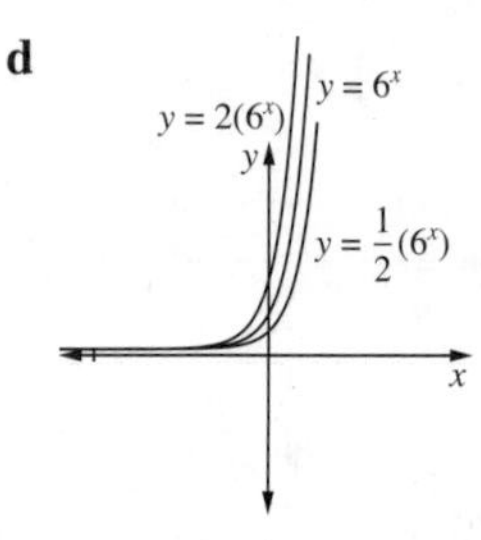

Answers

2

x	−1	0	1	2	3
3^x	$\frac{1}{3}$	1	3	9	27
3^{-x}	3	1	$\frac{1}{3}$	$\frac{1}{9}$	$\frac{1}{27}$
$\frac{3^x + 3^{-x}}{2}$	1.7	1	1.7	4.6	13.5

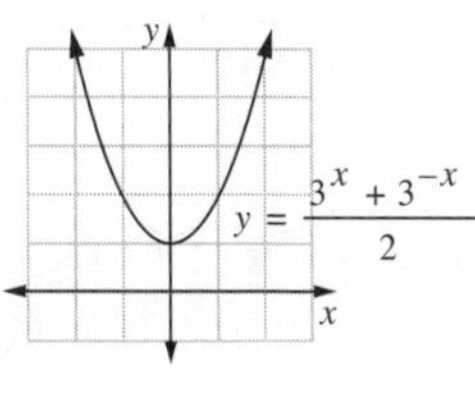

Page 52 **1 a**

x	−2	$-1\frac{1}{2}$	−1	$-\frac{1}{2}$	0	$\frac{1}{2}$	1	$1\frac{1}{2}$	2
y	−8	$-3\frac{3}{8}$	−1	$-\frac{1}{8}$	0	$\frac{1}{8}$	1	$3\frac{3}{8}$	8

b

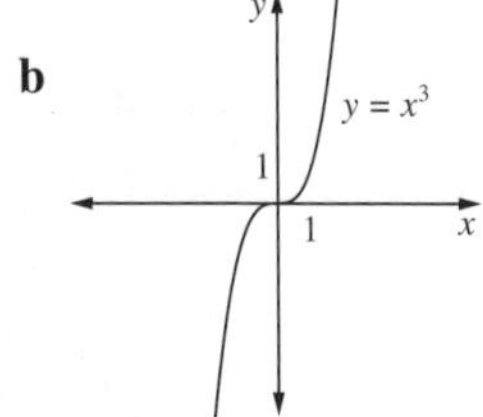

2 a

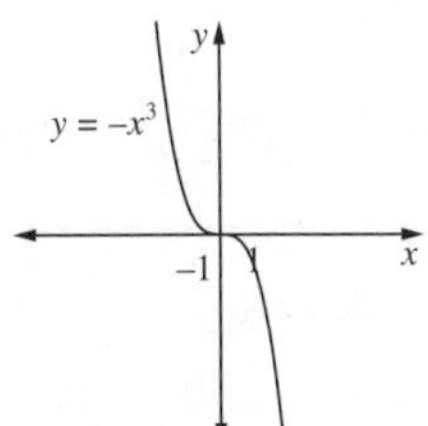

b

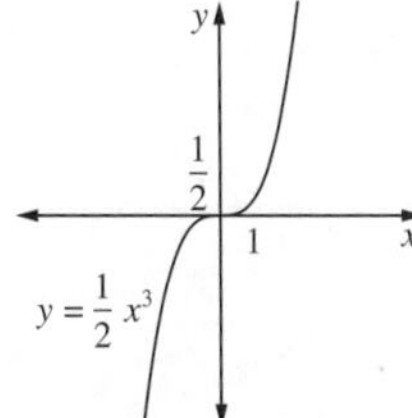

c

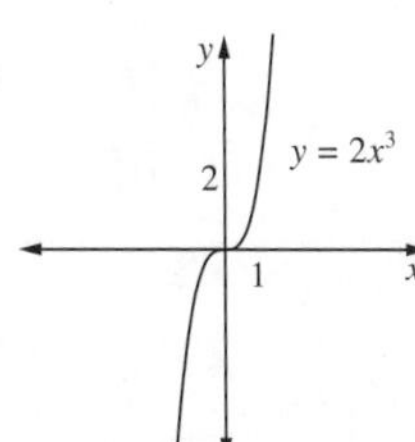

3 A negative constant reflects the graph in the x-axis. The magnitude of the constant determines the width of the curve.

4 a

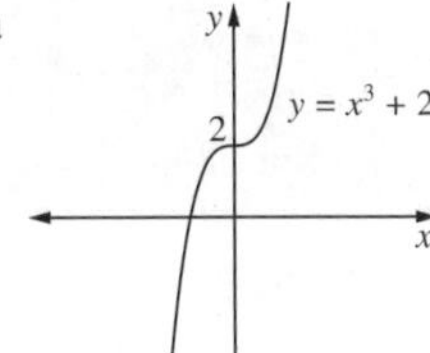

b

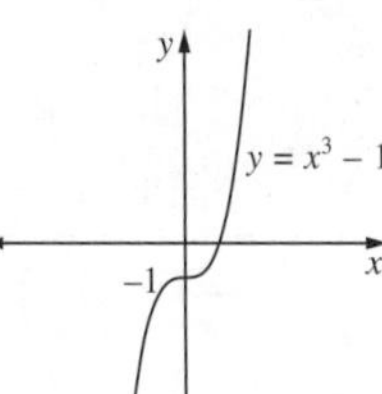

c

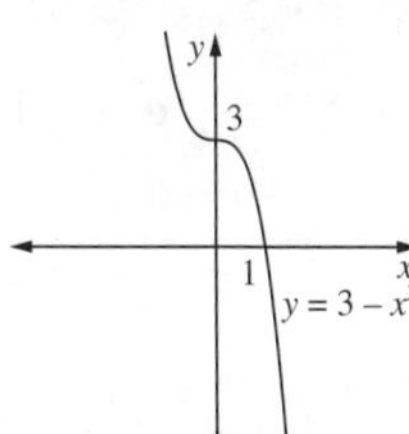

5 It moves the curve up or down the y-axis.

Page 53 **1 a**

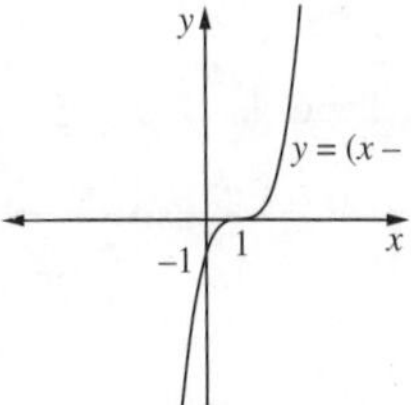

b

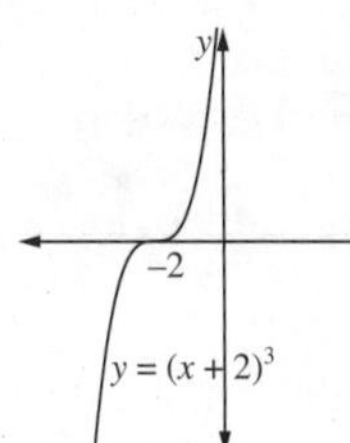

c

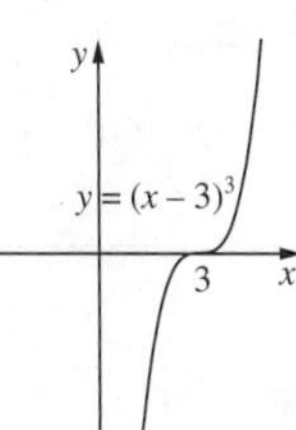

2 The curve moves to the right or left. The x-intercept is r.
3 a 6 **b** It gets very large.
c It gets very large negatively.
d $x = 1, -2$ or 3

e

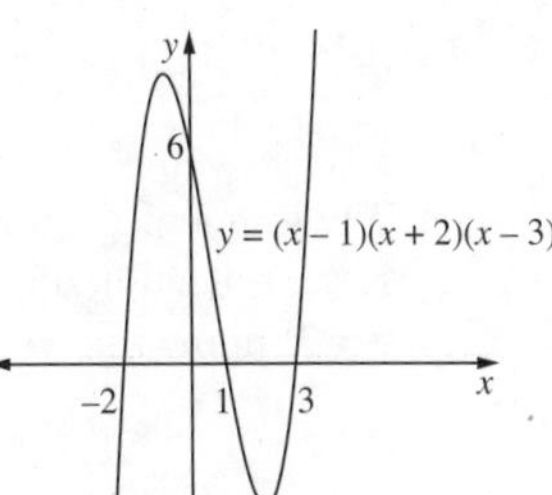

4 a

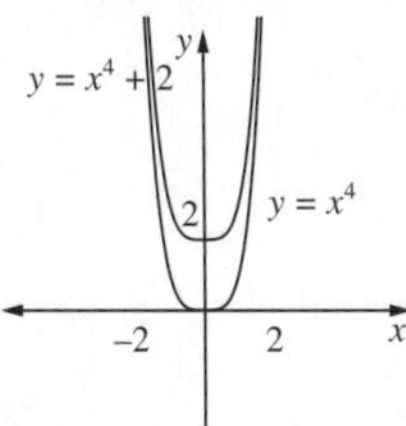

b

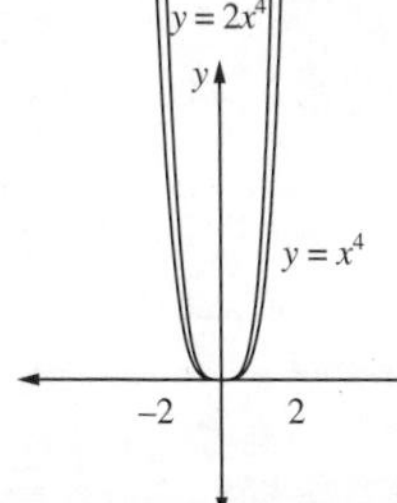

c

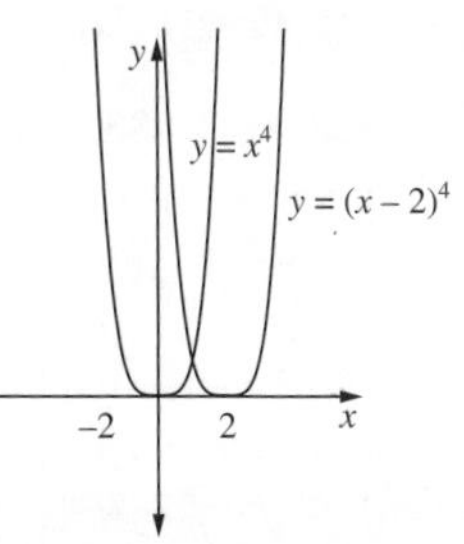

5 a true **b** false **c** true **d** false **e** true **f** false **g** false **h** true

Page 54 **1 a** straight line **b** hyperbola **c** straight line **d** parabola **e** parabola **f** exponential **g** parabola **h** hyperbola **i** none of these **j** circle **k** exponential **l** circle **2 a** D **b** H **c** F **d** G **e** I **f** C **g** A **h** B **i** E **j** L **k** J **l** K

3 a

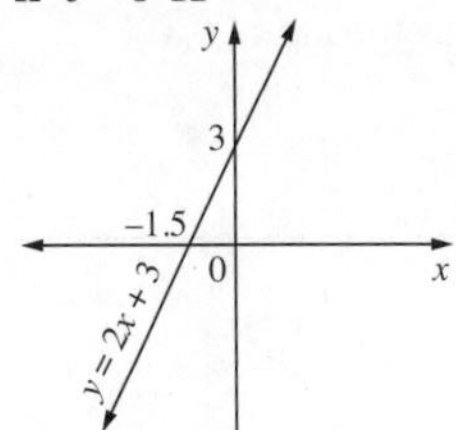

b

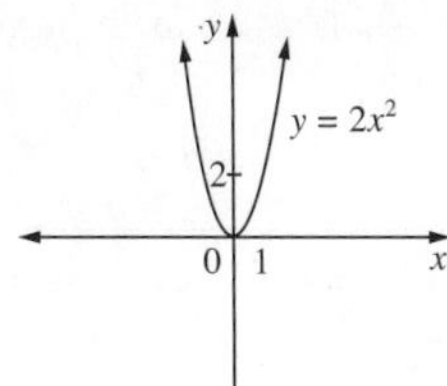

c

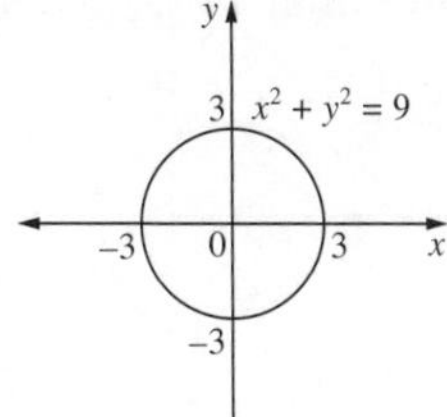

Answers

PAGE 55 **1 a** parabola **b** straight line **c** parabola **d** circle **e** exponential curve **f** hyperbola **g** cubic curve **h** circle **i** hyperbola **j** straight line **k** cubic curve **l** parabola **2 a** F **b** C **c** D **d** A **e** I **f** B **g** H **h** G **i** E **3 a** $(x-3)^2+(y+1)^2=16$ **b** $y=x^2-6x+5$ **c** $y=\dfrac{12}{x-3}+1$

PAGE 56 **1** C **2** A **3** C **4** A **5** D **6** B **7** B **8** A **9** B **10** A

PAGE 57 **1 a** 7 **b** 1, 7 **c** $x=4$ **d** $(4,-9)$ **e**

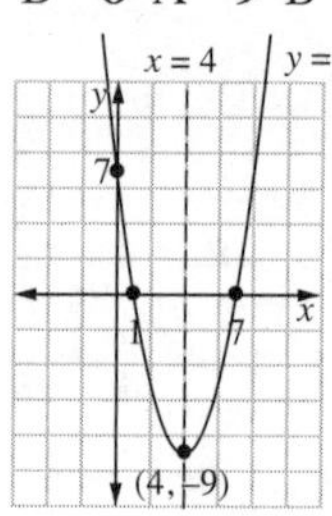

2 a $(3,-2)$
b 7 units
c $(3,-9)$ and $(-4,-2)$
3 a 1 **b** 3 **c** $x=-3, y=-1$

d

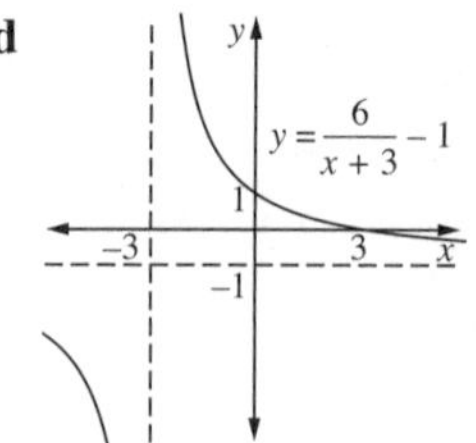

4 a

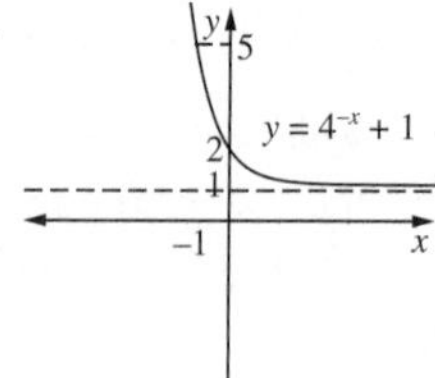

b

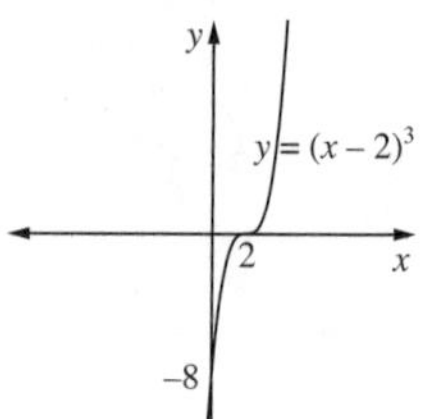

CHAPTER 4 – Logarithms, exponentials and functions

PAGE 58 **1 a** $2=\log_3 9$ **b** $3=\log_4 64$ **c** $3=\log_5 125$ **d** $5=\log_2 32$ **e** $4=\log_3 81$ **f** $-3=\log_2\left(\frac{1}{8}\right)$ **g** $6=\log_2 64$ **h** $3=\log_7 343$ **i** $-2=\log_3\left(\frac{1}{9}\right)$ **j** $-1=\log_3\left(\frac{1}{3}\right)$ **k** $-\frac{1}{2}=\log_{25}\left(\frac{1}{5}\right)$ **l** $\frac{2}{3}=\log_8 4$ **2 a** $2^1=2$ **b** $3^3=27$ **c** $5^{\frac{1}{2}}=\sqrt{5}$ **d** $3^2=9$ **e** $8^{\frac{1}{3}}=2$ **f** $27^{\frac{2}{3}}=9$ **g** $2^5=32$ **h** $2^4=16$ **i** $3^0=1$ **j** $4^3=64$ **k** $2^{\frac{1}{2}}=\sqrt{2}$ **l** $2^7=128$ **3 a** 2 **b** 3 **c** 4 **d** 2 **e** 6 **f** 2 **g** 2 **h** 3 **i** 4 **j** 3 **k** 3 **l** 3 **4 a** 8 **b** 81 **c** 25 **d** $\frac{1}{6}$ **e** 3 **f** 3 **g** 1 **h** $\frac{1}{2}$ **i** 3 **j** 8 **k** 3 **l** 5

PAGE 59 **1 a** xy **b** $\frac{x}{y}$ **c** n **d** x **e** 1 **f** 0 **g** – (negative) **2 a** 0 **b** 0 **c** 0 **d** 1 **e** 1 **f** 1 **g** 3 **h** 4 **i** 5 **j** 8 **k** 25 **l** 81 **m** 2 **n** 3 **o** 2 **3 a** $\log_a 15$ **b** $\log_n 4$ **c** $\log_m 24$ **d** $\log_x 3$ **e** $\log_e 50$ **f** $\log_b 162$ **4 a** 2 **b** 1 **c** 2 **d** $\frac{5}{3}$ **e** 3 **f** 0 **g** 2 **h** 4

PAGE 60 **1 a** 7 **b** 2 **c** 2 **d** 1 **e** 1 **f** 3 **2 a** $\log_a x+\log_a y$ **b** $\log_a x+2\log_a y-\log_a z$ **c** $\log_a 2+\log_a x-\log_a(x-1)$ **d** $3\log_a x+\frac{1}{2}\log_a y$ **3 a** $\log_a\left(\frac{xy}{z^2}\right)$ **b** $\log_a\left(\frac{x^3}{y^2}\right)$ **c** $\log_a\left(\sqrt{x}y^2\right)$ **d** $\log_a\left(\frac{x^2\sqrt{z}}{y^3}\right)$ **4 a** 2.914 **b** 3.4638 **c** 1.8831 **d** 3.7662 **e** 2.6116 **f** –0.7285 **g** 0.5773 **h** 2.1546

PAGE 61 **1 a** –3, –2, –1, 0, 1, 2, 3 **b**

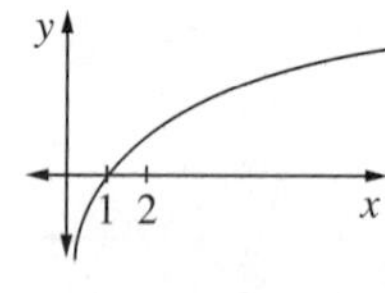

2 a and b

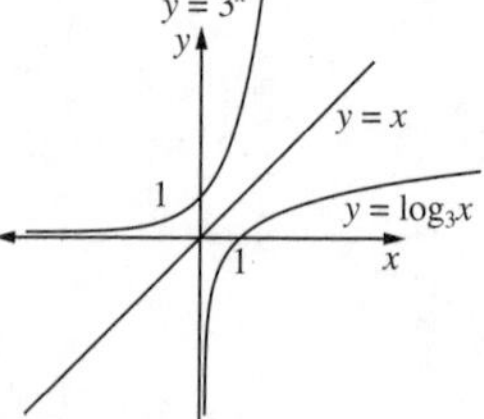

c The graph of $y=\log_3 x$ is the reflection of $y=3^x$ in the line $y=x$.

3 a

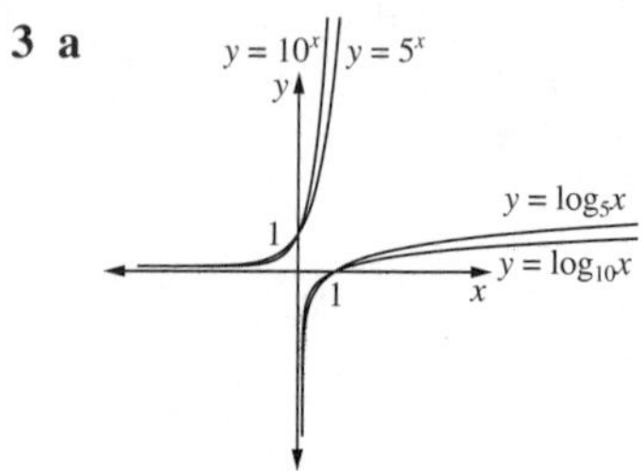

b The exponential graphs have a similar shape. They both pass through (0, 1) and have the x-axis as an asymptote. The log graphs are reflections of the exponential graphs in the line $y=x$. Both log graphs have the y-axis as an asymptote and pass through (1, 0).

Answers

PAGE 62 **1 a** $x=5$ **b** $x=6$ **c** $x=9$ **d** $x=2$ **e** $x=4$ **f** $x=5$ **g** $x=4$ **h** $x=4$ **i** $x=3$ **j** $x=4$ **k** $x=4$ **l** $x=3$ **2 a** $x=-2$ **b** $x=-3$ **c** $x=-4$ **d** $x=\frac{1}{2}$ **e** $x=\frac{1}{2}$ **f** $x=\frac{1}{7}$ **g** $x=-\frac{1}{3}$ **h** $x=-2$ **i** $x=-\frac{3}{4}$ **j** $x=\frac{1}{2}$ **k** $x=-\frac{1}{2}$ **l** $x=-1\frac{1}{2}$ **m** $x=-\frac{1}{2}$ **n** $x=\frac{2}{3}$ **o** $x=\frac{3}{4}$ **p** $x=1\frac{1}{2}$ **q** $x=\frac{3}{2}$ **r** $x=\frac{1}{4}$ **s** $x=\frac{3}{2}$ **t** $x=-\frac{1}{4}$ **3 a** 2.8 **b** 1.5 **c** 0.6 **d** −3.7

PAGE 63 **1 a** $x=512$ **b** $x=15$ **c** $x=1$ **d** $a=\frac{1}{5}$ **e** $x=2$ **f** $m=\frac{1}{81}$ **g** $x=4$ **h** $x=6$ **i** $x=5$ **j** $x=\frac{1}{3}$ **k** $n=\frac{1}{2}$ **l** $x=\frac{1}{2}$ **m** $a=-1$ **n** $x=1\frac{1}{2}$ **o** $x=\frac{1}{4}$ **2 a** $x=3$ **b** $x=4$ **c** $x=-6$ **d** $x=6$ **e** $x=-3$ **f** $x=1$ **g** $x=-3$ **h** $x=-4$ **i** $x=-3\frac{1}{2}$ **j** $x=1\frac{5}{6}$ **k** $x=-2$ **l** $x=5$ **m** $x=3$ **n** $x=\frac{3}{4}$ **o** $x=1\frac{2}{7}$ **p** $n=-\frac{1}{4}$ **q** $m=-\frac{1}{2}$ **r** $a=-\frac{11}{12}$

PAGE 64 **1 a** function **b** function **c** not function **d** function **2 a** 3 **b** 11 **c** 1 **d** 5 **e** 7 **f** −1 **3 a** 6 **b** 9 **c** 21 **d** $5\frac{1}{2}$ **e** $5\frac{1}{4}$ **f** 13 **4 a** 4 **b** 7 **5 a** $5-9a$ **b** $5-\frac{3}{2a}$ **6 a** $(-a)^2=a^2$ **b** $(-a)^3=-a^3$ **7** A, B

PAGE 65 **1 a** all real x, all real y **b** all real x, $y \geq 2$ **c** all real x except $x=0$, all real y except $y=0$ **d** all real x, $y \leq 9$ **e** all real x, all real y **f** $x>0$, all real y **2 a** all real x **b** all real y **c** $y=\frac{1}{2}x-2$ **d**

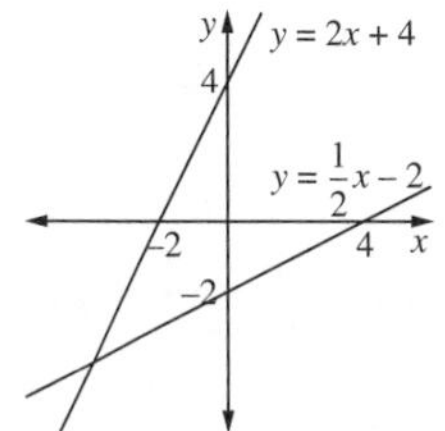

3 a Because for every value of y there must be one, and only one, value of x. **b** $y=\sqrt{x}$

c

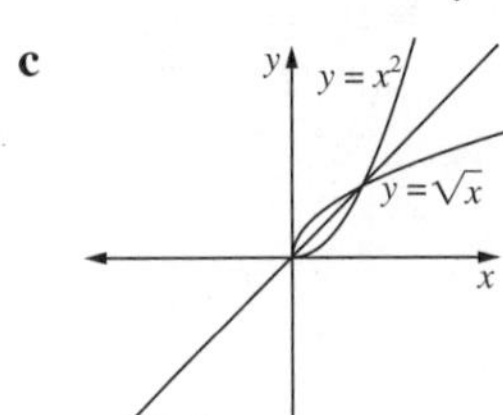

d The inverse function is the reflection of the function in the line $y=x$.

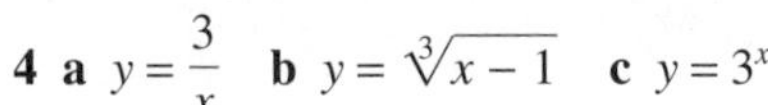

PAGE 66 **1 a** −1 **b** 11 **c** 2 **d** 8 **e** −10 **f** 5 **g** $3a-4$ **h** $-3a-4$ **i** $\frac{3}{a}-4$

2 $y=2^x$

x	−2	−1	0	1	2
y	$\frac{1}{4}$	$\frac{1}{2}$	1	2	4

$y=2^{2x}$

x	−2	−1	0	1	2
y	$\frac{1}{16}$	$\frac{1}{4}$	1	4	16

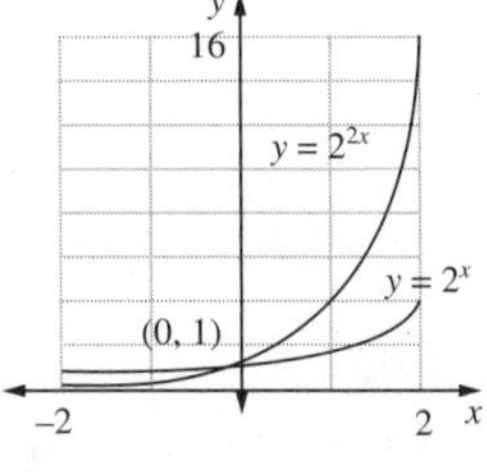

3 a $5=\log_2 32$ **b** $3=\log_3 27$ **c** $5=\log_4 1024$ **d** $4=\log_5 625$ **e** $3=\log_6 216$ **f** $3=\log_7 343$ **g** $3=\log_9 729$ **h** $4=\log_{10} 10\,000$ **i** $7=\log_3 2187$

4 a $3^3=27$

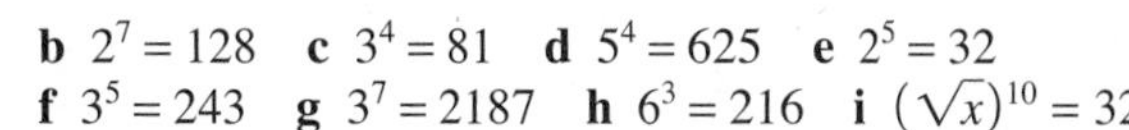

5 a 5 **b** 1 **c** $\log_a x$ **d** −1 **e** 2 **f** 1 **6 a** $x=6$ **b** $x=-4$ **c** $x=2$

PAGE 67 **1** A **2** C **3** B **4** A **5** B **6** C **7** B **8** B **9** C **10** C

PAGE 68 **1 a** 3 **b** −1 **2 a** 3 **b** $\log m$ **3** $b=\log_a x$ **4 a** $x=4$ **b** $x=1$ **c** $x=9$ **d** $x=5$ **e** $x=3\frac{1}{2}$ **f** $x=-2$ **5 a** $x>0$ **b** all real y **c** $y=4^x$ **d**

$y=\log_4 x$

x	$\frac{1}{16}$	$\frac{1}{4}$	1	4	16
y	−2	−1	0	1	2

$y=4^x$

x	−2	−1	0	1	2
y	$\frac{1}{16}$	$\frac{1}{4}$	1	4	16

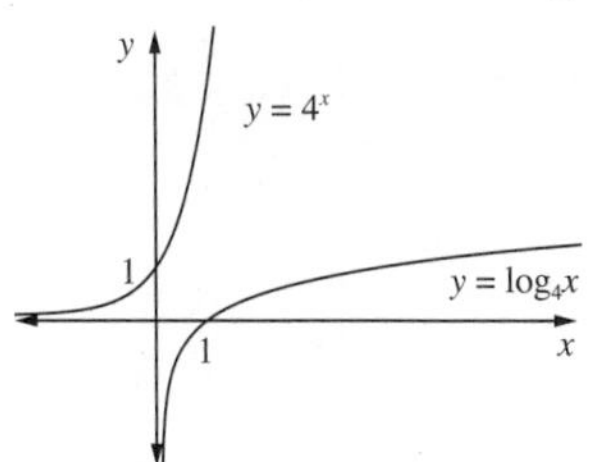

Answers

Chapter 5 – Polynomials and curve sketching

Page 69 **1** C, E, F **2 a** $3, 6x^3, 6, 7$ **b** $3, 4x^3, 4, -2$ **c** $5, 8x^5, 8, 0$ **d** $8, 5x^8, 5, -3$ **e** $3, 9x^3, 9, 5$ **f** $4, x^4, 1, -5$ **g** $2, 9x^2, 9, -8$ **h** $3, x^3, 1, 2$ **i** $5, 6x^5, 6, 9$ **3** B, C, F, G, H **4 a i** 1 **ii** 11 **iii** 5 **b i** 3 **ii** 4 **iii** 2 **c i** 6 **ii** 4 **iii** 18 **d i** 0 **ii** 2 **iii** 12 **5 a** 5 **b** 5

Page 70 **1 a** $5x^3+3x^2+2x-8$ **b** $4x^3+2x^2+14x-9$ **c** $6x^5+12x^4+x^2+9x$ **d** $3x^6+9x^2+11x-3$ **e** x^5+8x^2+7x-9 **f** $x^4+9x^3-8x^2-2x+16$ **2 a** x^3+11x^2-7x+2 **b** $3x^5+x^4+x^3+6x^2+x+11$ **c** $2x^3+9x^2+3x-5$ **d** $3x^5+x^4+8x^2-9x+18$ **e** $3x^5+x^4-x^3+3x^2+2x+6$ **f** $3x^3+14x^2-8x+7$ **3 a** $2x^4+x^3+2x^2+9x-15$, degree = 4 **b** $x^4+x^3+10x^2+x-4$, degree = 4 **c** $5x^3+2x^2-10$, degree = 3 **d** $6x^3+7x^2-x-9$, degree = 3 **e** $x^4+x^3-6x^2+16x$, degree = 4 **f** $5x^3+14x^2-x-2$, degree = 3 **g** $8x^4+3x^3+2x^2+2x+6$, degree = 4

Page 71 **1 a** $x^3-3x^2-8x+10$ **b** $3x^3-2x^2+6x-4$ **c** $6x^5-5x^4-15x^2+3x$ **d** $2x^6-5x^4-2x^3-x$ **e** $8x^2+8x+4$ **f** $-x^4+5x^3-7x^2+3x+13$ **2 a** $-2x^3-x^2+11x-16$ **b** $-7x^5-x^4+4x^3+7x^2-x-2$ **c** $-3x^3-2x^2+5x-7$ **d** $-7x^5-x^4+3x^3+6x^2-7x+7$ **e** $-7x^5-x^4+x^3+5x^2+4x-9$ **f** $-x^3-x^2-6x+9$ **3 a** $-3x^4+x^3-4x-1$, degree = 4 **b** $x^5+4x^2+2x-10$, degree = 5 **c** $8x^3-11x^2+7x+2$, degree = 3 **d** $x^4-4x^3+11x^2-7x$, degree = 4 **e** $x^5-3x^3-9x^2+13x-3$, degree = 5 **f** $6x^3-2x^2-3x-1$, degree = 3 **g** $8x^3-x^2+8x-11$, degree = 3

Page 72 **1 a** x^3+6x^2+7x+2 **b** x^3-x^2-2x+8 **c** x^3-5x^2+x+10 **d** $x^3+10x^2+13x-24$ **e** $2x^3-9x^2-11x-3$ **f** $3x^4+x^3-5x^2-4x+4$ **2 a** $x^4+7x^3-12x^2-21x+27$ **b** $x^4+4x^3+5x^2+2x$ **c** $6x^5+12x^4-25x^3+10x^2+21x-28$ **d** $32x^5-12x^4+4x^3+6x^2-10x$ **e** $6x^5-16x^4+2x^3-9x^2+24x-3$ **f** $9x^4+12x^3+10x^2+4x+1$ **3 a** $20x^3+17x^2-33x-9$, degree = 3 **b** $2x^5+3x^4-5x^3-6x^2+2x$, degree = 5 **c** $6x^5+3x^4-10x^3-5x^2+6x+3$, degree = 5

Page 73 **1 a**

$$\begin{array}{r} x+2 \\ x+1\overline{)x^2+3x+5} \\ \underline{x^2+x} \\ 2x+5 \\ \underline{2x+2} \\ 3 \end{array}$$

b

$$\begin{array}{r} x^2+4x+6 \\ x-1\overline{)x^3+3x^2+2x-1} \\ \underline{x^3-x^2} \\ 4x^2+2x \\ \underline{4x^2-4x} \\ 6x-1 \\ \underline{6x-6} \\ 5 \end{array}$$

c

$$\begin{array}{r} 2x^2+5x-3 \\ 3x-2\overline{)6x^3+11x^2-19x+2} \\ \underline{6x^3-4x^2} \\ 15x^2-19x \\ \underline{15x^2-10x} \\ -9x+2 \\ \underline{-9x+6} \\ -4 \end{array}$$

2 a $5x^2+6x+3=(x+2)(5x-4)+11$ **b** $x^2-5x+16=(x-1)(x-4)+12$ **c** $6x^3+4x^2-7x+5=(x+1)(6x^2-2x-5)+10$ **d** $x^4-3x^3-5x^2+7x-9=(x-4)(x^3+x^2-x+3)+3$ **e** $x^3+2x^2-8=(x+5)(x^2-3x+15)-83$ **f** $x^4+3x^2+2x=(x+1)(x^3-x^2+4x-2)+2$

Page 74 **1 a** 11 **b** 2 **c** −1 **d** −9 **e** 84 **f** 5 **2 a** −12 **b** −13 **c** 50 **d** 14 **e** 11 **f** 53 **3 a** $p=1$ **b** $m=4$

Page 75 **1 a** $(x+1)(5x+3)$ **b** $(x-1)^2(x+3)$ **c** $(x-2)(x-3)(2x-1)$ **d** $(x+3)(x+4)(x+5)$ **2 a** $(x+1)(x+2)(x+3)$ **b** $(x-2)(x+5)(2x-1)$ **c** $(2x+3)(x^2+x+2)$ **d** $(x-1)(x+1)(2x-3)$ **3 a** $(x-5)(x+1)(4x-1)$ **b** $(x-1)(x+2)(3x+1)$ **c** $(x+1)(x+3)(2x-1)$ **d** $(x-2)(x-5)(4x+1)$ **4** $p=-2$

Page 76 **1 a** $P(1)=0$ **b**

$$\begin{array}{r} x^2-2x-8 \\ x-1\overline{)x^3-3x^2-6x+8} \\ \underline{x^3-x^2} \\ -2x^2-6x \\ \underline{-2x^2+2x} \\ -8x+8 \\ \underline{-8x+8} \\ 0 \end{array}$$

c $P(x)=(x-1)(x+2)(x-4)$ **d** 3

2 a $P(3)=0$ **b** $Q(x)=x^3+5x^2+2x-8$ **c** $Q(-2)=0$ **d**

$$\begin{array}{r} x^2+3x-4 \\ x+2\overline{)x^3+5x^2+2x-8} \\ \underline{x^3+2x^2} \\ 3x^2+2x \\ \underline{3x^2+6x} \\ -4x-8 \\ \underline{-4x-8} \\ 0 \end{array}$$

e $P(x)=(x-3)(x+2)(x+4)(x-1)$ **f** 4 **3** n **4 a** $1, -2, -3$ **b** $-1, -3, -4$ **c** $-1, -2, 6$ **d** $0, -3, -2, 2$

Answers

PAGE 77 **1 a** $x = 1, x = -3, x = -7$ **b** $x = 0, x = -1, x = -2, x = 8$ **2 a** $P(3) = 0$ **b** $x = 3, x = -1, x = -2$ **3** $x = 2, x = 3$ **4 a** $x = 1, -2, -5$ **b** $x = -2, x = 3, x = 5$ **c** $x = 1, x = 1.5$ or $x = -3$ **d** $x = 0, -1, -2$ or 4

PAGE 78 **1 a**

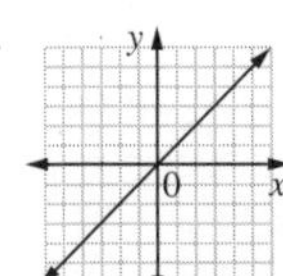

b

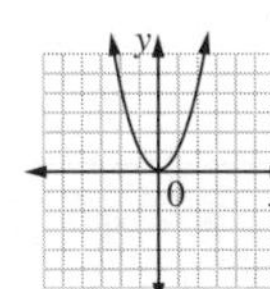

c

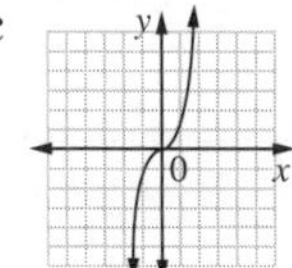

d

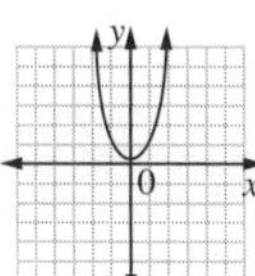

2 a

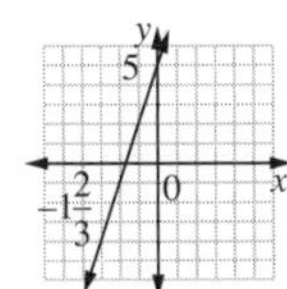

b

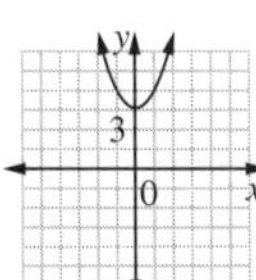

c

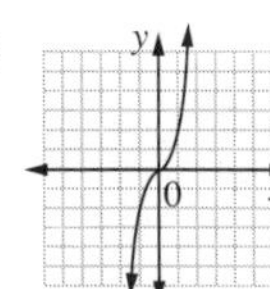

d

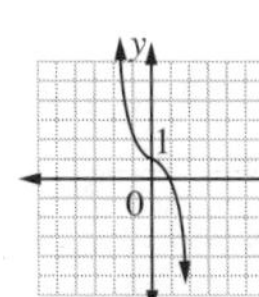

3 a

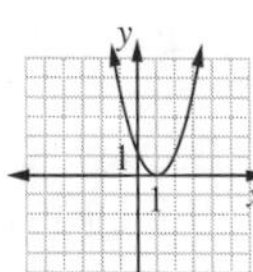

b

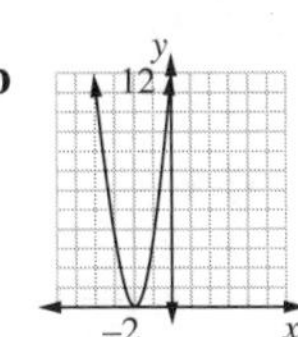

c

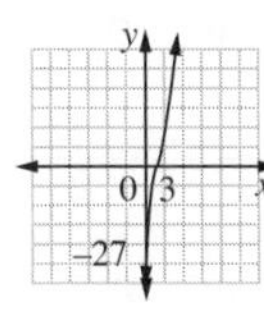

d

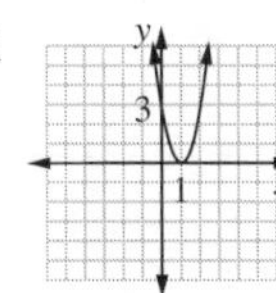

4 a

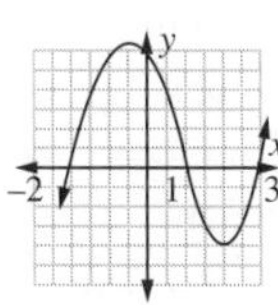

b

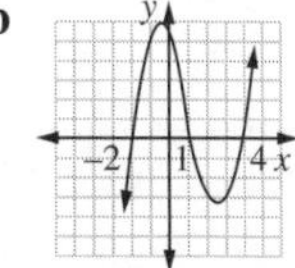

PAGE 79 **1 a** -8 **b** $-4, -1$ and 2 **c** gets very large positively **d** gets very large negatively **e** [see below] **2 a** [see below] **b** [below] **3 a** $(x-1)(x+2)(x+6)$ **b** [below]

1e

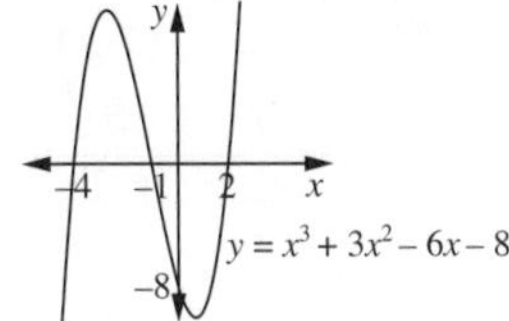

2a

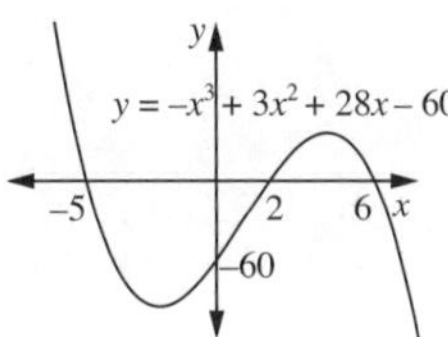

2b

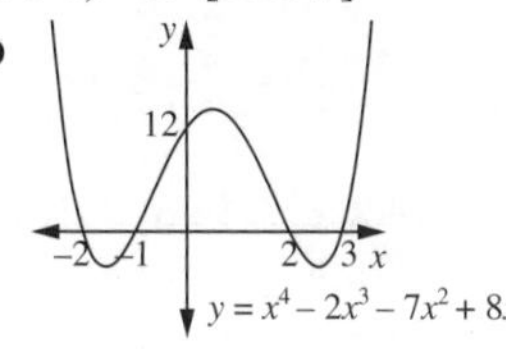

3b

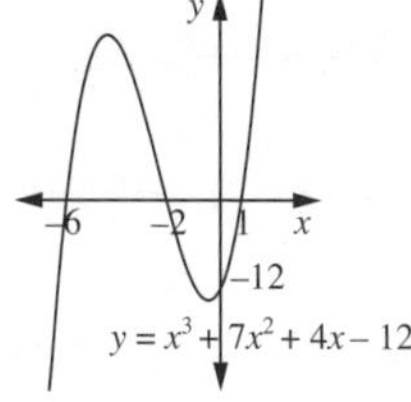

PAGE 80 **1**

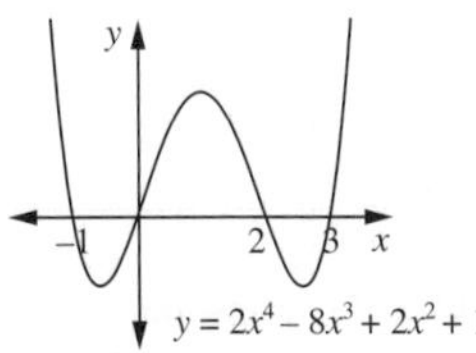

2 a $(x-1)^2(x+3)$ **b** $x = 1$ or $x = -3$ **c** $x = 1$ **d** 3 **e** 5 **f** 1

g

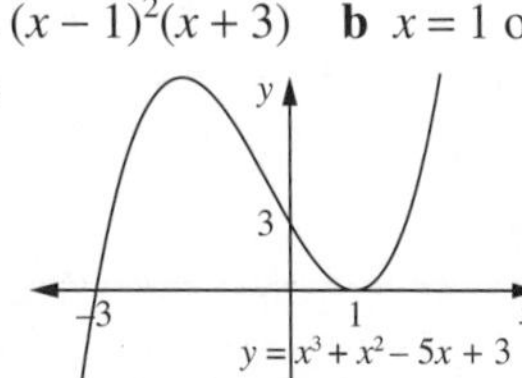

3 a $x = 1$ **b** -3 **c** 5 **d** Yes

e

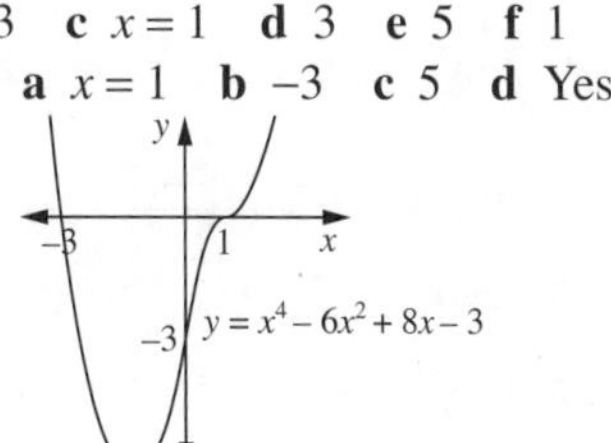

PAGE 81 **1** [see below]

a

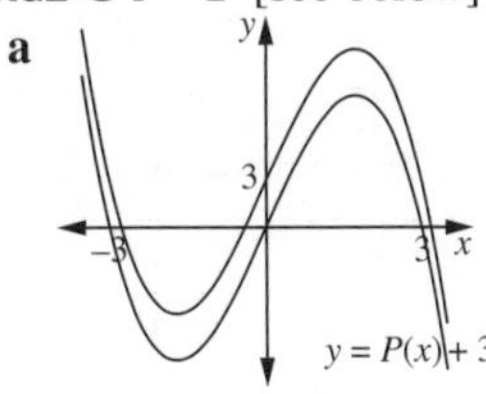

b

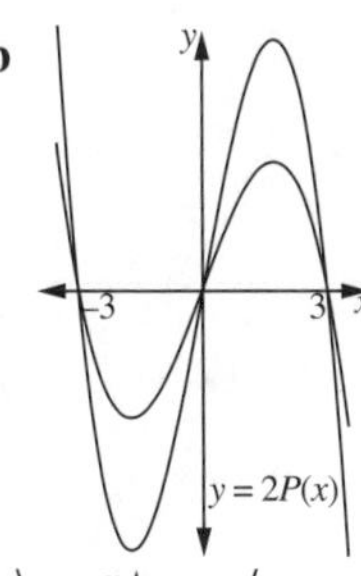

c

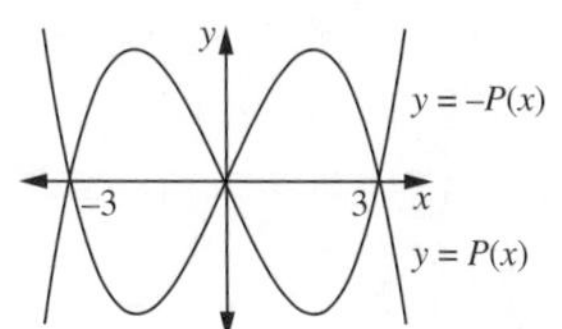

2 a

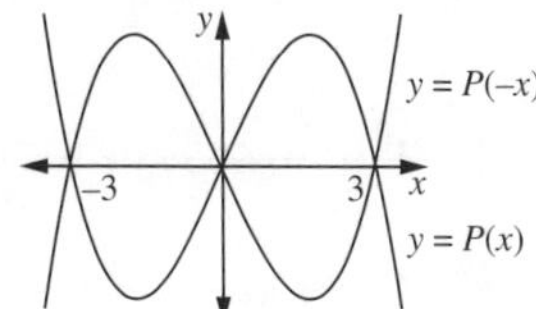

b

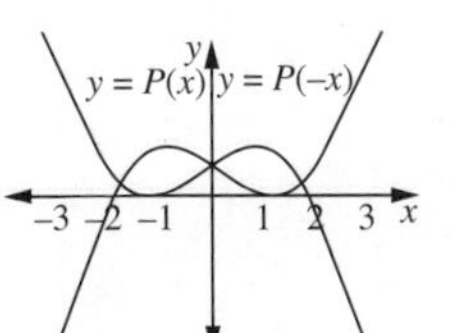

c

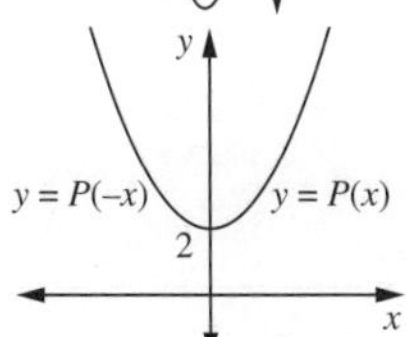

d

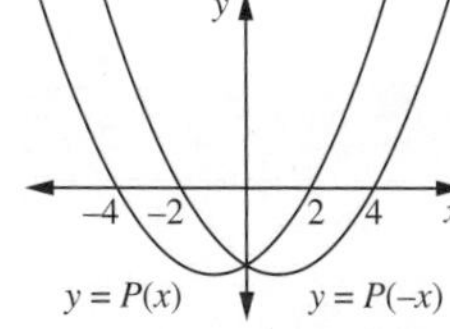

3 a moves the curve up or down **b** zeros remain the same, curve becomes longer and thinner (or shorter and wider) between the zeros **c** inverts the curve, reflected in the x-axis **d** reflected in the y-axis

PAGE 82 **1** D **2** D **3** A **4** D **5** A **6** D **7** C **8** B **9** B **10** A

PAGE 83 **1 a** 3 **b** 1 **c** 4 **d** -9 **2 a** $6x^3 + 2x^2 - 7x + 3$ **b** $2x^2 - 15x - 21$ **c** $x^3 - 3x^2 - 4x + 12$

3 a $x^3 + 7x^2 - 2x - 4 = (x-2)(x^2 + 9x + 16) + 28$

b $x^4 + 5x^3 - 9x^2 + 34x - 4 = (x+7)(x^3 - 2x^2 + 5x - 1) + 3$

4 a 8 **b** -5 **5 a** $x(x+2)(x-5)$ **b** 0 **c** 0, -2 and -5 **d**

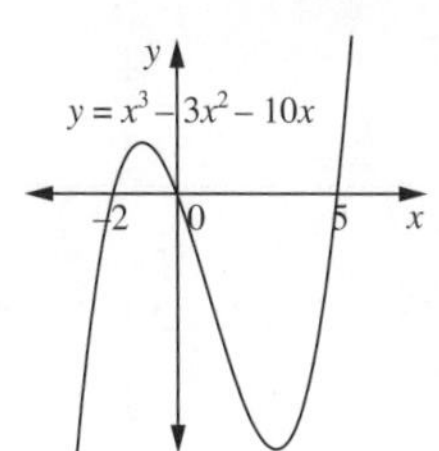

Answers

Chapter 6 – Surface area and volume

Page 84 **1 a** 654 cm^2 **b** 1734 cm^2 **c** 4990 cm^2 **2 a** 660 cm^2 **b** 1240 cm^2 **c** 690 cm^2 **3 a** 428 m^2 **b** 313.6 m^2 **c** 1120 m^2 **4 a** 1099.6 cm^2 **b** 1193.8 cm^2 **c** 1281.8 cm^2

Page 85 **1** 89 mm **2** 23.1 cm **3 a** 13 cm **b** 20 cm **4 a** 8 m **b** 17 m **5** 28 cm **6** 9.6 m, 3 m

Page 86 **1 a** 504.0 cm^2 **b** 236.3 cm^2 **c** 288.7 cm^2 **2 a** 1144 cm^2 **b** 445.1 cm^2 **c** 855.0 cm^2

Page 87 **1 a** 301.6 cm^2 **b** 339.3 cm^2 **c** 364.0 cm^2 **d** 1617.2 cm^2 **2 a** 264π cm^2 **b** 800π cm^2 **3** 23 cm **4** 8 cm

Page 88 **1 a** 196π cm^2 **b** 324π cm^2 **c** 3136π cm^2 **d** 1764π cm^2 **e** 275.56π cm^2 **f** 571.21π cm^2 **2 a** 576π cm^2 **b** 1024π cm^2 **3 a** 942.48 cm^2 **b** 1847.26 cm^2 **4 a** 1260 cm^2 **b** 2890 cm^2 **c** 1230 cm^2 **5** 5.35 cm

Page 89 **1 a** 565 cm^2 **b** 761 cm^2 **c** 104 cm^2 **d** 10 720 cm^2 **e** 5429 cm^2 **f** 5070 cm^2

Page 90 **1 a** 614.1 cm^3 **b** 461.9 cm^3 **c** 663.3 cm^3 **2 a** 2167 cm^3 **b** 471.0 cm^3 **c** 3348 cm^3 **3 a** 4477.20 cm^3 **b** 336.00 cm^3 **c** 2483.02 cm^3 **4 a** 1170.0 m^3 **b** 2470.0 m^3 **c** 77 584.8 cm^3

Page 91 **1 a** 361.6 cm^3 **b** 174.1 cm^3 **c** 216.2 cm^3 **2 a** 306.05 cm^3 **b** 2.33 m^3 **c** 776.83 cm^3 **3 a** 1230 cm^3 **b** 392 cm^3 **4 a** 1.9 m^3 **b** 326.5 cm^3 **c** 396.8 cm^3

Page 92 **1 a** 121.5 cm^3 **b** 223.3 cm^3 **c** 1838.6 cm^3 **d** 1766.9 cm^3 **e** 72804.1 cm^3 **f** 37.7 m^3 **2 a** 1005.3 cm^3 **b** 55.9 cm^3 **3 a** 73.45 m^3 **b** 86.86 m^3 **c** 49.18 m^3

Page 93 **1 a** 3053.6 cm^3 **b** 4188.8 cm^3 **c** 113 097.3 cm^3 **d** 22 449.3 cm^3 **e** 15 002.5 cm^3 **f** 91 952.3 cm^3 **2 a** 4188.8 cm^3 **b** 150 532.6 cm^3 **c** 17 157.3 cm^3 **d** 102 160.4 cm^3 **3 a** 1526.8 cm^3 **b** 15 529.7 cm^3

Page 94 **1 a** 20 910.4 cm^3 **b** 1392.8 cm^3 **c** 2382.4 cm^3 **d** 753.98 cm^3 **e** 9600 cm^3 **f** 358.8 cm^3 **g** 26 900.4 cm^3 **h** 20 357.5 cm^3

Page 95 **1 a** 8 : 27 **b** 1 : 8 **c** 8 : 125 **2 a** 31.9 cm^3 **b** 13.0 cm^3 **c** 5.1 cm^3 **3 a** 125 : 64 **b i** 8 : 7 **ii** 512 : 343 **c** 544 cm^3

Page 96 **1 a** 5.147×10^8 km^2 **b** 1.098×10^{12} km^3 **2 a** 261.3 m^2 **b** 397.18 m^3 **3** 61.26 m^2 **4 a** 377 cm^3 **b** 377 mL **5 a** 22 560 cm^2 **b** 204 800 cm^3

Page 97 **1** D **2** B **3** C **4** B **5** C **6** A **7** B **8** B **9** B **10** D

Page 98 **1 a** 57.2 cm **b** 13896.4 cm^2 **c** 184 506.5 cm^3 **2 a** 1018 cm^2 **b** 3054 cm^3 **3 a** 360 cm^2 **b** 400 cm^3 **4 a** 301 cm^2 **b** 301 cm^3 **5 a** 7304.2 cm^2 **b** 45945.8 cm^3 **6 a** 20 cm **b** 14 cm **c** 1144 cm^2 **d** 2240 cm^3

Chapter 7 – Trigonometry

Page 99 **1 a** 0.966 **b** 0.675 **c** –0.5 **d** 2.156 **e** –0.269 **f** 0.336 **2 a** 30° **b** 43°12′ **c** 69°47′ **d** 27°8′ **e** 53°40′ **f** 15° **3 a** $\sin\theta = \frac{4}{5}, \cos\theta = \frac{3}{5}, \tan\theta = \frac{4}{3}$ **b** $\sin\theta = \frac{5}{13}, \cos\theta = \frac{12}{13}, \tan\theta = \frac{5}{12}$ **c** $\sin\theta = \frac{15}{17}, \cos\theta = \frac{8}{17}, \tan\theta = \frac{15}{8}$ **4 a** 29.5 cm **b** 12.3 cm **c** 9.5 cm **d** 21.7 cm **e** 13.6 cm **f** 9.4 m **5 a** 36°52′ **b** 67°23′ **c** 61°56′

Page 100 **1 a** 164.85 m **b** 20° **c** 78.32 m **2 a** 9.5 cm **b** 36°52′ **c** 15 cm **3** $p = 7.5$ cm, $q = 8.7$ cm **4 a** 452 m **b** 370 m

Page 101 **1 a** 130° **b** 072° **c** 250° **d** 140° **e** 270° **f** 046° **g** 030° **h** 122° **2** 89 km **3 a**

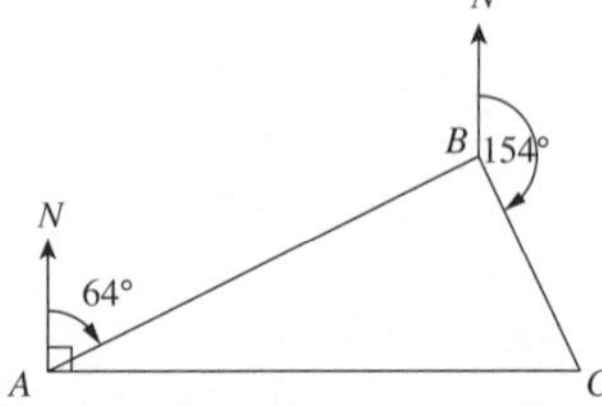

b 90° **c** 26° **d** 125.7 km **e** 270° **4** 122°

Page 102 **1 a** 17 cm **b** $\sqrt{325}$ cm **c** 19°26′ **2** 45°14′ **3** 39°31′

Page 103 **1 a** 12.1 m **b** 8.3 m **c** 14.7 m **2 a** 14.0 m **b** 17.9 m **c** 34.9 m **3 a** 62.2 m **b** 42 m **4** 197 m **5** 72 m

Page 104 **1 a** 1 unit **b** 1 unit **c i** y **ii** x **iii** $\frac{y}{x}$ **2 a i** y **ii** x **iii** $\frac{y}{x}$ **b** positive, negative, negative **3 a i** y **ii** x **iii** $\frac{y}{x}$ **b** negative, negative, positive **4 a i** y **ii** x **iii** $\frac{y}{x}$ **b** negative, positive, negative **5 a** $\frac{\sin\theta}{\cos\theta}$ **b** 0.75 **6 a** 0.8192 **b** –0.9397 **c** –0.8391 **d** –0.7660 **e** –11.4301 **f** –0.9272 **g** 0.2679 **h** –0.8480 **i** 0.9976

Answers

Page 105 **1 a**

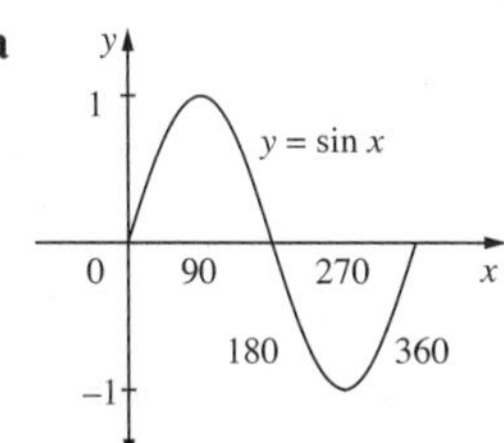

b

c

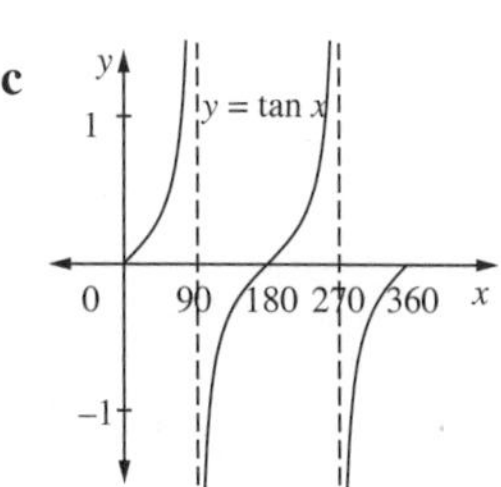

2 a 360° **b** 360° **c** 180° **3 a** 1 **b** −1 **4 a** 1 **b** −1 **5 a** no **b** as $x°$ gets closer to 90°, tan $x°$ gets larger and larger. Tan $x°$ is undefined whenever cos $x°$ = 0.
6 all the curves are periodic. $y = \cos x°$ is the same curve as $y = \sin x°$ but moved across by 90°.

Page 106 **1 a** positive **b** negative **c** negative **d** negative **e** positive **f** negative **2 a** 0.9063 **b** −0.9744 **c** −3.7321 **d** −0.2126 **e** 0.9986 **f** −0.5150 **3 a** sin 70° **b** −tan 65° **c** sin 22° **d** −cos 50° **e** −cos 68° **f** −tan 52° **g** −cos 63° **h** sin 82° **4 a** 145° **b** 102° **c** 119° **d** 134° **5 a** 30°, 150° **b** 51°, 129° **c** 63°, 117° **d** 14°, 166° **6 a** 7°11′, 172°49′ **b** 21°6′ **c** 73°48′ **d** 107°27′ **e** 23°35′, 156°25′ **f** 153°26′ **g** 36°52′, 143°8′ **h** 152°52′ **7 a** 56° **b** 117°

Page 107 **1 a** $\frac{\sqrt{3}}{2}$ **b** $\frac{\sqrt{3}}{2}$ **c** $\frac{1}{\sqrt{3}}$ **d** $\sqrt{3}$ **e** $\frac{1}{2}$ **f** $\frac{1}{2}$ **2 a** 1 **b** $\frac{1}{\sqrt{2}}$ **c** $\frac{1}{\sqrt{2}}$ **3 a** 0 **b** 1 **c** 0 **d** 1 **e** 0 **4 a** $\frac{1}{2}$ **b** $-\frac{1}{\sqrt{2}}$ **c** $-\sqrt{3}$ **d** $-\frac{1}{\sqrt{3}}$ **e** $-\frac{1}{2}$ **f** $\frac{1}{\sqrt{2}}$ **g** $\frac{\sqrt{3}}{2}$ **h** −1 **i** $-\frac{\sqrt{3}}{2}$ **5 a** $\sin 30° = \frac{1}{2}$; $\cos 60° = \frac{1}{2}$; $\frac{1}{2} \times \frac{1}{2} = \frac{1}{4}$ **b** $\sin 60° = \frac{\sqrt{3}}{2}$; $\cos 30° = \frac{\sqrt{3}}{2}$; $\frac{\sqrt{3}}{2} + \frac{\sqrt{3}}{2} = \sqrt{3}$ **c** $\sin 45° = \frac{1}{\sqrt{2}}$; $\cos 45° = \frac{1}{\sqrt{2}}$; $\frac{1}{\sqrt{2}} \div \frac{1}{\sqrt{2}} = 1 = \tan 45°$
6 a $3\sqrt{2}$ **b** $4\sqrt{3}$ **c** $8\sqrt{3}$

Page 108 **1 a** 23°35′, 156°25′ **b** 16°42′ **c** 60° **d** 44°26′, 135°34′ **e** 158°12′ **f** 38°19′, 141°41′ **g** 33°54′ **h** 26°34′ **i** 104°29′ **2 a** 60° **b** 90° **c** 45° **d** 30° **e** 45° **f** 30° **g** 60° **h** 30° **i** 45° **3 a** 120° **b** 120° **c** 180° **4 a** 0°, 180° **b** 60°, 120° **c** 45°, 135° **5 a** 150° **b** 135° **c** 30°, 150° **6 a** cos **b** 40° **c** 62° **d** 75°

Page 109 **1 a** 11.65 **b** 23.68 **c** 18.17 **d** 39.36 **e** 11.85 **f** 3.47 **2 a i** 77° **ii** 8.2 km **b i** 40° **ii** 7.9 km **c i** 105° **ii** 11.8 m

Page 110 **1 a** 47° **b** 39° **c** 19° **2 a** 129° **b** 125° **c** 100° **3 a** 57° or 123° **b** 68° or 112° **c** 36° or 144° **4 a** 25° (155° ✘) **b** 48° (132° ✘)

Page 111 **1 a** 83° or 97° **b** 42° **c** 69° or 111° **2 a** 31° **b** 31° **3** 41° **4** a 27° **b** 85° **c** 20 km

Page 112 **1 a** 5.3 **b** 7.6 **c** 17.4 **d** 24.4 **e** 7.4 **f** 13.9 **2 a** 20 m **b** 53 m **3 a** 10.4 m **b** 10.2 m

Page 113 **1 a** 82° **b** 103° **c** 29° **d** 49° **e** 123° **f** 52° **2 a** 35°26′ **b** 96°23′ **c** 48°11′ **d** yes

Page 114 **1 a** 121.7 **b** 44.6 **c** 14.2 **d** 23.3 **e** 45.8 **f** 46.8 **2 a** 90° **b** 30° **c** 84°, 96° **d** 42° **e** 55°, 125° **f** 115°

Page 115 **1 a** 25 m^2 **b** 70 m^2 **c** 2925 m^2 **d** 2971 m^2 **e** 17 m^2 **f** 686 m^2 **2 a** 33 cm^2 **b** 169 cm^2 **c** 77 cm^2 **3 a** 24 **b** 64°47′

Page 116 **1 a** 29° **b** 2039 m^2 **2 a** 125 km **b** 191.5 km **c** 295° **3** a 115° **b** 9.3 m **c** 698 km

Page 117 **1** a 95° **b** 51° **c** 014° **2 a** 111.8 ha **b** 1317 m **3 a** 15° **b** 243.15 m **c** 197 m

Page 118 **1** B **2** C **3** C **4** D **5** A **6** B **7** B **8** D **9** A **10** D

Page 119 **1 a** 130° **b** 30° **2 a** B **b** $\frac{1}{\sqrt{2}}$ **3 a** 506.1 m **b** 377.7 m **c** 631.5 m **4 a** 4.8 **b** 11.1 **5 a** 2 m^2 **b** 28 m^2 **6** 83° or 97° **7 a** 32° **b** 238°

Chapter 8 – Circle geometry

Page 120 **1 a** centre **b** radius **c** diameter **d** circumference **e** chord **f** arc **g** tangent **h** secant **i** semi-circle **j** quadrant **k** sector **l** segment **2 a** $\angle ACD$ **b** $\angle FBC$ **c** $\angle BAD, \angle BFD$ **d** $\angle CEA, \angle CFA$ **e** $\angle EAD, \angle ECD$ **f** $\angle DBF, \angle DCF$ **3 a** 1 **b** cyclic **c** $\angle POQ$ (or $\angle QOP$)

Page 121 **1 a** equal angles **b** equidistant **c** perpendicular **d** bisects **e** centre **f** bisects, right angles **g** centre **2 a** $x = 9$ cm **b** $y = 18$ cm **c** $a = 6.3$ cm **d** $x = 90$ **e** $y = 90$ **f** $a = 90$ **g** $x = 85$ **h** $y = 50$ **i** $a = 120$ **j** $x = 5$ cm **k** $y = 7$ cm **l** $a = 12$ cm **3 a** 6 cm **b** 9 cm **c** 21 cm

Page 122 **1 a** circumference, arc **b** right angle **c** circumference, arc, equal **2 a** $a = 40$ **b** $m = 70$ **c** $x = 30$ **d** $y = 90$ **e** $m = 90$ **f** $a = 30$ **g** $x = 110$ **h** $n = 90$ **3 a** $m = n = 35$ **b** $x = y = 15$ **c** $a = b = 40$ **d** $x = 15, y = 35$ **e** $x = y = 90$ **f** $a = 60, b = 120$ **g** $a = 35, b = 55$ **h** $x = 60, y = 30$

Page 123 **1 a** opposite, supplementary **b** exterior, equal **2 a** $x = 92$ **b** $y = 70$ **c** $a = 115$ **d** $x = 100$ **3 a** $p = 75, q = 120$ **b** $m = 95, n = 89$ **c** $a = 94, b = 93$ **d** $x = 120, y = 60$ **e** $a = 85, b = 95$ **f** $m = 110, n = 70$ **g** $a = 105, b = 75$ **h** $x = 75, y = 83$ **i** $a = 91, x = 115, y = 65$ **j** $a = 80, b = 100, c = 105$ **k** $m = 93, n = 87, p = 83$ **l** $a = 80, c = 100, b = 160$

Answers

PAGE 124 **1 a** perpendicular, radius **b** equal **c** collinear **2 a** $x = 90$ **b** $y = 90$ **c** $a = 90$ **d** $a = 50$ **e** $x = 35$ **f** $m = 30$ **g** $a = 14$ cm **h** $b = 21$ cm **3 a** $c = 9$ cm **b** $x = 10$ cm, $p = 65°$ **c** $y = 7$ cm, $m = 60°$ **d** $a = 20$ cm, $n = 50°$ **e** $a = b = 90, c = 140$ **f** $m = l = 90, n = 130$ **g** $x = y = 90, p = 120$ **h** $x = 180$

PAGE 125 **1** equal, alternate segment **2 a** $m = 50$ **b** $a = 45$ **c** $x = 30$ **d** $x = 120$ **e** $x = 60$ **f** $x = 48, y = 60$ **g** $x = 50, y = 70$ **h** $x = 135, y = 25$ **i** $a = 65, b = 65, c = 65$ **j** $x = 70, y = 80$ **k** $a = 25, b = 75$ **l** $m = 35, n = 70, p = 75$ **m** $x = 35$ **n** $x = 70$ **o** $a = 55, b = 110$ **p** $m = n = 40$

PAGE 126 **1 a** product, equal **b** product, equal **c** square, equals, secant **2 a** $x = 2$ **b** $y = 1$ **c** $a = 5$ **d** $m = 4$ **3 a** $x = 7$ **b** $x = 1$ **c** $a = 3$ **d** $x = 3$ **e** $y = 1$ **f** $m = 4$ **4 a** $x = 6$ **b** $a = 22.9$ **c** $x = 7$

PAGE 127 **1 a** $x = 8$ cm **b** $a = 90$ **c** $x = 85$ **d** $m = 48$ **e** $x = 90$ **f** $a = 90$ **g** $y = 20$ **h** $x = 9$ cm **i** $x = 9$ **j** $x = 4$ cm **k** $x = 40$ **l** $x = 3$ **2 a** $a = b = 50$ **b** $m = 70, n = 115$ **c** $x = 70, y = 110$ **d** $m = n = 85$ **e** $x = 30, y = 75$ **f** $x = 90, y = 7$ cm **g** $x = 12$ **h** $x = 12$

PAGE 128 **1 a** $a = 20$ **b** $x = 40$ **c** $a = 52$ **d** $m = 101$ **2 a** $a = 140, b = 40$ **b** $x = 30, y = 60$ **c** $a = 37, b = 68, c = 75$ **d** $x = 35, y = 35$

PAGE 129 **1 a** 120°, reflex $\angle AOC = 240°$, angle at centre is twice angle at circumference **b** 35°, $DCE = 110°$ (vertically opposite), $\angle CDE = \angle ABC$ (alternate $AB \parallel DE$), $\angle CED = \angle ABC$ (angles in same segment) so $\angle CDE = \angle CED$ **c** 50°, $\angle BAC = 65°$ (angles in same segment) but $AB = BC$ so $\angle BAC = \angle BCA$ **d** 36°, $\angle AOD = \angle ODC$ (alternate, $AB \parallel DC$), $\angle ACD = \frac{1}{2}\angle AOD$ (angle on circumference) **2** 3.9 cm **3 a** 20 cm **c** 12 cm **4 a** 120°

PAGE 130 **6 b** 3α

PAGE 131 **1** B **2** C **3** C **4** A **5** C **6** C **7** C **8** D **9** A **10** B

PAGE 132 **1 a** 120° **b** 90° **2 a** 46° **b** 111° **3 a** angle in semi-circle **b** interior opposite angle **4 a** angle between tangent and chord = angle in alt. segment **6 b** 15 cm **d** 7.2 cm

CHAPTER 9 – Statistics and probability

(Note for students: When asked to comment or explain the answers given are suggested solutions. Other answers could adequately answer the question. If in doubt check with your teacher.)

PAGE 133 **1 a** $\frac{4}{25}$ **b** $\frac{1}{10}$ **2** rolling one die **3** $\frac{1}{4}$ **4** $\frac{1}{4}$ **5 a**

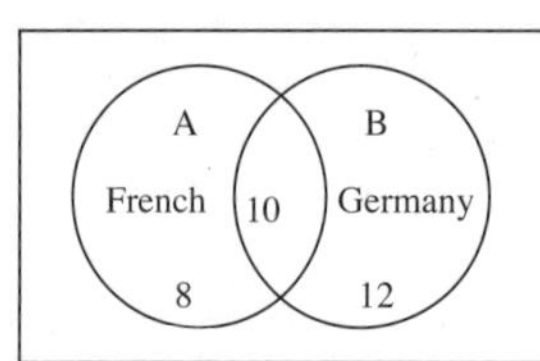

b 10 **c** $\frac{4}{15}$ **6 a** $\frac{1}{50}$ **b** $\frac{1}{4950}$ **c** $\frac{4753}{4950}$

7 a

1st child	2nd child	
H	A	HA
	J	HJ
	G	HG
A	H	AH
	J	AJ
	G	AG
J	H	JH
	A	JA
	G	JG
G	H	GH
	A	GA
	J	GJ

b $P(H) = \frac{1}{2}$, $P(GH) = \frac{1}{3}$

PAGE 134 **1 a** 8.25 **b** 7 **c** 8 **d** 7 **e** 2.5 **2** 55 kg **3 a i** \$62 **ii** \$64 **b i** \$46 **ii** \$48 **c** \$46 **d** The median and range are very similar, both are just a little higher for females. The distribution for females is more skewed. **4 a** A(3, 6, 9, 14, 30); B(4, 7, 11, 22, 28)

b

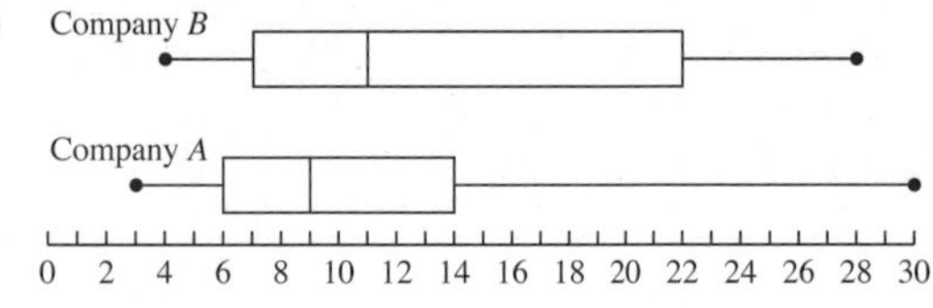

c both distributions are positively skewed. Although company A has the larger range, the interquartile range for company B is much larger. Company B has the larger median.

PAGE 135 **1 a** 7 **b** 6 **c**

x	5	7	11	2	8	6	3
$x - \bar{x}$	–1	1	5	–4	2	0	–3
$(x - \bar{x})^2$	1	1	25	16	4	0	9

d 56 **e** 2.8 (1 d.p.) **2 a** 3.9 **b** 8.0 **c** 2.6 **d** 3.9 **3 a** $\bar{x} = 6.9, \sigma_n = 2.2$ **b** $\bar{x} = 37.6, \sigma_n = 4.4$ **c** $\bar{x} = 13.6, \sigma_n = 4.4$ **d** $\bar{x} = 20.7, \sigma_n = 8.4$

Answers

PAGE 136 **1 a** 18 **b** 2.6 **2 a** stay the same, the score being added is equal to the mean **b** decrease, the spread of all the scores relative to the mean has decreased **3 a** decrease, the score being added is less than the mean **b** increase, the spread of scores relative to the mean has increased **4 a** 23 **b** 2.6 **5 a** all the scores have increased by 5 **b** mean has increased by 5 because all the scores have increased **c** standard deviation does not change because the spread of marks relative to the mean does not change. **6 a** 36 **b** 5.3 **7 a** scores have doubled **b** mean has doubled because all the scores have doubled c standard deviation has doubled because the spread of marks relative to the mean has doubled

PAGE 137 **1 a** 78, 14.7 **b** 78, 12.5 **c** 78, 12.9 **d** all have the same mean but algebra has the largest standard deviation. The standard deviations of geometry and statistics are similar, with statistics being slightly higher **e** The range is largest for statistics so that has the largest spread of actual marks but algebra has the largest spread of marks in relation to the mean. **2 a** Science $\bar{x} = 61.8$, SD = 4.62; Mathematics: $\bar{x} = 78.6$, SD = 5.57 **b** Science **c** Science **3 a** Tom: $\bar{x} = 16, \sigma_n = 2.966$ Theo: $\bar{x} = 14.1, \sigma_n = 4.679$ **b** Tom, he has the smaller standard deviation.

PAGE 138 **1 a** 9 **b** 5 **c** 2.8 **d** the range gives the total spread of all the scores, the interquartile range gives the spread of the middle 50% of scores. The standard deviation shows which scores are most consistent. **2 a** 32 **b** 6 **c** 7.6 **d** range **e** interquartile range **f** The range is affected the most because it is the difference between the highest and lowest score. The interquartile range is affected the least because it is concerned with the middle 50% of scores. **3 a i** 43 **ii** 46 **b i** 23.5 **ii** 17 **c i** 13.5 **ii** 12.6 **d** The theory marks have the slightly higher range but lower interquartile range and standard deviation. This is reflected in the shape of the distribution where the theory marks have a more even distribution about the mean.

PAGE 139 **1** The greater the number of hours of study, the higher the marks. The greatest increase in marks occurred for the lower hours of study.

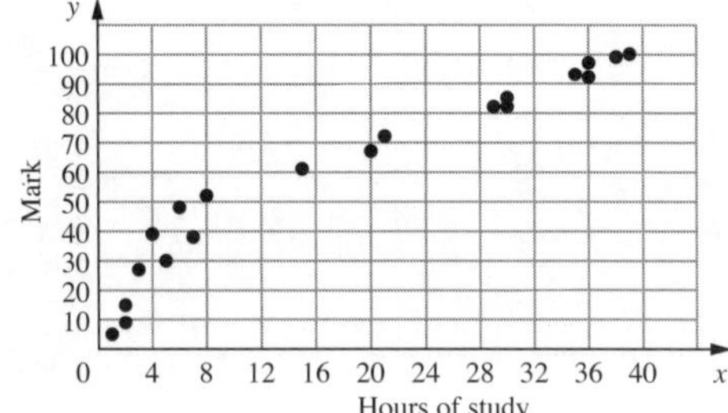

2 The graph shows older cars cost less than newer cars. There is a linearly decreasing trend in the price.

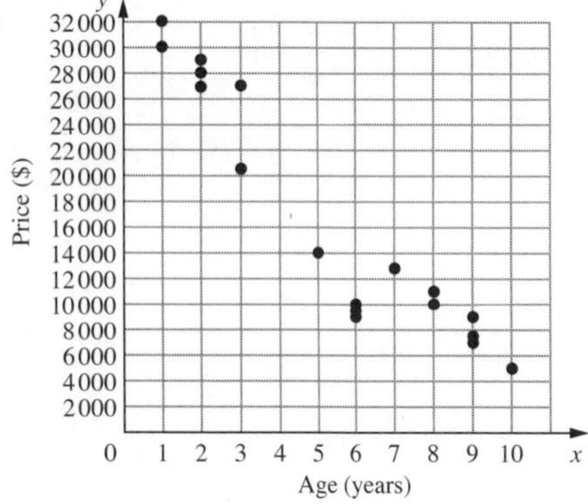

3

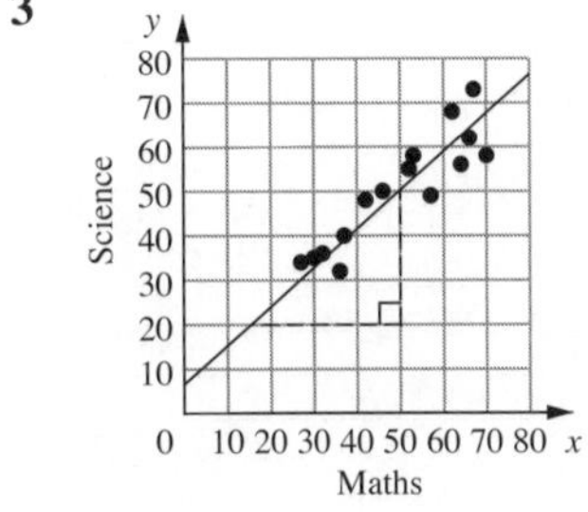

PAGE 140 **1 a** and **b** **c** $y = x + 10$

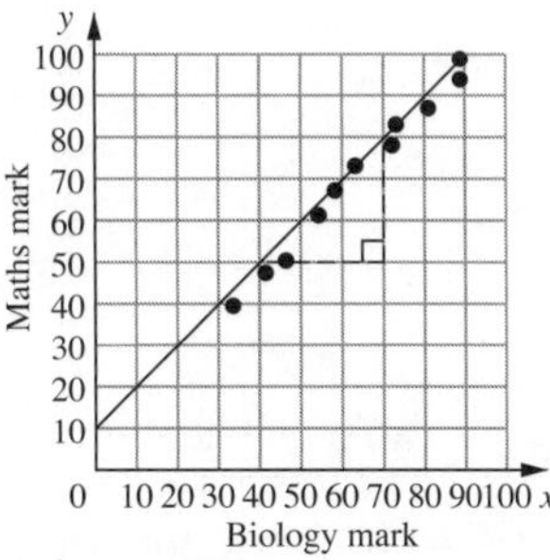

2 a and **b** **c (i)** 100 **(ii)** 25 **d** $y = x + 5$

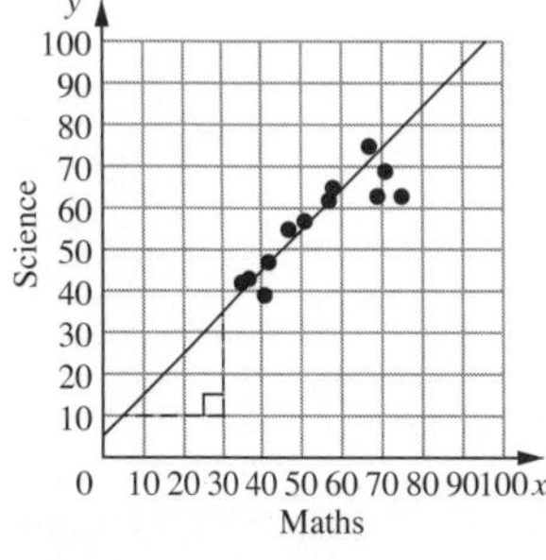

PAGE 141 **1 a** $h = 10n$ **b (i)** 85 cm **(ii)** 130 cm **c (i)** $5\frac{1}{2}$ weeks **(ii)** 16 weeks

2 a $h = 7n + 20$ **b (i)** 79.5 cm **(ii)** 111 cm **c (i)** 5 weeks **(ii)** 20 weeks **d** There is not a lot of difference for small values of n but as n gets larger the difference between the predictions increases.

3 a $V = 25\,000 - 2000a$ **b** \$15 000 **c** 10 years **d** After $12\frac{1}{2}$ years the equation will give a value that is negative.

PAGE 142 (Some possible responses) **1** No, we don't have any information about the sample so we cannot tell whether it is representative of all students. **2** The size of the study, the age and sex of the participants, where it was conducted, who paid for it. **3** They were trying to introduce doubt into voters minds by implying that the politician might change her mind. **4** The results might be biased because of the content of the program shown. Viewers of that program might not be representative of the general public. **5** The previous driving record of the drivers and the number of previous claims made. **6** The need to provide facilities for young families such as child care and schools.

Answers

Page 143 1 C 2 D 3 D 4 C 5 A 6 B 7 C 8 A 9 D 10 A

Page 144 1 All the cans lie within 2 standard deviations of the mean. **2 a** $\bar{x} = 77.3, \sigma_n = 7.86$ **b** $\bar{x} = 74.9$ $\sigma_n = 11.05$ **c** English as it has the higher standard deviation. **d** English, as the maths mark is more than one standard deviation below the mean while the English mark is less than one standard deviation below the mean.

3 a

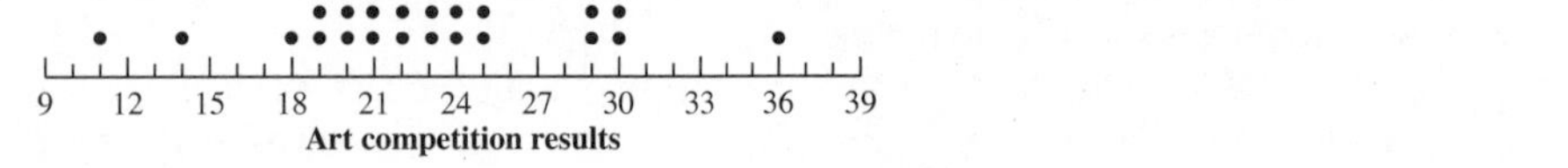

b 25 **c** 5 **d** 4.9

e the range covers all scores whereas the interquartile range only covers the middle 50% of scores. The standard deviation measures the spread from the mean. **4** (Different lines, with different equations are possible.) **a**

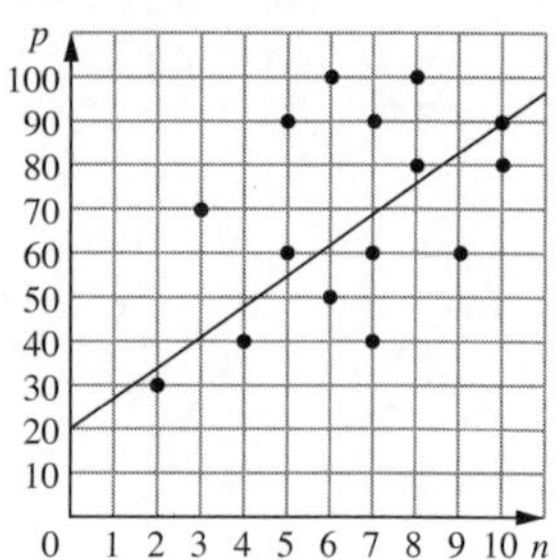

b $p = 7n + 20$ **c** \$139 **d** 40 toys

5 Probably not, as this sample is too small to be significant. There is no mention, also of the 'recommended' daily fruit amount in terms of some quantity nor how many students ate close to this amount.

Exam Papers

Page 145 1 C 2 D 3 A 4 D 5 C 6 A 7 C 8 C 9 A 10 B

Page 146–147 1 16 **2** 3 **3** (2, –3) **4** $\frac{1}{2}$ **5** 30 m^3 **6** $-\frac{\sqrt{3}}{2}$ **7** 30 **8** 3.5 **9** 0 **10** $\sqrt{3} + 3\sqrt{5}$ **11** 022° **12** 59 **13** 0.75 **14** $x = 0, y = 0$ **15** 85.7%

Page 148 1 a $6\sqrt{3}$ **b** $21 - 12\sqrt{3}$ **c** $\frac{\sqrt{21}}{2}$ **2 a** $x = 0$ or 16 **b** $x = -3$ or 5 **c** $x = -\frac{1}{4}$ **3 a** $x = \frac{1}{4}$ **b** $\left(\frac{1}{4}, -6\frac{1}{8}\right)$ **c** (0, –6) **d** $(-1\frac{1}{2}, 0)$ and (2, 0) **e**

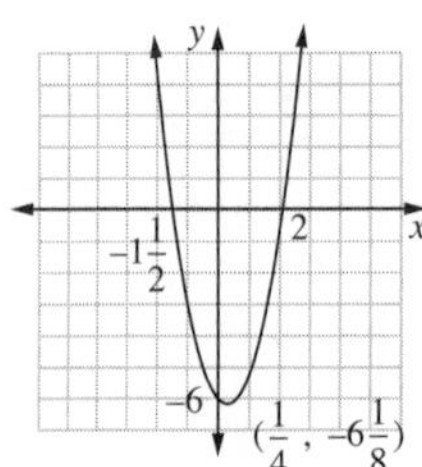

4 $(x - 2)(x^2 + 3x - 2) + 5$

Page 149 5 a 41° **b** 108° **6 a** 1.02 m **b** 1.01 m^3 **c** 5.58 m^2 **7**

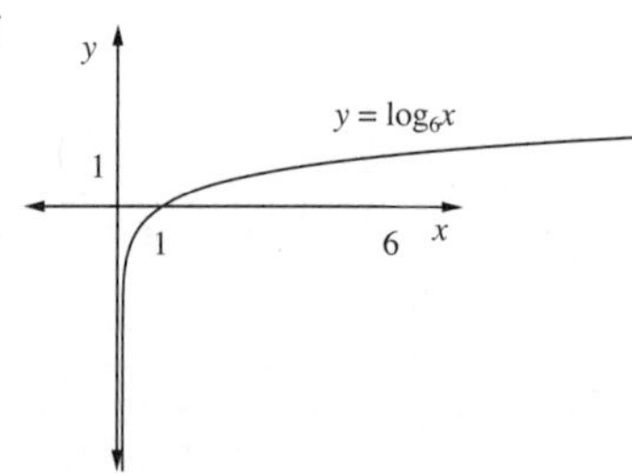

Page 150 8 a 8.1 **b** 46° **9 a**

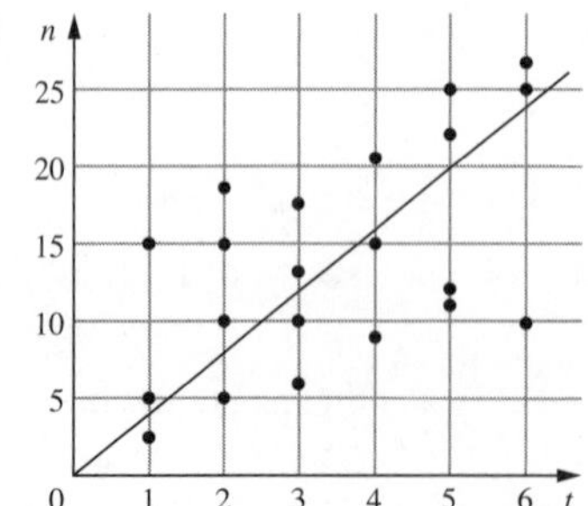

(Different lines are possible and will produce different answers for the following parts)
b $n = 4t$ **c** 40 **10 a** 118.5 m **b** 134 m

Page 151 1 D 2 D 3 B 4 C 5 A 6 D 7 B 8 D 9 A 10 A

Answers

PAGES 152–153 **1** 9 **2** 8 **3** 3.05 **4** $\frac{19}{55}$ **5** 3 **6** $\frac{\sqrt{3}}{2}$ **7** $\angle ABC$ **8** $x = 256$ **9** 703 **10** $6\sqrt{3} - 6$

11

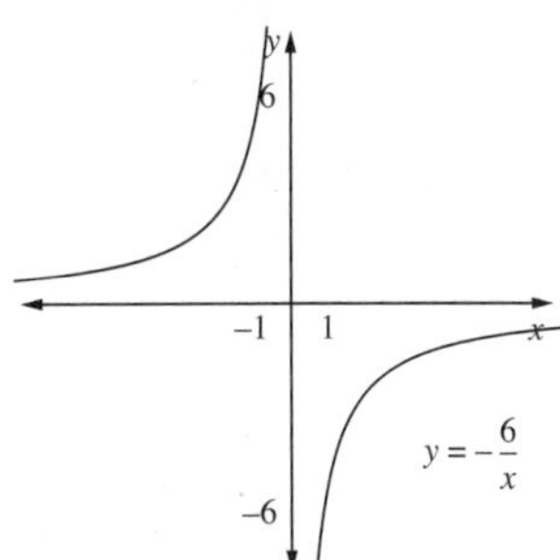

12 25.1 m^3 **13** n = 1.5 **14** 5.6 **15** $x > 0$

PAGE 154 **1 a** $2x^5 - 13x^4 + 27x^3 - 34x^2 + 20x + 6$ **b** $(3x - 2)(2x + 5)$ **2 a** $x = 1\frac{1}{2}$ or -2 **b** $x = 1.541$ or -4.541

3 a

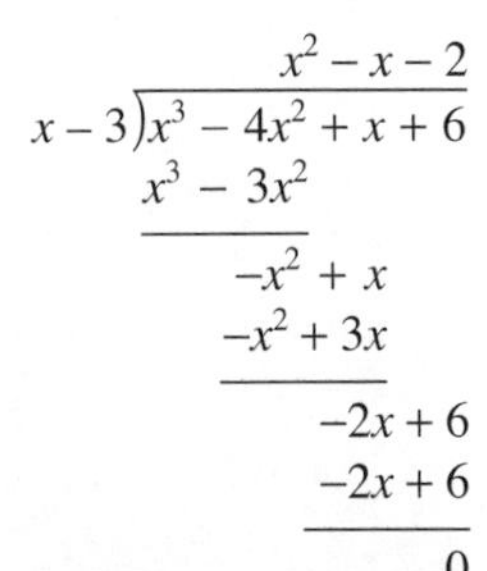

b $(x - 3)(x - 2)(x + 1)$ **c**

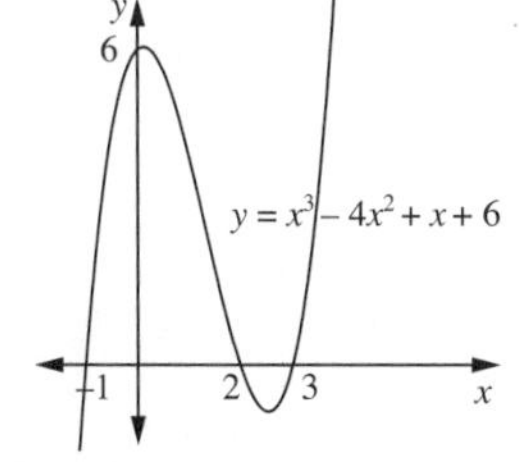

4 a 0.178 **b** 1.486

PAGE 155 **5 a** 13 units **b** $(-3, 4)$ **c** on **6 a** 720 cm^3 **b** 15 cm **c** 13 cm **d** 564 cm^2 **7** 109°

PAGE 156 **8** 135° **9 a** 73.9 **b** 9.83 **c** 34 **d** 14 **e** The second mark was slightly better. The mark in the second exam was slightly more than 1 standard deviation above the mean while in the first exam the mark is slightly less than 1 standard deviation above the mean.

PAGE 157 **1** C **2** B **3** C **4** A **5** C **6** A **7** A **8** D **9** B **10** D

PAGES 158–159 **1** $x = 147$ **2** 59°42′ **3** 64° **4** $x = 39°$ **5** 3 **6** 1 **7** 2.47 **8** $(3x + 1)(x - 1)$ **9** $5\sqrt{10}$ **10** $(4, -11)$ **11** 6.5 m **12** $p = -5$ **13** 75 cm^2 **14** It is biased. It implies the current uniform is not good.

15

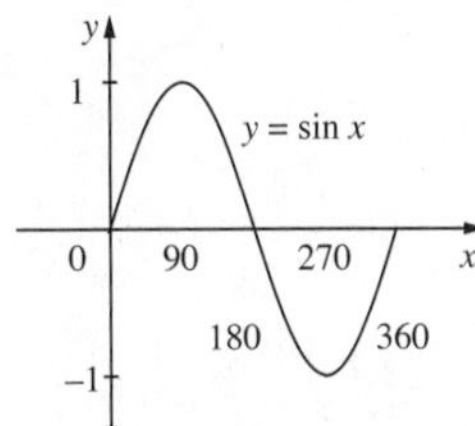

PAGE 160 **1 a** $3 + \sqrt{3}$ **b** $\frac{1}{\sqrt[3]{x^2}}$ **2 a** $x = \frac{5 \pm \sqrt{53}}{14}$ **b** $x = \pm 2$ or ± 3 **3 a** 4.26 **b** $m = 12$

4 a

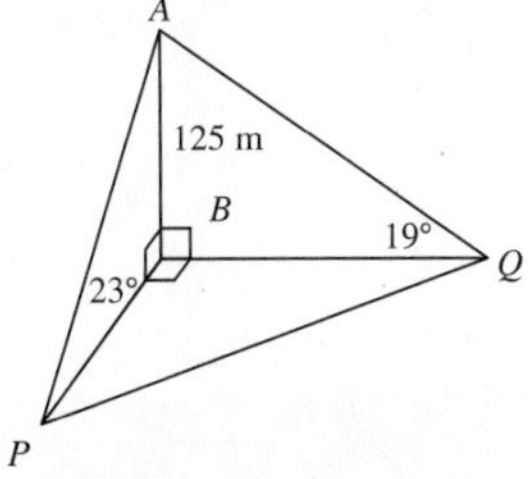

b 294.5 m **c** 363 m **d** 467 m

Answers

PAGE 161 **5 a** 34°6′ **b** 32.8 m^2 **6 a**

$$\begin{array}{r} x^2+8x+16 \\ x-2\overline{)x^3+6x^2+0x-32} \\ x^3-2x^2 \qquad\qquad \\ \hline 8x^2+0x \qquad \\ 8x^2-16x \qquad \\ \hline 16x-32 \\ 16x-32 \\ \hline 0 \end{array}$$

b $P(x) = (x-2)(x+4)^2$ **c**

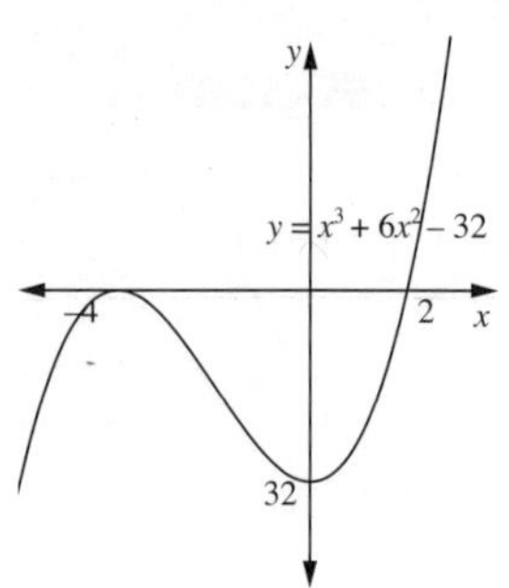

7 a 4.8 cm **b** 23 206.4 cm^3 **c** 4748.8 cm^2

PAGE 162 **8 a** $N = 17 - \frac{w}{2}$ **b** 18 weeks **c** it will give a negative number of minutes

9 a

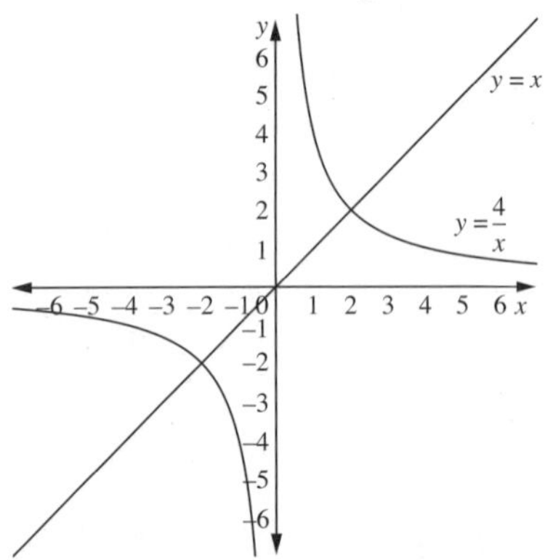

b (2, 2) and (−2, −2) **10 a** 8 **b** Yes, Pythagoras' theorem is satisfied for $\triangle OPT$ where O is the centre of the circle.

PAGE 163 **1** C **2** B **3** A **4** B **5** D **6** B **7** C **8** B **9** C **10** C

PAGES 164–165 **1** $13\sqrt{2}$ **2** $x = \pm 5$ **3** $\frac{2}{9}$ **4** $x = \frac{1}{4}, x = -2$ **5** $x^2 + y^2 = 25$ **6** 84.82 m^2 **7** $x = 2$ **8** $a = 28°$ **9** 1 **10** $x = 5\sqrt{2}$ **11** 14 m^2 **12** $(2x+9)(2x-3)$ **13** $x = \frac{1}{729}$ **14** $17 - 4\sqrt{15}$ **15** $y = 1$

PAGE 166 **1 a** $y = 4 \pm \sqrt{11}$ **b** $x = \frac{7 \pm \sqrt{29}}{10}$ **c** $x = 9$ or $x = -1$ **d** $x = -2, y = -3$ and $x = 3, y = 2$ **2 a** 49 **b** 17.5 **c** $75.\dot{6}$ **d** 12.0 **3 a** $x = 3$ **b** No, the vertex is at (3, 4) and the parabola is concave up

PAGE 167 **4 a** 55 cm **b** 45 945.8 cm^3 **c** 46 L **5** 75.4 cm^2 **6 a** $p = 3, q = -11$ **b** $P(x) = (x-2)(x+3)(2x+1)$

c

y

$y = 2x^3 + 3x^2 - 11x - 6$

−3 $-\frac{1}{2}$ 2 x

PAGE 168 **7 a** 65° **b** 35° **c** 024° **d** 39 km **8 a**

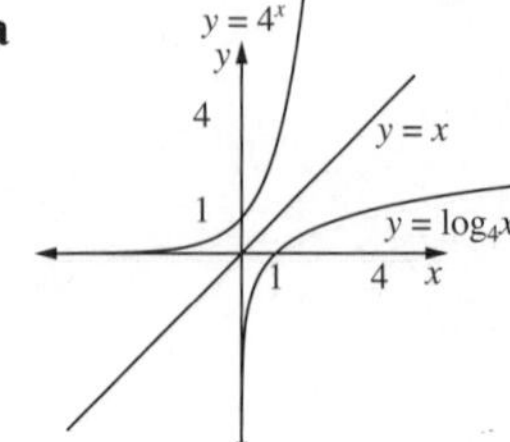

b 15.5 units